T0338675

13C-NMR of Natural Products

Volume 2
Diterpenes

^{13}C-NMR of Natural Products

Volume 2

Diterpenes

Atta-ur-Rahman
and
Viqar Uddin Ahmad

H.E.J. Research Institute of Chemistry
University of Karachi
Karachi, Pakistan

Springer Science+Business Media, LLC

Library of Congress Cataloging-in-Publication Data

13C-NMR of natural products / Atta-ur-Rahman and Viqar Uddin
Ahmad.
    p.   cm.
  Includes bibliographical references and index.
  Contents: v. 1. Monoterpenes and sesquiterpenes -- v.
2. Diterpenes.
  ISBN 978-0-306-43898-1   ISBN 978-1-4615-3288-0 (eBook)
  DOI 10.1007/978-1-4615-3288-0
  1. Natural products--Analysis. 2. Nuclear magnetic resonance
spectroscopy. I. Rahman, Atta-ur-, 1942- II. Ahmad, Viqar Uddin.
III. Title: Carbon-thirteen NMR of natural products.
QD415.A267   1991
547.3'0877--dc20   91-10877
                                                       CIP

ISBN 978-0-306-43898-1

© 1992. Springer Science+Business Media New York
Originally published by Plenum Press, New York in 1992

All rights reserved

No part of this book may be reproduced, stored in a retrieval system, or transmitted
in any form or by any means, electronic, mechanical, photocopying, microfilming,
recording, or otherwise, without written permission from the Publisher

PREFACE

Natural product chemistry presents a vast and exciting area which has attracted the attention of some of the foremost organic chemists. With the development of modern spectroscopic techniques, the number of new natural products continues to grow exponentially thereby placing at the hands of phytochemists and pharmacologists a vast array of novel and interesting compounds with potential for drug development. The need for books which present spectral data of such compounds has accordingly grown.

The present volume, which is the second of this new series, covers the ^{13}C-NMR data of diterpenes reported till late 1989. The compounds are arranged according to their structural types in various subclasses. Within each subclass, they are arranged in order of increasing molecular weight. Compounds in the same subclass and having the same molecular weight are arranged in order to increasing carbon number while those having the same molecular formula are arranged alphabetically. In case the ^{13}C-chemical shifts were not assigned by the authors, then these are presented unassigned below the structures. If the chemical shifts have been reported in several solvents, then the values presented are those in CDCl$_3$, but references have been included to their data in other solvents. We have avoided giving long chemical names of compounds and used trivial names but where trivial names were not given by the authors, then the compounds have been designated by alphabets (representing the first alphabets of the plant genus and species) followed by a number. The data given is for natural compounds and the data for simple derivatives has only been given where the data of the parent natural compounds was not available. The symbols *, **, ***, +, ++, +++, #, @, \$, ¢, &, have been employed to indicate interchangeability of assigned values.

The numbering systems used are given at the beginning of each subclass. Compound name, molecular formula, molecular weight, source and compounds-type indices have been included, which should add to the utility of the book.

The authors wish to thank the following research fellows for their help in the literature survey: Miss Anis Fatima, Miss Jahan Ara, Miss Nazli Rasool, Miss Rehana Perveen, Miss Rubeena Saleem, Miss Saleha Ishrat, Miss Seema Iqbal, Miss Shahida Begum and Miss Zahida Begum. We are grateful to Mr. Kamran Faisal Khan, Mr. Mohammad Afzal Butt, Mr. Wajih-ul-Hasnain, Mr, Mohammad Shakir and Mr. Asif Mehmood Raja for typing of the manuscript. Our thanks go to Mr. Mehmood Alam for the secretarial work and to Mr. Abdul Hafeez for drawing the structures. We are indebted to Zelin Private Limited (Karachi), Pakistan Atomic Energy Commission (Islamabad), Sumitomo Corporation (Tokyo) and Rotary Club (Karachi) for financial support.

Finally we wish to express our sincere gratitude to Prof. K. Antonakis, Prof. F. Arcamone, Prof. Y. Asakawa, Dr. Fatima Z. Basha, Prof. Jan Bergman, Prof. H.J. Bestmann, Prof. H. Budzikiewicz, Prof. L. Castedo, Prof. M. Chierichetti, Prof. J.D. Connolly, Prof. B. Danieli, Prof. S. Dev, Prof. H. Duddeck, Mr. E.B. Gapit, Prof. B.Q. Guevara, Prof. T. Higa, Dr. H. Hirota, Miss J. Hudson, Dr. Byung-Hoon Han, Prof. S. Ito, Prof. P. Joseph-Nathan, Prof. J. Jurczak, Dr. L. Kenne, Prof. S.V. Kessar, Prof. W. Kraus, Prof. J.P. Kutney, Dr. J. McLeod, Prof. P.W. Le Quesne, Prof. J. Levy, Mr. K. Mahmood, Dr. S. Malik, Prof. K. Mori, Dr. T. Nakano, Prof. Nanjun-Sun, Prof. I. Ninomiya, Prof. M. Nishizawa, Prof. C. Oehlschlager, Prof. K.A. Parker, Dr. A. Panthong, Prof. J.H. Pazur, Prof. G.D. Prestwich, Prof. A.V. Rama Rao, Prof. V. Reutrakul, Prof. M.V. Sargent, Prof. P. Sinay, Prof. Y. Shizuri, Prof. G. Snatzke, Prof. W. Steglich, Prof. C. Szantay, Dr. F. Tillequin, Prof. A. Vasella, Prof. R. Verpoorte, Prof. R. Vlahov, Prof. W. Voelter, Prof. D. Walton, Prof. Z.J. Witczak. Prof. Xu.Xing-Xiang, Prof. X.X. Xu, Prof. Yong-Long Liu, Prof. A. Zamojski and Prof. K.P. Zeller for supply of copies of various publications.

Atta-ur-Rahman
Ph.D. (Cantab.), Sc.D. (Cantab).
Viqar Uddin Ahmad
Ph.D. (Kar.), Dr.rer.nat. (Bonn)

CONTENTS

Diterpenes

Indexes

DITERPENES

LINEAR

Basic skeleton:

β-SPRINGENE

Source: *Antidorcas marsupialis*
Mol. formula: $C_{20}H_{32}$
Mol. wt.: 272
Solvent: $CDCl_3$

B.V. Burger, M. Le Roux, H.S.C. Spies, V. Truter and R.C. Bigalke, <u>Tetrahedron Lett.</u>, (52), 5221 (1978).

(5E,9E)-13,14-EPOXY-6,10,14-TRIMETHYLPENTADECA-5,9-DIEN-2-ONE

Source: *Cystophora moniliformis*

Mol. formula: $C_{18}H_{30}O_2$

Mol. wt.: 278

Solvent: $CDCl_3$

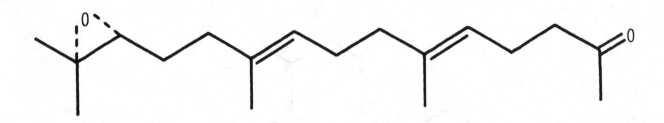

CH_3: 15.8, 15.8, 18.6, 27.3 and 29.8

CH_2: 22.3, 24.8, 26.3, 36.2, 39.5 and 43.6

CH: 64.1, 122.7 and 124.7

C: 58.2, 134.2 and 136.3

C=O: 208.8

I.A. van Altena, <u>Aust. J. Chem.</u>, 41 (1), 49 (1988).

(2E,6E,10E,13E)-3,7,11,15-TETRAMETHYL-HEXADECA-2,6,10, 13,15-PENTENOL

Source: *Bifurcaria bifurcata*

Mol. formula: $C_{20}H_{32}O$

Mol. wt.: 288

Solvent: $CDCl_3$

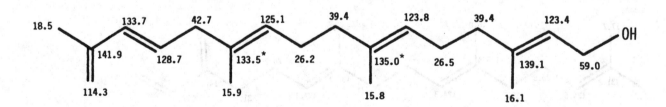

L. Semmak, A. Zerzouf, R. Valls, B. Banaigs, G. Jeanty and C. Francisco, <u>Phyto-chemistry</u>, 27(7), 2347 (1988).

GERANYLGERANIOL

Source: *Bixa orellana*

Mol. formula: $C_{20}H_{34}O$

Mol. wt.: 290

Solvent: $CDCl_3$

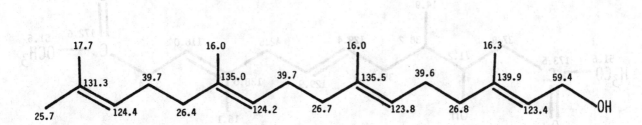

I.J.O. Jondiko and G. Pattenden, <u>Phytochemistry</u>, **28** (11), 3159 (1989).

||||||||||||||||||||||||||||||||||||||||||||||||||||||||||||||||||||||||||||||||||||||||||||||||||||||||||||||||||||||||||||||||||||||

PHYTOL

Source: *Gracilaria andersoniana*

Mol. formula: $C_{20}H_{40}O$

Mol. wt.: 296

Solvent: $CDCl_3$

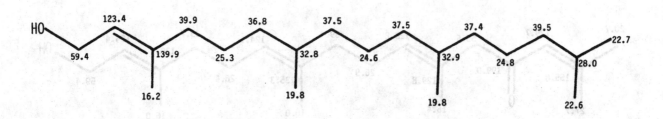

1. R.A. Goodman, E. Oldfield and A. Allerhand, <u>J. Am. Chem. Soc.</u>, **95** (23), 7553 (1973).

2. J.J. Sims and J.A. Pettus, Jr., <u>Phytochemistry</u>, **15** (6), 1076 (1976).

3ε-HYDROXY-4ε,9-DIMETHYL-6<u>E</u>,9<u>E</u>-DODECADIENEDIOIC ACID (METHYL ESTER)

Source: *Nicotianum tobaccum*
Mol. formula: $C_{16}H_{26}O_5$
Mol. wt.: 298

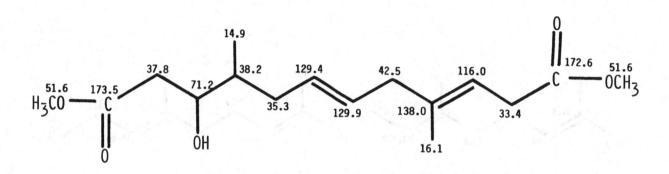

T. Chuman, M. Noguchi, A. Ohkubo and S. Toda, <u>Tetrahedron Lett.</u>, 19(35), 3045 (1977).

ELEGANOLONE

Source: *Cystoseira elegans*
Mol. formula: $C_{20}H_{32}O_2$
Mol. wt.: 304
Solvent: $CDCl_3$

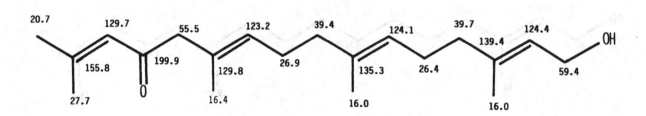

V. Amico, G. Oriente, M. Piattelli, G. Ruberto and C. Tringali, <u>Phytochemistry</u>, 20(5), 1085 (1981).

ELEGANEDIOL

Source: *Bifurcaria bifurcata* Ross
Mol. formula: $C_{20}H_{34}O_2$
Mol. wt.: 306
Solvent: $CDCl_3$

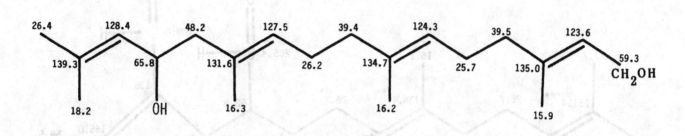

J.F. Biard, J.F. Verbist, R. Floch and Y. Letourneux, Tetrahedron Lett:, 21(19), 1849 (1980).

|||||||||||||||||||||||||||||||||||||||||||||||||||||||||||||||||||||||||||||||||||||||||||||||||||||||||||||||||||||||||||||||||||||||||||||||||||

12-(S)-HYDROXYGERANYLGERANIOL

Source: *Bifurcaria bifurcata*
Mol. formula: $C_{20}H_{34}O_2$
Mol. wt.: 306
Solvent: $CDCl_3$

R. Valls, B. Banaigs, C. Francisco, L. Codomier and A. Cave, Phytochemistry, 25(3), 751 (1986).

Source: *Udotea flabellum* (Ellis and Solander) Lamouroux
Mol. formula: $C_{20}H_{30}O_3$
Mol. wt.: 318
Solvent: $CDCl_3$

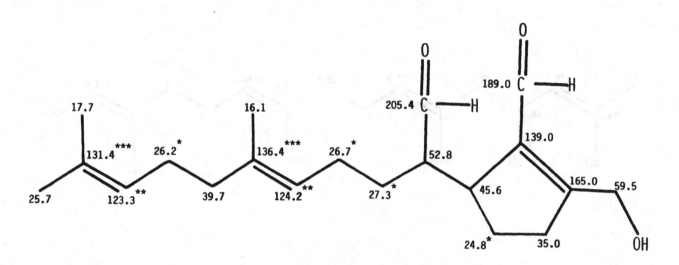

V.J. Paul and W. Fenical, <u>Tetrahedron</u>, 40(15), 2913 (1984).

||||||||||||||||||||||||||||||||||||||||||||||||||||||||||||||||||||||||||||||||||||||||||||||||||||||||

(2<u>E</u>,10<u>E</u>)-1-HYDROXY-6,13-DIKETO-7-METHYLENE-3,11,15-TRIMETHYLHEXADECA-2,10,14-TRIENE

Source: *Cystoseira crinita*
Mol. formula: $C_{20}H_{30}O_3$
Mol. wt.: 318
Solvent: $CDCl_3$

V. Amico, G. Oriente, M. Piattelli, G. Ruberto and C. Tringali, <u>Phytochemistry</u>, **20** (5), 1085 (1981).

FLEXILIN

Source: *Caulerpa flexilis*

Mol. formula: $C_{19}H_{28}O_4$

Mol. wt.: 320

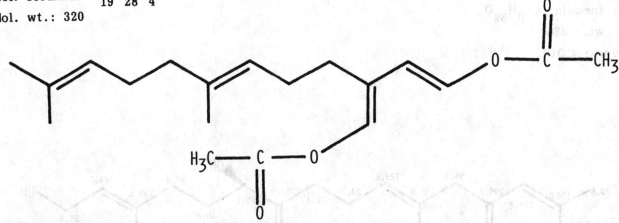

CH$_3$: 16.0, 17.7, 20.7, 20.7 and 25.7

CH$_2$: 25.5, 26.6, 26.8 and 39.7

CH: 113.4, 122.2, 124.3, 134.3 and 135.6

C: 121.2, 131.3 and 136.0

C=O: 167.4 and 167.8

A.J. Blackman and R.J. Wells, <u>Tetrahedron Lett.</u>, (33), 3063 (1978).

||||||||||||||||||||||||||||||||||||||||||||||||||||||||||||||||||||||||||||||||||||||||||||||||||||||||||||||||||||||||||||||||||||||||||||||

(2$\underline{E}$,5$\underline{E}$,10$\underline{E}$)-1,7-DIHYDROXY-13-KETO-3,7,11,15-TETRAMETHYLHEXADECA-2,6,10,14-TETRAENE

Source: *Cystoseira crinita* Bory

Mol. formula: $C_{20}H_{32}O_3$

Mol. wt.: 320

Solvent: $CDCl_3$

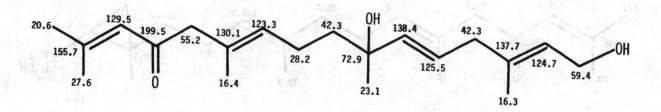

V. Amico, G. Oriente, M. Piattelli, G. Ruberto and C. Tringali, <u>Phytochemistry</u>, 20(5), 1085 (1981).

(2E,10E)-1,6-DIHYDROXY-7-METHYLENE-13-KETO-3,11,15-TRIMETHYLHEXADECA-2,10,14-TRIENE

Source: *Cystoseira crinita* Bory

Mol. formula: $C_{20}H_{32}O_3$

Mol. wt.: 320

Solvent: $CDCl_3$

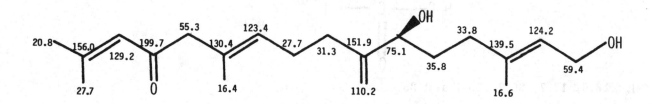

V. Amico, G. Oriente, M. Piattelli, G. Ruberto and C. Tringali, Phytochemistry, **20**(5), 1085 (1981).

||||||||||||||||||||||||||||||||||||||||||||||||||||||||||||||||||||||||||||||||||||||||||||||||||||||||||||||||||||||||||||||

EPOXYELEGANOLONE

Source: *Bifurcaria bifurcata* Ross

Mol. formula: $C_{20}H_{32}O_3$

Mol. wt.: 320

Solvent: $CDCl_3$

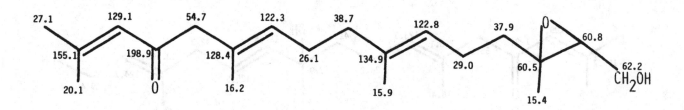

J.F. Biard, J.F. Verbist, R. Floch and Y. Letourneux, Tetrahedron Lett., **21**(19), 1849 (1980).

(2E,6E,10E)-12(S)-HYDROXY-3,7,11,15-TETRAMETHYL-HEXADECA-2,6,10,14-TETRANOIC ACID

Source: *Bifurcaria bifurcata*

Mol. formula: $C_{20}H_{32}O_3$

Mol. wt.: 320

Solvent: $CDCl_3$

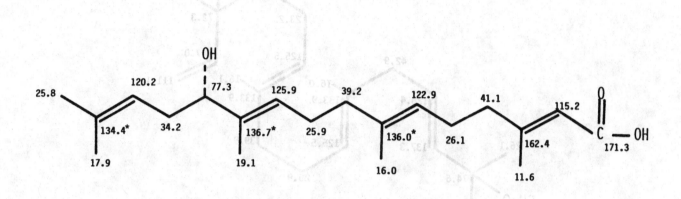

L. Semmak, A. Zerzouf, R. Valls, B. Banaigs, G. Jeanty and C. Francisco, Phytochemistry, 27(7), 2347 (1988).

13-METHOXYGERANYLLINALOOL

Source: *Arachniodes maximowiczii* Ohwi

Mol. formula: $C_{21}H_{36}O_2$

Mol. wt.: 320

Solvent: C_5D_5N

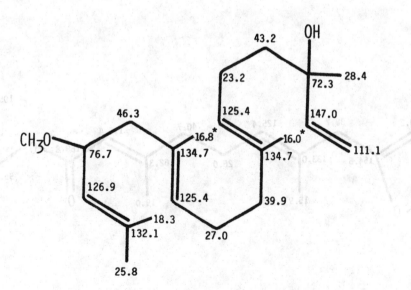

N. Tanaka, H. Sakai, T. Murakami, Y. Saiki, C.-M. Chen and Y. Iitaka, Chem. Pharm. Bull., (Tokyo), 34(3), 1015 (1986).

15-METHOXY-3,7,11,15-TETRAMETHYLHEXADECA-1,6(E), 10(E),13(E)-TETRAENE

Source: *Arachniodes maximowiczii* Ohwi

Mol. formula: $C_{21}H_{36}O_2$

Mol. wt.: 320

Solvent: C_5D_5N

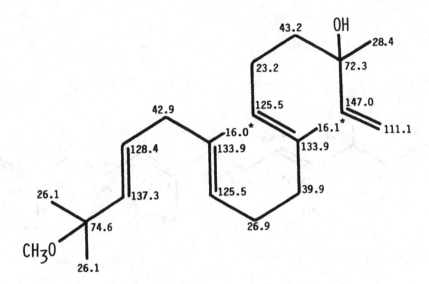

N. Tanaka, H. Sakai, T. Murakami, Y. Saiki, C.-M. Chen and Y. Iitaka, <u>Chem. Pharm. Bull.</u>, (Tokyo), 34(3), 1015 (1986).

|||||||||||||||||||||||||||||||||||||||||||||||||||||||||||||||||||||||||||||||||||||||||||||||||||||||||||||||||||

MARISLIN

Source: *Chromodoris marislae* Bertsch

Mol. formula: $C_{20}H_{24}O_4$

Mol. wt.: 328

Solvent: C_6D_6

J.E. Hochlowski and D.J. Faulkner, <u>Tetrahedron Lett.</u>, 22(4), 271 (1981).

(E̲,E̲,E̲)-3-HYDROXYMETHYL-7,11-DIMETHYL-2,6,10-
HEXADECATRIEN-1,14,15-TRIOL

Source: *Tithonia pedunculata*

Mol. formula: $C_{20}H_{36}O_4$

Mol. wt.: 340

Solvent: $CDCl_3$

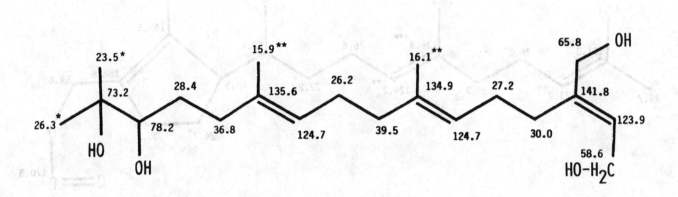

A.-L. Perez, A. Ortega and A.R. De Vivar, <u>Phytochemistry</u>, 27(12), 3897 (1988).

||||||||||||||||||||||||||||||||||||||||||||||||||||||||||||||||||||||||||||||||||||||||||||||||||||||||||||||||||||

THYMIFODIOIC ACID

Source: *Baccharis thymifolia* H. et A.

Mol. formula: $C_{20}H_{26}O_5$

Mol. wt.: 346

Solvent: $CDCl_3$

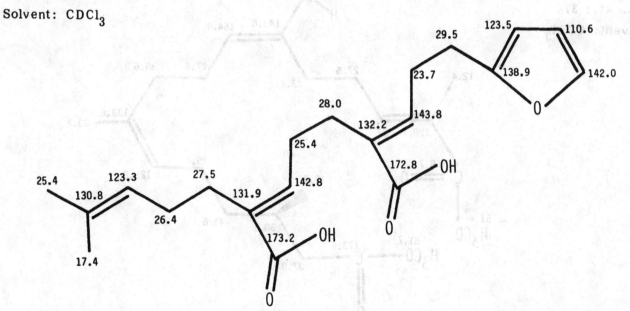

J.R. Saad, M.J. Pestchanker and O.S. Giordano, <u>Phytochemistry</u>, **26** (11), 3033 (1987).

DITERPENOID UF-VIII (ACETATE)

Source: *Udotea flabellum* (Ellis and Solander) Lamouroux

Mol. formula: $C_{22}H_{32}O_4$

Mol. wt.: 360

Solvent: $CDCl_3$

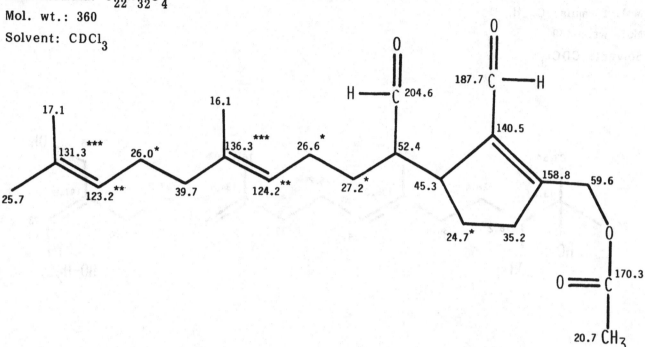

V.J. Paul and W. Fenical, Tetrahedron, 40(15), 2913 (1984).

||||i||||||||||||||||||||||||||||||||||||||||||||||||||||||||||||||||||||||||||||||

DIMETHYL(14S)-(2E,6E,10Z)-6-FORMYL-2,10,14-TRIMETHYLHEXADECA-2,6,10-TRIENEDIOIC ACID (DIMETHYL ESTER)

Source: *Eremophila glutinosa* Chinnock

Mol. formula: $C_{22}H_{34}O_5$

Mol. wt.: 378

Solvent: $CDCl_3$

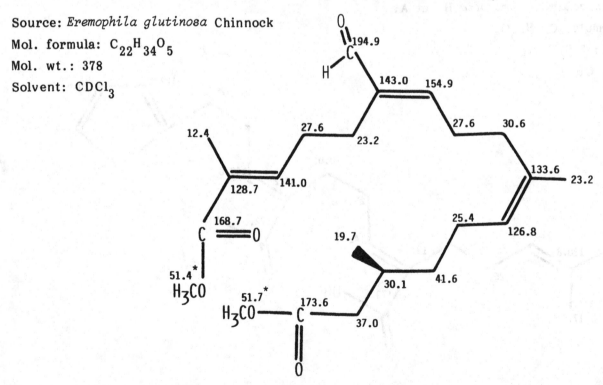

E.L. Ghisalberti, P.R. Jefferies and G.M. Proudfoot, Aust. J. Chem., 34 (7), 1491 (1981).

DIMETHYL (2E,6Z)-2-[3'Z,7'E)-9'-HYDROXY-4',8'-DIMETHYLNONA-3',7'-DIENYL]-6-METHYLOCTA-2,6-DIENEDIOATE

Source: *Eremophila exilifolia* F. Muell.

Mol. formula: $C_{22}H_{34}O_5$

Mol. wt.: 378

Solvent: $CDCl_3$

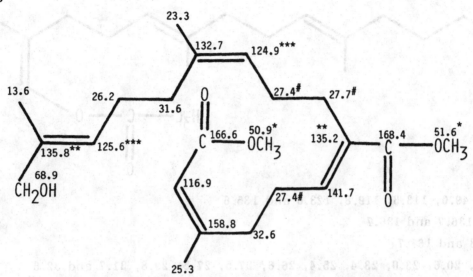

E.L. Ghisalberti, P.R. Jefferies and G.M. Proudfoot, Aust. J. Chem., **34** (7), 1491 (1981).

||||||||||||||||||||||||||||||||||||||||||||||||||||||||||||||||||||||||||||||||||||||||||||||||||||||||||||||||||||||||||||||||

DIMETHYL(2E,6Z)-2-[3'Z,7'Z)-9'-HYDROXY-4',8'-DIMETHYLNONA-3',7'-DIENYL]-6-METHYLOCTA-2,6-DIENEDIOIC ACID (METHYL ESTER)

Source: *Eremophila exilifolia* F. Muell.

Mol. formula: $C_{22}H_{34}O_5$

Mol. wt.: 378

Solvent: $CDCl_3$

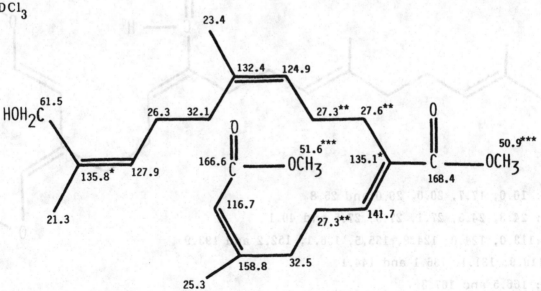

E.L. Ghisalberti, P.R. Jefferies and G.M. Proudfoot, Aust. J. Chem., **34** (7), 1491 (1981).

TRIFARIN

Source: *Caulerpa trifaria*

Mol. formula: $C_{24}H_{38}O_4$

Mol. wt.: 390

CH_2: 40.6

CH: 34.0, 49.0, 113.3, 119.8, 123.0 and 135.6

C: 121.2, 136.7 and 136.7

C=O: 167.3 and 167.7

16.1, 20.6, 20.6, 23.0, 23.4, 25.4, 26.6, 27.5, 27.5, 29.8, 31.7 and 32.6

A.J. Blackman and R.J. Wells, <u>Tetrahedron Lett.</u>, (33), 3063 (1978).

||||||||||||||||||||||||||||||||||||||||||||||||||||||||||||||||||||||||||||||||||||||||||||||||||||||||||||||||||||

UDOTEAL

Source: *Udotea flabellum* (Ellis and Solander).

Mol. formula: $C_{24}H_{34}O_5$

Mol.wt.: 402

Solvent: C_6D_6

CH_3: 16.0, 17.7, 20.0, 20.0 and 25.8

CH_2: 24.3, 24.5, 27.1, 27.5, 27.5 and 40.1

CH: 113.0, 124.0, 124.8, 135.5, 136.1, 152.2 and 193.9

C: 119.9, 131.1, 136.1 and 144.1

C=O: 166.5 and 167.3

V.J. Paul, H.H. Sun and W. Fenical, <u>Phytochemistry</u>, 21(2), 468 (1982).

DIMETHYL(2Z,6E,10Z,14E)-7-METHOXYCARBONYL-3,11,15-TRIMETHYLHEXADECA-2,6,10,14-TETRAENEDIOIC ACID (DIMETHYL ESTER)

Source: *Eremophila exilifolia* F. Muell.

Mol. formula: $C_{23}H_{34}O_6$

Mol. wt.: 406

Solvent: $CDCl_3$

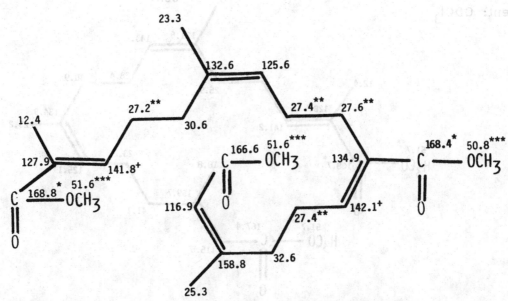

E.L. Ghisalberti, P.R. Jefferies and G.M. Proudfoot, <u>Aust. J. Chem.</u>, 34 (7), 1491 (1981).

DIMETHYL (2E,6E,10Z,14Z)-6-METHOXYCARBONYL-2,10,14-TRIMETHYLHEXADECA-2,6,10,14-TETRAENEDIOIC ACID (DIMETHYL ESTER)

Source: *Eremophila glutinosa* Chinnock

Mol. formula: $C_{23}H_{34}O_6$

Mol. wt.: 406

Solvent: $CDCl_3$

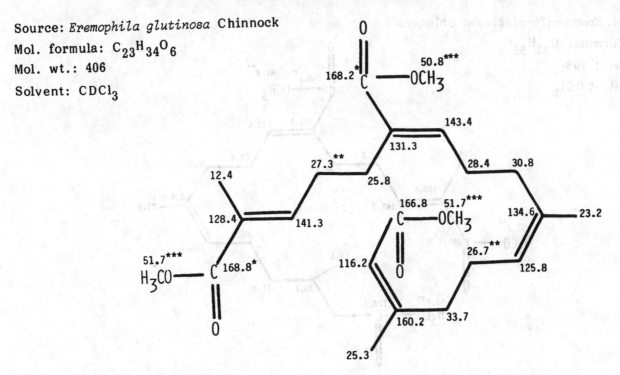

E.L. Ghisalberti, P.R. Jefferies and G.M. Proudfoot, <u>Aust. J. Chem.</u>, 34 (7), 1491 (1981).

DIMETHYL (2E,6E,10Z,14E)-6-METHOXYCARBONYL-2,10,14-TRIMETHYLHEXADECA-2,6,10,14-TETRAENEDIOATE (TRIMETHYL ESTER)

Source: *Eremophila glutinosa* Chinnock

Mol. formula: $C_{23}H_{34}O_6$

Mol. wt.: 406

Solvent: $CDCl_3$

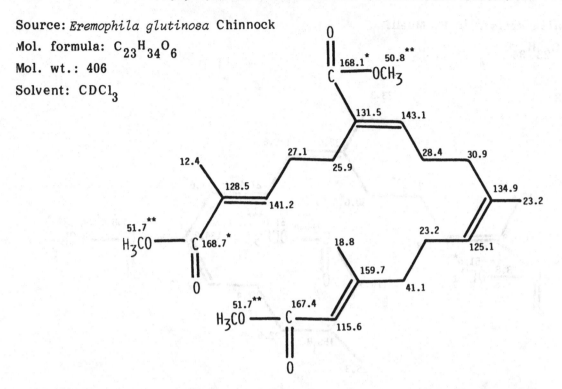

E.L. Ghisalberti, P.R. Jefferies and G.M. Proudfoot, Aust. J. Chem., **34**(7), 1491 (1981).

||||||||||||||||||||||||||||||||||||||||||||||||||||||||||||||||||||||||||||||||||||||||||||||

DIMETHYL(14S)-(2E,6E,10Z)-6-METHOXYCARBONYL-2,10,14-TRIMETHYLHEXADECA-2,6,10-TRIENEDIOIC ACID (DIMETHYL ESTER)

Source: *Eremophila glutinosa* Chinnock

Mol. formula: $C_{23}H_{36}O_6$

Mol. wt.: 408

Solvent: $CDCl_3$

E.L. Ghisalberti, P.R. Jefferies and G.M. Proudfoot, Aust. J. Chem., **34** (7), 1491 (1981).

17-ACETOXYTHYMIFODIOIC ACID (DIMETHYL ESTER)

Source: *Baccharis thymifolia* Hook and Arn

Mol. formula: $C_{24}H_{32}O_7$

Mol. wt.: 432

Solvent: $CDCl_3$

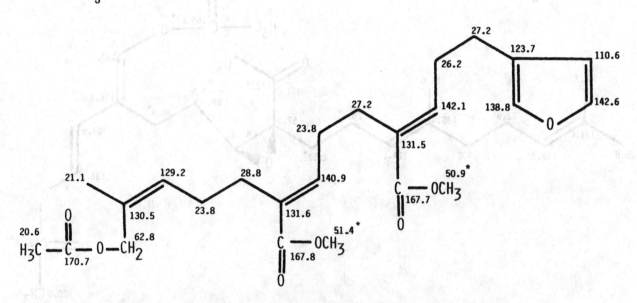

J.R. Saad, M.J. Pestchanker and O.S. Giordano, <u>Phytochemistry</u>, **26** (11), 3033 (1987).

DITERPENE PF-1

Source: *Pseudochlorodesmis furcellata* (Zanard.) Boerg

Mol. formula: $C_{24}H_{32}O_7$

Mol. wt.: 432

Solvent: $CDCl_3$

V.J. Paul, P. Ciminiello and W. Fenical, <u>Phytochemistry</u>, **27** (4), 1011 (1988).

DITERPENE UA-I

Source: *Udotea argentea* Zanard
Mol. formula: $C_{24}H_{32}O_7$
Mol. wt.: 432
Solvent: $CDCl_3$

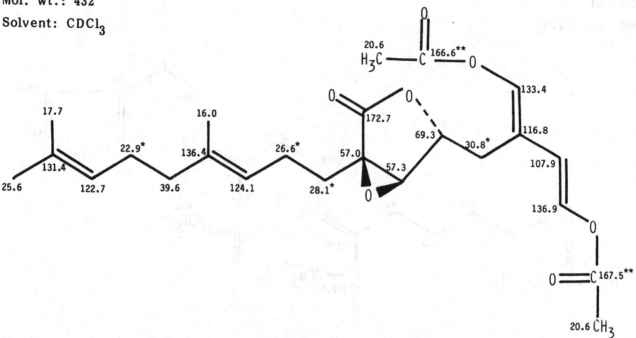

V.J. Paul and W. Fenical, <u>Phytochemistry</u>, 24(10), 2239 (1985).

|||||||||||||||||||||||||||||||||||||||||||||||||||||||||||||||||||||||||||||||||||||||||||||||||||||||||||||||||||||||||||||

DITERPENE PD-III

Source: *Penicillus dumetosus*
Mol. formula: $C_{26}H_{38}O_6$
Mol. wt.: 446
Solvent: $CDCl_3$

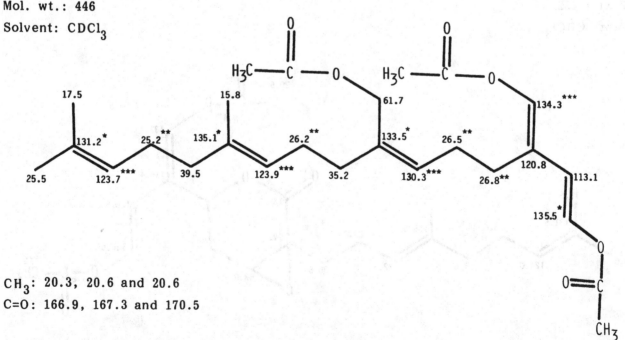

CH_3: 20.3, 20.6 and 20.6
C=O: 166.9, 167.3 and 170.5

V.J. Paul and W. Fenical, <u>Tetrahedron</u>, 40(15), 2913 (1984).

DITERPENE CE-III

Source: *Cystoseira elegans*

Mol. formula: $C_{28}H_{42}O_5$

Mol. wt.: 458

Solvent: $CDCl_3$

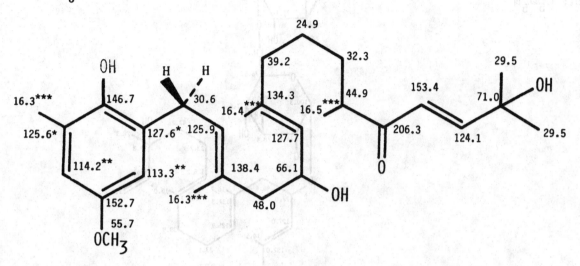

B. Banaigs, C. Francisco, E. Gonzalez, L. Codomier and W. Fenical, <u>Tetrahedron Lett.</u>, **23** (32), 3271 (1982).

||||||||||||||||||||||||||||||||||||||||||||||||||||||||||||||||||||||||||||||||||||||||||||||||||||||||||||||||||||||||||||||||||||||||||||||||

HALIMEDATRIAL-IV

Source: *Halimeda* sp.

Mol. formula: $C_{28}H_{38}O_9$

Mol. wt.: 518

Solvent: $CDCl_3$

CH_3: 20.6, 20.6, 20.8 and 21.2

C=O: 166.7, 167.6, 167.7 and 169.9

V.J. Paul and W. Fenical, <u>Tetrahedron</u>, 40(16), 3053 (1984).

13-HYDROXYGERANYLLINALOOL-13-O-β-D-(6'-O-β-L-FUCOPYRANOSYL)-GLUCOPYRANOSIDE

Source: *Arachniodes maximowiczii* Ohwi

Mol. formula: $C_{32}H_{54}O_{11}$

Mol. wt.: 614

Solvent: C_5D_5N

N. Tanaka, H. Sakai, T. Murakami, Y. Saiki, C.-M. Chen and Y. Iitaka, Chem. Pharm. Bull., (Tokyo), 34(3), 1015 (1986).

||||||||||||||||||||||||||||||||||||||||||||||||||||||||||||||||||||||||||||||||||||||||||||||||||

13-HYDROXYGERANYLLINALOOL-3,13-O-β-D-DIGLUCOSIDE

Source: *Arachniodes maximowiczii* Ohwi

Mol. formula: $C_{32}H_{54}O_{12}$

Mol. wt.: 630

Solvent: C_5D_5N

N. Tanaka, H. Sakai, T. Murakami, Y. Saiki, C.-M. Chen and Y. Iitaka, Chem. Pharm. Bull., (Tokyo), 34(3), 1015 (1986).

DIFURAN TERPENOID

Basic skeleton:

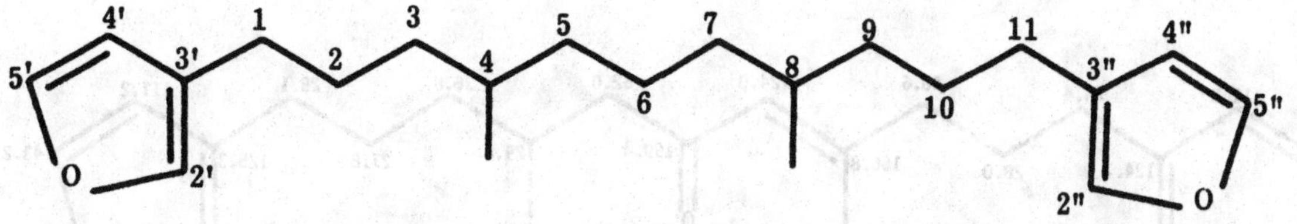

CACOSPONGIENONE A

Source: *Cacospongia scalaris* Schmidt
Mol. formula: $C_{21}H_{28}O_3$
Mol. wt.: 328
Solvent: C_6D_6

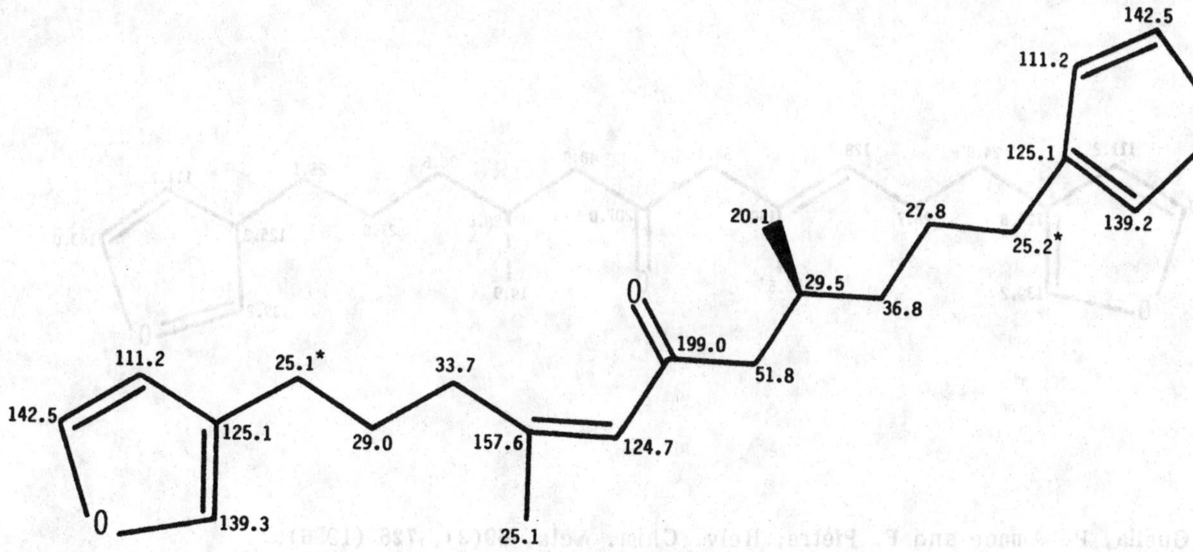

G. Guella, P. Amade and F. Pietra, Helv. Chim. Acta, **69** (3), 726 (1986).

CACOSPONGIENONE B

Source: *Cacospongia scalaris* Schmidt

Mol. formula: $C_{21}H_{28}O_3$

Mol. wt.: 328

Solvent: C_6D_6

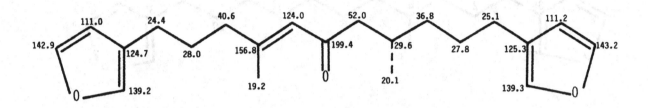

G. Guella, P. Amade and F. Pietra, Helv. Chim. Acta, **69** (3), 726 (1986).

DIHYDROFUROSPONGIN-2

Source: *Cacospongia scalaris* Schmidt

Mol. formula: $C_{21}H_{28}O_3$

Mol. wt.: 328

Solvent: C_6D_6

G. Guella, P. Amade and F. Pietra, Helv. Chim. Acta, **69**(3), 726 (1986).

CACOSPONGIONE A

Source: *Cacospongia scalaris* Schmidt
Mol. formula: $C_{21}H_{30}O_3$
Mol. wt.: 330
Solvent: C_6D_6

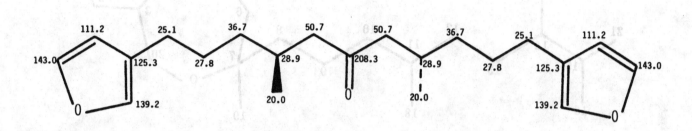

G. Guella, P. Amade and F. Pietra, Helv. Chim. Acta, **69** (3), 726 (1986).

FUROSPONGIN-1

Source: *Cacospongia scalaris* Schmidt
Mol. formula: $C_{21}H_{30}O_3$
Mol. wt.: 330
Solvent: C_6D_6

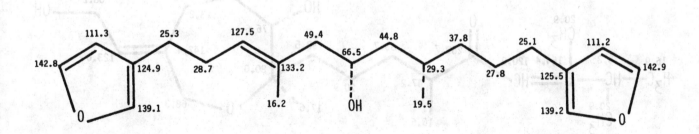

G. Guella, P. Amade and F. Pietra, Helv. Chim. Acta, **69** (3), 726 (1986).

OXEPANE

Basic skeleton:

||||||||||||||||||||||||||||||||||||||||||||||||||||||||||||||||||||||||||||||||||||||||||||||||||||||||||||||||||||||||||||||||||||||||||

MONTANOL

Source: *Montanoa tomentosa* Cerv.
Mol. formula: $C_{21}H_{36}O_4$
Mol. wt.: 352
Solvent: $CDCl_3$

Y. Oshima, G.A. Cordell and H.H.S. Fong, <u>Phytochemistry</u>, 25 (11), 2567 (1986).

TOMENTANOL

Source: *Montanoa tomentosa* Cerv.
Mol. formula: $C_{21}H_{36}O_4$
Mol. wt.: 352
Solvent: $CDCl_3$

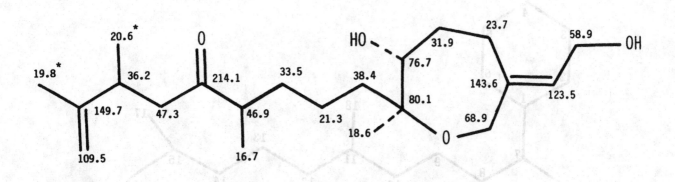

Y. Oshima, G.A. Cordell and H.H.S. Fong, <u>Phytochemistry</u>, 25(11), 2567 (1986).

TOMENTOL

Source: *Montanoa tomentosa* Cerv.
Mol. formula: $C_{21}H_{36}O_4$
Mol. wt.: 352
Solvent: $CDCl_3$

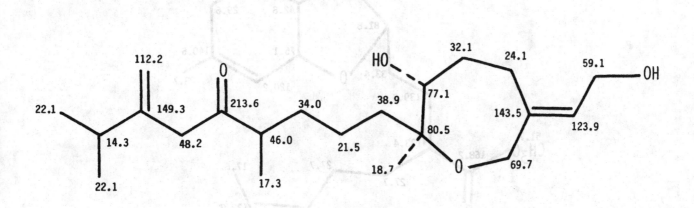

Y. Oshima, G.A. Cordell and H.H.S. Fong, <u>Phytochemistry</u>, 25(11), 2567 (1986).

Basic skeleton:

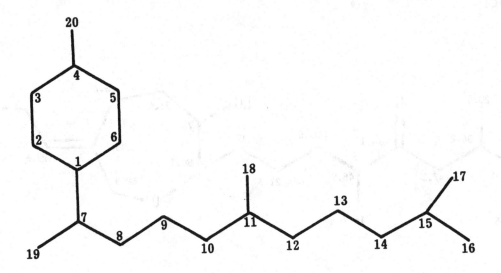

TETRAHYDROFURAN

Source: *Eremophila foliosissima* Kraenzlin.

Mol. formula: $C_{21}H_{32}O_3$

Mol. wt.: 332

Solvent: $CDCl_3$

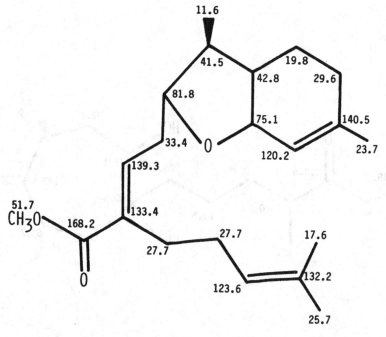

P.G. Forster, E.L. Ghisalberti and P.R. Jefferies, <u>Tetrahedron</u>, **43** (13), 2999 (1987).

TETRAHYDROPYRAN

Source: *Eremophila foliosissima* Kraenzlin

Mol. formula: $C_{21}H_{32}O_3$

Mol. wt.: 332

Solvent: $CDCl_3$

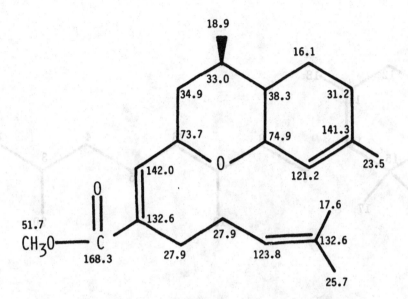

P.G. Forster, E.L. Ghisalberti and P.R. Jefferies, <u>Tetrahedron</u>, 43 (13), 2999 (1987).

||||||||||||||||||||||||||||||||||||||||||||||||||||||||||||||||||||||||||||||||||||||||||||||||||||||||||||

DITERPENE EF-1(METHYL ESTER)

Source: *Eremophila foliosissima* Kraenzlin

Mol. formula: $C_{23}H_{36}O_5$

Mol. wt.: 392

Solvent: $CDCl_3$

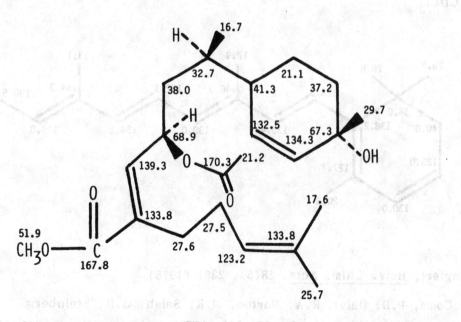

P.G. Forster, E.L. Ghisalberti and P.R. Jefferies, <u>Tetrahedron</u>, 43 (13), 2999 (1987).

RETINANE

Basic skeleton:

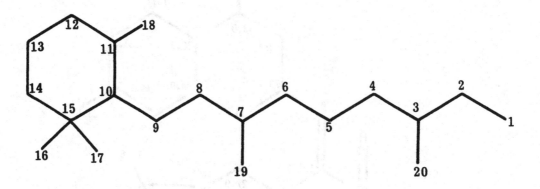

||||||||||||||||||||||||||||||||||||||||||||||||||||||||||||||||||||||||||||||||||||||||||||||

3-DEHYDRORETINAL

Source: Retinas of fresh water fishes and ling-cod

Mol. formula: $C_{20}H_{26}O$

Mol. wt.: 282

Solvent: $CDCl_3$

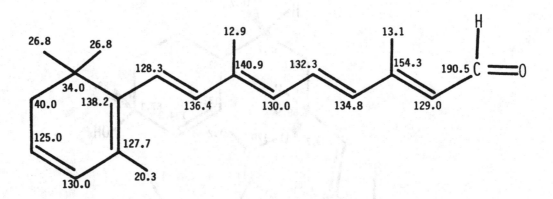

1. G. Englert, Helv. Chim. Acta, **58**(8), 2367 (1975).

2. H.R. Coma, P.D. Dalvi, R.A. Morton, M.K. Salah, G.R. Steinberg and A.L. Stubbs, Biochem. J., **52**, 535 (1952).

3-DEHYDRORETINOL

Source: *Ophiodon elongatus* and *Esox lucius*

Mol. formula: $C_{20}H_{28}O$

Mol. wt.: 284

Solvent: $CDCl_3$

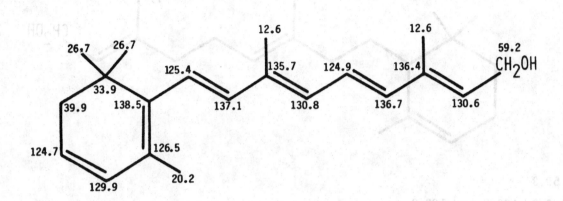

1. G. Englert, Helv. Chim. Acta, **58** (8), 2367 (1975).

2. H.R. Cama, P.D. Dalvi, R.A. Morton, M.K. Salah, G.R. Steinberg and A.L. Stubbs, Biochem. J., **52**, 535 (1952).

||||||||||||||||||||||||||||||||||||||||||||||||||||||||||||||||||||||||||||||||||||||||||||||||||||||||||||

RETINOL

Source: *Stereolepis gigas* and *S. ishinagi*

Mol. formula: $C_{20}H_{30}O$

Mol. wt.: 286

Solvent: $CDCl_3$

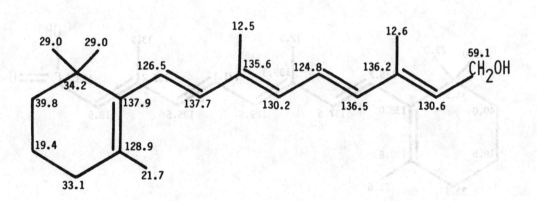

1. G. Englert, Helv. Chim. Acta, **58** (8), 2367 (1975).

2. J.G. Baxter and C.D. Robeson, J. Am. Chem. Soc., **64** (10), 2411 (1942).

CAULERPOL

Source: *Caulerpa brownii*
Mol. formula: $C_{20}H_{34}O$
Mol. wt.: 290

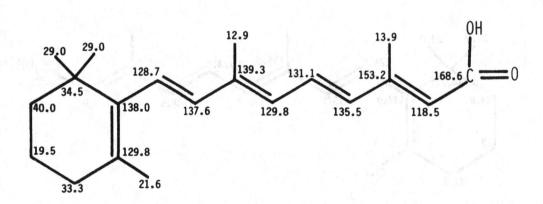

CH$_2$: 59.3
CH: 120.0, 123.8 and 123.8
C: 136.1, 136.9 and 139.3
16.2, 23.2, 26.5, 27.5, 30.0, 31.8, 32.7, 39.7, 40.6 and 49.1

A.J. Blackman and R.J. Wells, <u>Tetrahedron Lett.</u>, (31), 2729 (1976).

||||||||||||||||||||||||||||||||||||||||||||||||||||||||||||||||||||||||||||||||||||||||||||||||||||||||

RETINOIC ACID

Source: Rat liver
Mol. formula: $C_{20}H_{28}O_2$
Mol. wt.: 300
Solvent: $CDCl_3$+Dioxane 1:1

1. G. Englert, <u>Helv. Chim. Acta</u>, 58 (8), 2367 (1975).

2. S. Mahadevan, S.K. Murthy and J. Ganguly, <u>Biochem. J.</u>, 85, 326 (1962).

AGELASINE E

Source: *Agelas nakamurai* Hoshino

Mol. formula: $C_{26}H_{40}ClN_5$

Mol. wt.: 457

Solvent: $CDCl_3$

CH$_3$: 15.9, 17.2, 26.0 and 28.2

CH$_2$: 23.5, 24.6, 26.0, 32.3, 36.1, 38.0, 39.4, 48.4 and 108.6

CH: 53.5, 115.2 and 122.7

C: 34.6, 136.2, 146.8 and 149.0

H. Wu, H. Nakamura, J.I. Kobayashi, Y. Ohizumi and Y. Hirata, <u>Tetrahedron Lett.</u>, 25 (34), 3719 (1984).

||||||||||||||||||||||||||||||||||||||||||||||||||||||||||||||||||||||||||||||||||||||||||||||||||||||||||||||||

LABDANE

Basic skeleton:

E-15,16-BISNORLABDA-8(17),11-DIENE-13-ONE

Source: *Alpinia formosana*
Mol. formula: $C_{18}H_{28}O$
Mol. wt.: 260
Solvent: $CDCl_3$

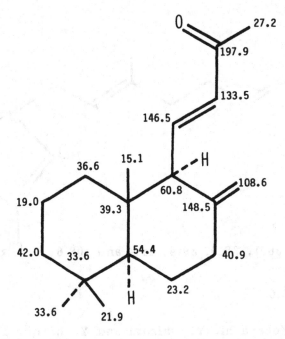

H. Itokawa, S. Yoshimoto and H. Morita, <u>Phytochemistry</u>, 27(2), 435 (1988).

12,13 E-BIFORMEN

Source: *Palafoxia rosea* (Bush.) Cory
Mol. formula: $C_{20}H_{32}$
Mol. wt.: 272
Solvent: $CDCl_3$

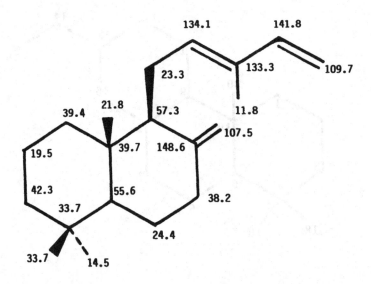

F. Bohlmann and H. Czerson, <u>Phytochemistry</u>, 18(1), 115 (1979).

Source: *Hedychium coronarium*

Mol. formula: $C_{20}H_{28}O$

Mol. wt.: 284

Solvent: $CDCl_3$

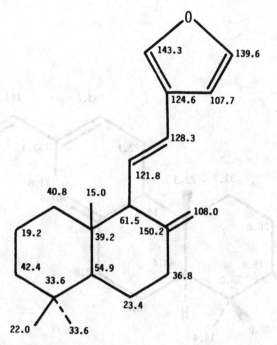

H. Itokawa, H. Morita, K. Takeya and M. Motidome, <u>Chem. Pharm. Bull.</u>, (Tokyo), 36(7), 2682 (1988).

||||||||||||||||||||||||||||||||||||||||||||||||||||||||||||||||||||||||||||||||||||||||||||||||||||||||||||

3-OXO-12,11E-BIFORMEN

Source: *Palafoxia rosea* (Bush.) Cory

Mol. formula: $C_{20}H_{30}O$

Mol. wt.: 286

Solvent: $CDCl_3$

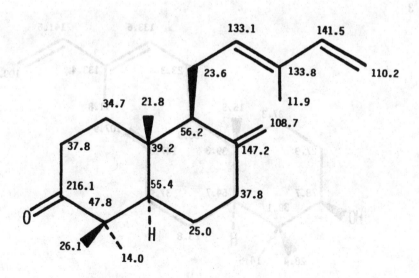

F. Bohlmann and H. Czerson, <u>Phytochemistry</u>, 18 (1), 115 (1979).

3α-HYDROXY-12,13E-BIFORMEN

Source: *Palafoxia rosea* (Bush.) Cory

Mol. formula: $C_{20}H_{32}O$

Mol. wt.: 288

Solvent: $CDCl_3$

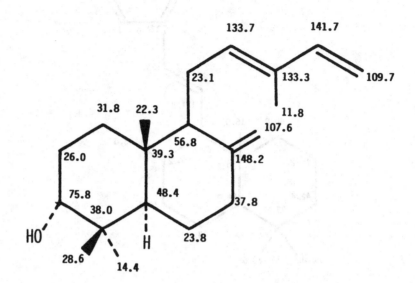

F. Bohlmann and H. Czerson, Phytochemistry, **18** (1), 115 (1979).

3β-HYDROXY-12,13E-BIFORMEN

Source: *Palafoxia rosea* (Bush.) Cory

Mol. formula: $C_{20}H_{32}O$

Mol. wt.: 288

Solvent: $CDCl_3$

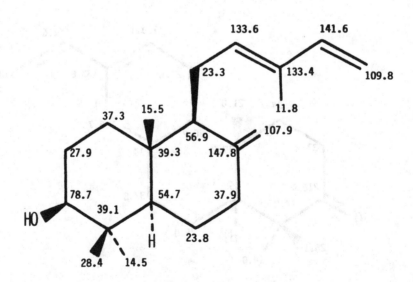

F. Bohlmann and H. Czerson, Phytochemistry, **18** (1), 115 (1979).

ISORAIMONOL

Source: *Nicotiana raimondii*

Mol. formula: $C_{20}H_{32}O$

Mol. wt.: 288

Solvent: $CDCl_3$

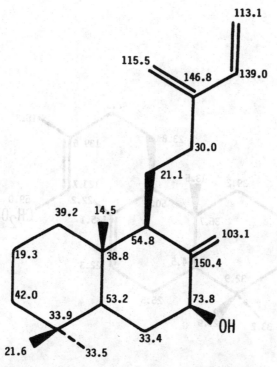

M. Noma, F. Suzuki, K. Gamou and N. Kawashima, <u>Phytochemistry</u>, 21(2), 395 (1982).

||||||||||||||||||||||||||||||||||||||||||||||||||||||||||||||||||||||||||||||||||||||||||||||||||||||||||||||||||||||||||

RAIMONOL

Source: *Nicotiana raimondii*

Mol. formula: $C_{20}H_{32}O$

Mol. wt.: 288

Solvent: $CDCl_3$

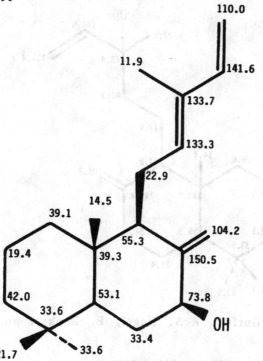

M. Noma, F. Suzuki, K. Gamou and N. Kawashima, <u>Phytochemistry</u>, 21(2), 395 (1982).

LABDA-7,13E-DIEN-15-OL

Source: *Nicotiana setchellii*

Mol. formula: $C_{20}H_{34}O$

Mol. wt.: 290

Solvent: $CDCl_3$

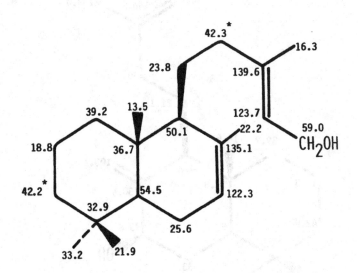

H. Suzuki, M. Noma and N. Kawashima, <u>Phytochemistry</u>, 22(5), 1294 (1983).

MANOOL

Source: *Dacrydium biforme* Pilg and *Cupressus sempervirens* L.

Mol. formula: $C_{20}H_{34}O$

Mol. wt.: 290

Solvent: $CDCl_3$

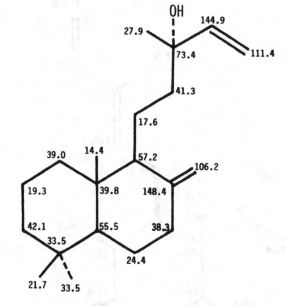

1. B.L. Buckwalter, I.R. Burfitt, A.A. Nagel, E. Wenkert and F. Naf, <u>Helv. Chim. Acta.</u>, 58(6), 1567 (1975).

2. G. Ohloff, <u>Helv. Chim. Acta.</u>, 41(3), 845 (1958).

HAVARDIOL

Source: *Grindelia havardii* Steyerm.

Mol. formula: $C_{19}H_{32}O_2$

Mol. wt.: 292

Solvent: $CDCl_3$

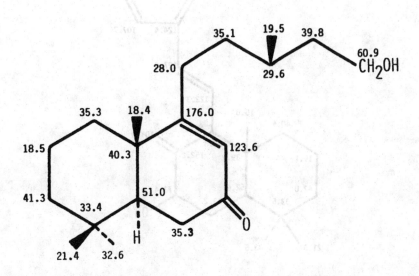

S.D. Jolad, B.N. Timmermann, J.J. Hoffmann, R.B. Bates and T.J. Siahaan, <u>Phytochemistry</u>, 26(2), 483 (1987).

||||||||||||||||||||||||||||||||||||||||||||||||||||||||||||||||||||||||||||||||||||||||||||||||||||||||||||||||||||

STEREBIN D

Source: *Stevia rebaudiana* Bertoni

Mol. formula; $C_{18}H_{30}O_3$

Mol. wt.: 294

Solvent: $CDCl_3$

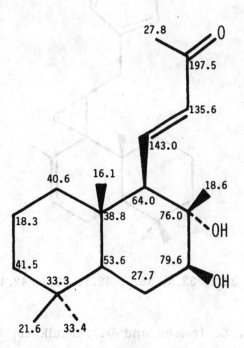

Y. Oshima, J.-I. Saito and H. Hikino, <u>Tetrahedron</u>, 42 (23), 6443 (1986).

CORONARIN A

Source: *Hedychium coronarium*

Mol. formula: $C_{20}H_{28}O_2$

Mol. wt.: 300

Solvent: $CDCl_3$

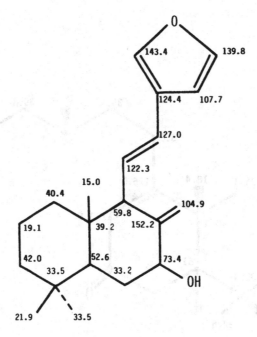

H. Itokawa, H. Morita, K. Takeya and M.Motidome, <u>Chem. Pharm. Bull.</u> (Tokyo), **36**(7), 2682 (1988).

GHISELININ

Source: *Hypselodoris ghiselini*

Mol. formula: $C_{20}H_{28}O_2$

Mol. wt.: 300

Solvent: C_6D_6

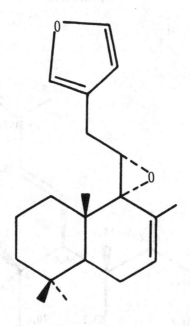

14.6, 19.1, 22.4, 23.8, 25.2, 32.1, 33.5, 35.1, 38.1, 40.2, 42.4, 50.2, 58.4, 70.0, 111.7, 122.7, 140.4 and 143.1

J.E. Hochlowski, R.P. Walker, C. Ireland and D.J. Faulkner, <u>J. Org. Chem.</u>, **47** (1), 88 (1982).

(E)-LABDA-8(17),12-DIENE-15,16-DIAL

Source: *Curcuma heyneana* Val. and V. Zijp
Mol. formula: $C_{20}H_{30}O_2$
Mol. wt.: 302
Solvent: $CDCl_3$

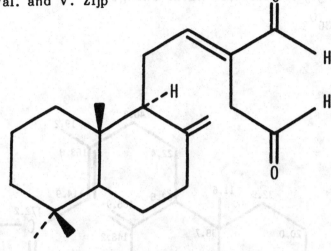

CH$_3$: 14.3, 21.6 and 33.5
CH$_2$: 19.2, 24.0, 24.6, 37.7, 39.1, 39.2, 41.9 and 107.8
CH: 55.3, 56.5 and 159.7
C: 33.5, 39.5, 134.8 and 147.9
C=O: 193.4 and 197.1

K. Firman, T. Kinoshita, A. Itai and U. Sankawa, <u>Phytochemistry</u>, 27(12), 3887 (1988).

||||||||||||||||||||||||||||||||||||||||||||||||||||||||||||||||||||||||||||||||||||||||||||||||||||||||

METASEQUOIC ACID A

Source: *Metasequoia glyptostroboides* Hu et Cheng
Mol. formula: $C_{20}H_{30}O_2$
Mol. wt.: 302
Solvent: $CDCl_3$

F. Sakan, T. Iwashita and N. Hamanaka, <u>Chem. Lett.</u>, (1), 123 (1988).

Source: *Metasequoia glyptostroboides* Hu et Cheng.

Mol. formula: $C_{20}H_{30}O_2$

Mol. wt.: 302

Solvent: $CDCl_3$

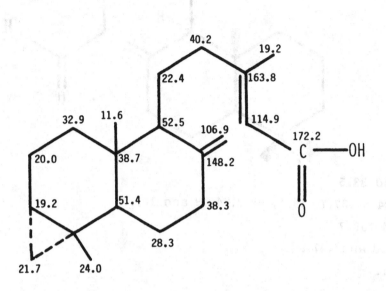

F. Sakan, T. Iwashita and N. Hamanaka, <u>Chem. Lett.</u>, (1), 123 (1988).

PERROTTETIANAL A

Source: *Porella perrottetiana*

Mol. formula: $C_{20}H_{30}O_2$

Mol. wt.: 302

Solvent: $CDCl_3$

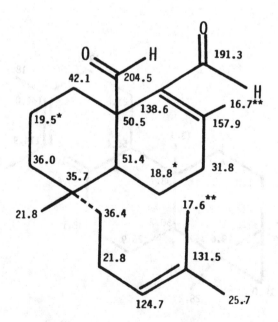

Y. Asakawa, M. Toyota and T. Takemoto, <u>Phytochemistry</u>, 18(10), 1681 (1979).

EPERUA-7,13-DIEN-15-OIC ACID

Source: *Hymeneae coubaril* L.

Mol. formula: $C_{20}H_{32}O_2$

Mol. wt.: 304

Solvent: $CDCl_3$

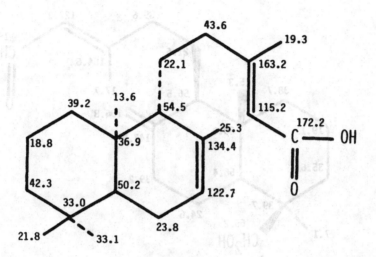

P.M. Imamura, A.J. Marsaioli, L.E.S. Barata and E.A. Ruveda, <u>Phytochemistry</u>, 16 (11), 1842 (1977).

||||||||||||||||||||||||||||||||||||||||||||||||||||||||||||||||||||||||||||||||||||||||||||||||||||||||||||||||||||||||

8,13-EPOXY-LABD-14-EN-11-ONE

Source: *Coleus forskohlii* Briq.

Mol. formula: $C_{20}H_{32}O_2$

Mol. wt.: 304

Solvent: $CDCl_3$

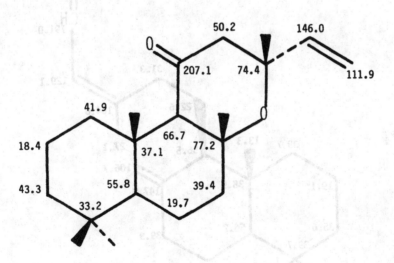

CH_3: 15.5, 21.6, 27.9, 31.2 and 33.5

B. Gabetta, G. Zini and B. Danieli, <u>Phytochemistry</u>, 28 (3), 859 (1989).

19-HYDROXYLABDA-8(17),13E-DIEN-15-AL

Source: *Juniperus thurifera*

Mol. formula: $C_{20}H_{32}O_2$

Mol. wt.: 304

Solvent: $CDCl_3$

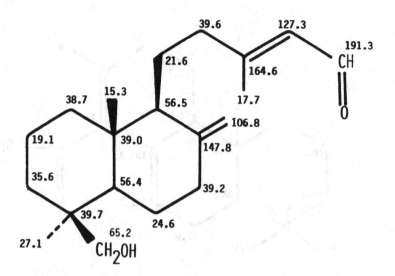

A. San Feliciano, M. Medarde, J.L. Lopez, J.M.M. Del Corral, P. Puebla and A.F. Barrero, Phytochemistry, **27** (7), 2241 (1988).

||||||||||||||||||||||||||||||||||||||||||||||||||||||||||||||||||||||||||||||

19-HYDROXYLABDA-8(17),13Z-DIEN-15-AL

Source: *Juniperus thurifera*

Mol. formula: $C_{20}H_{32}O_2$

Mol. wt.: 304

Solvent: $CDCl_3$

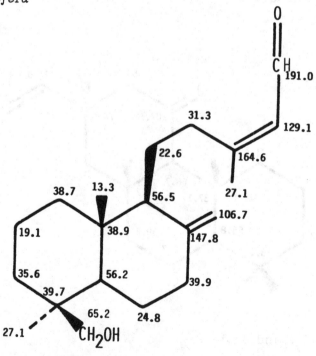

A. San Feliciano, M. Medarde, J.L. Lopez, J.M.M. Del Corral, P. Puebla and A.F. Barrero, Phytochemistry, 27(7), 2241 (1988).

ISOAGATHOLAL

Source: *Thujopsis dolabrata* Sieb et Zucc.
Mol. formula: $C_{20}H_{32}O_2$
Mol. wt.: 304
Solvent: $CDCl_3$

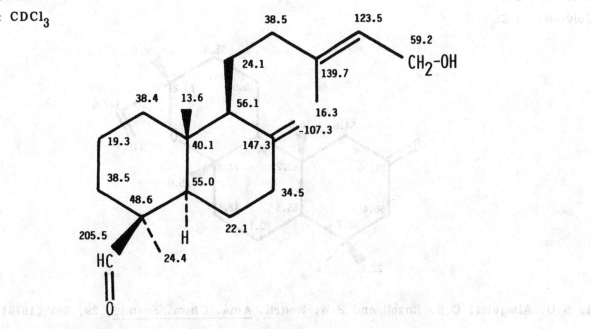

S. Hasegawa and Y. Hirose, <u>Phytochemistry</u>, **19**(11), 2479 (1980).

(E)-LABDA-8(17),12-DIENE-15-OL-16-AL

Source: *Alpinia formosana*
Mol. formula: $C_{20}H_{32}O_2$
Mol. wt.: 304
Solvent: $CDCl_3$

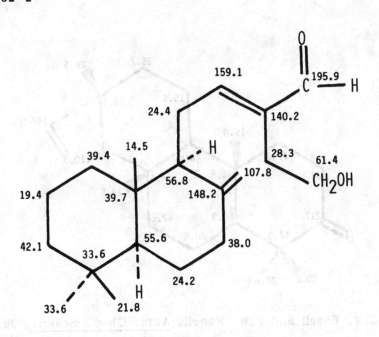

H. Itokawa, S. Yoshimoto and H. Morita, <u>Phytochemistry</u>, **27** (2), 435 (1988).

2-OXOMANOYL OXIDE

Source: *Dacrydium colensoi*
Mol. formula: $C_{20}H_{32}O_2$
Mol. wt.: 304
Solvent: $CDCl_3$

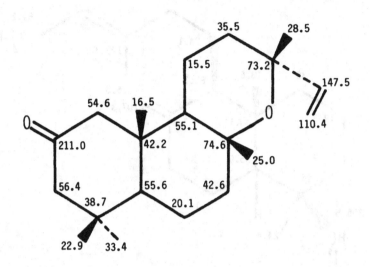

1. S.O. Almqvist, C.R. Enzell and F.W. Wehrli, Acta. Chem. Scand., 29, 695 (1975).

2. J.R. Hosking and C.W. Brandt, Chem. Ber., 68, 286 (1935).

3-OXOMANOYL OXIDE

Source: *Xylia dolabriformis*
Mol. formula: $C_{20}H_{32}O_2$
Mol. wt.: 304
Solvent: $CDCl_3$

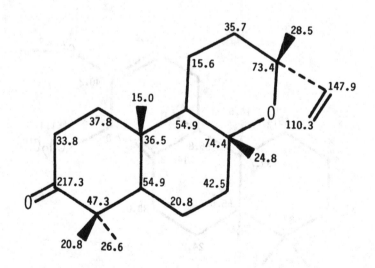

1. S.O. Almqvist, C.R. Enzell and F.W. Wehrli, Acta. Chem. Scand., 29, 695 (1975).

2. R.A. Laidlaw and J.W.W. Morgan, J. Chem. Soc., 644 (1963).

HAVARDIC ACID E (METHYL ESTER)

Source: *Grindelia havardii* Steyerm.

Mol. formula: $C_{19}H_{30}O_3$

Mol. wt.: 306

Solvent: $CDCl_3$

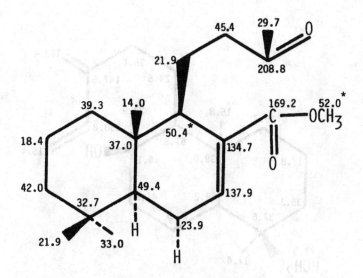

S.D. Jolad, B.N. Timmermann, J.J. Hoffmann, R.B. Bates and T.J. Siahaan, <u>Phytochemistry</u>, 26(2), 483 (1987).

||||||||||||||||||||||||||||||||||||||||||||||||||||||||||||||||||||||||||||||||||||||||||||||||||||

AGATADIOL

Source: *Juniperus thurifera*

Mol. formula: $C_{20}H_{34}O_2$

Mol. wt.: 306

Solvent: $CDCl_3$

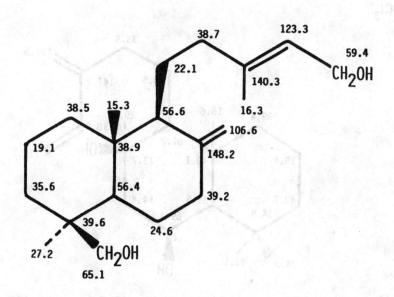

A.San Feliciano, M. Medarde, J.L. Lopez, J.M.M. Del Corral, P. Puebla and A.F. Barrero, <u>Phytochemistry</u>, 27(7), 2241 (1988).

Source: *Sideritis paulii* (Pau) P.W. Ball ex Heywood

Mol. formula: $C_{20}H_{34}O_2$

Mol. wt.: 306

Solvent: $CDCl_3$

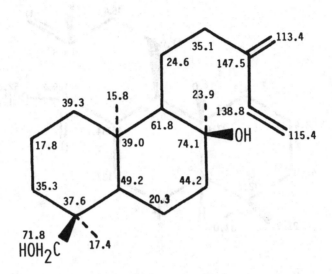

A. Garcia-Granados, A. Martinez and M.E. Onorato, <u>Phytochemistry</u>, **24**(3), 517 (1985).

ENT-6α,8α-DIHYDROXYLABDA-13(16),14-DIENE(18-DEOXYANDALUSOL)

Source: *Sideritis arborescens* Salzm. ex Bentham

Mol. formula: $C_{20}H_{34}O_2$

Mol. wt.: 306

Solvent: $CDCl_3$

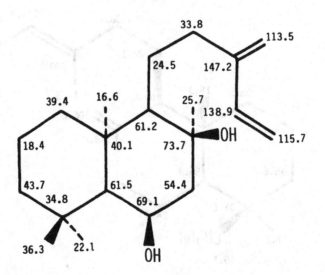

A. Garcia-Granados, A. Martinez and M.E. Onorato, <u>Phytochemistry</u>, **24**(3), 517 (1985).

DITERPENE PP-IX

Source: *Porella perrottetiana*
Mol. formula: $C_{20}H_{34}O_2$
Mol. wt.: 306
Solvent: $CDCl_3$

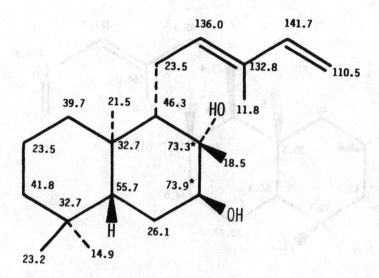

Y. Asakawa, M. Toyota and T. Takemoto, <u>Phytochemistry</u>, 18(10), 1681 (1979).

||||||||||||||||||||||||||||||||||||||||||||||||||||||||||||||||||||||||||||||||||||||||||||||||||||||||||

ENT-2α-HYDROXY-8,13β-EPOXYLABD-14-ENE

Source: *Sideritis perfoliata* L.
Mol. formula: $C_{20}H_{34}O_2$
Mol. wt.: 306
Solvent: $CDCl_3$

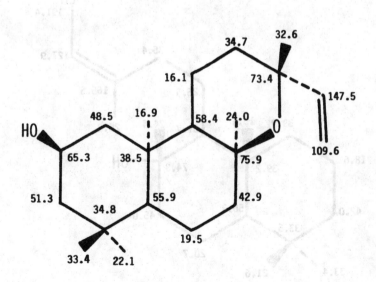

E. Sezik, N. Ezer, J.A. Hueso-Rodriguez and B. Rodriguez, <u>Phytochemistry</u>, 24(11), 2739 (1985).

8-HYDROXYLABD-13E-EN-15-AL

Source: *Juniperus thurifera*

Mol. formula: $C_{20}H_{34}O_2$

Mol. wt.: 306

Solvent: $CDCl_3$

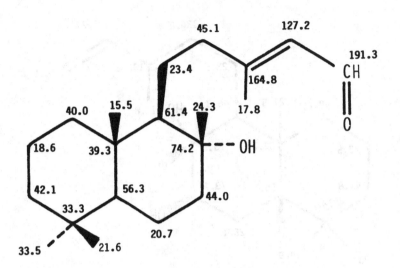

A. San Feliciano, M. Medarde, J.L. Lopez, J.M.M. Del Corral, P. Puebla and A.F. Barrero, <u>Phytochemistry</u>, 27(7), 2241 (1988).

8-HYDROXYLABD-13Z-EN-15-AL

Source: *Juniperus thurifera*

Mol. formula: $C_{20}H_{34}O_2$

Mol. wt.: 306

Solvent: $CDCl_3$

A. San Feliciano, M. Medarde, J.L. Lopez, J.M.M. Del Corral, P. Puebla and A.F. Barrero, <u>Phytochemistry</u>, 27(7), 2241 (1988).

2α-HYDROXYMANOYL OXIDE

Source: *Dacrydium colensoi*
Mol. formula: $C_{20}H_{34}O_2$
Mol. wt.: 306
Solvent: $CDCl_3$

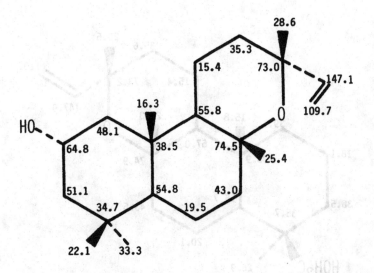

1. S.O. Almqvist, C.R. Enzell and F.W. Wehrli, <u>Acta. Chem. Scand.</u>, **29**, 695 (1975).

2. P.K. Grant, M.H.G. Munro and N.R. Hill, <u>J. Chem. Soc.</u>, 3846 (1965).

<u>ENT</u>-2α-HYDROXY-13-<u>EPI</u>-MANOYL OXIDE

Source: *Sideritis perfoliata* L.
Mol. formula: $C_{20}H_{34}O_2$
Mol. wt.: 306
Solvent: $CDCl_3$

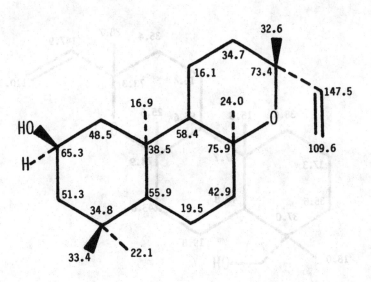

E. Sezik, N. Ezer, J.A. Hueso-Rodriguez and B. Rodriguez, <u>Phytochemistry</u>, **24** (11), 2739 (1985).

19-HYDROXYMANOYL OXIDE

Source: *Polemonium viscosum*

Mol. formula: $C_{20}H_{34}O_2$

Mol. wt.: 306

Solvent: $CDCl_3$

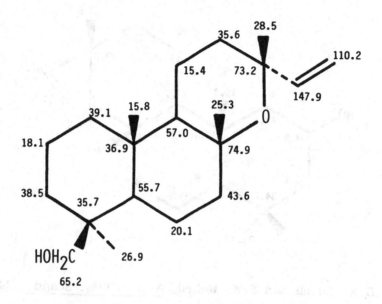

D.B. Stierle, A.A. Stierle and R.D. Larsen, <u>Phytochemistry</u>, 27(2), 517 (1988).

||||||||||||||||||||||||||||||||||||||||||||||||||||||||||||||||||||||||||||||||||||||||||||

JHANOL

Source: *Stevia rebaudiana*

Mol. formula: $C_{20}H_{34}O_2$

Mol. wt.: 306

Solvent: $CDCl_3$

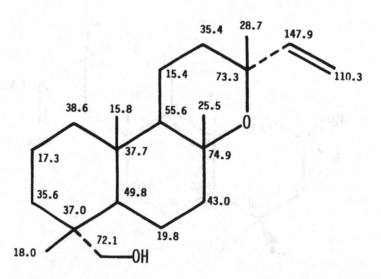

M. Sholichin, K. Yamasaki, R. Miyama, S. Yahara and O. Tanaka, <u>Phytochemistry</u>, 19(2), 326 (1980).

JUNGERMANOOL

Source: *Jungermannia torticalyx* Steph.
Mol. formula: $C_{20}H_{34}O_2$
Mol. wt.: 306
Solvent: $CDCl_3$

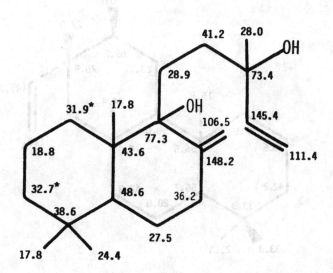

A. Matsuo, T. Nakamoto, M. Nakayama and S. Hayashi, Experientia, 32(8), 966 (1976).

RIBENOL

Source: *Sideritis varoi* Soc.
Mol. formula: $C_{20}H_{34}O_2$
Mol. wt.: 306
Solvent: $CDCl_3$

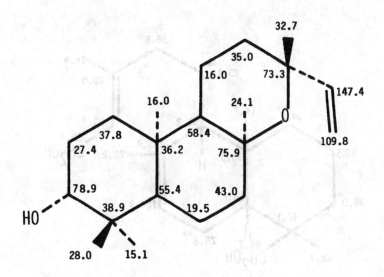

A. Garcia-Granados, A. Martinez, A. Molina, M.E. Onorato, M. Rico, A.S.De Buruaga and J.M.S.De Buruaga, Phytochemistry, 24(8), 1789 (1985).

VAROL

Source: *Sideritis varoi* Soc.

Mol. formula: $C_{20}H_{34}O_2$

Mol. wt.: 306

Solvent: $CDCl_3$

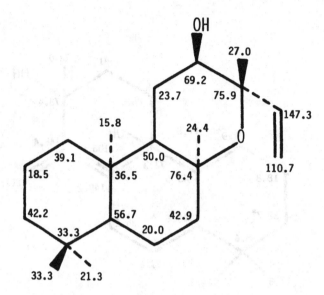

A. Garcia-Granados, A. Martinez, A. Molina, M.E. Onorato, M. Rico, A.S.De Buruaga and J.M.S.De Buruaga, <u>Phytochemistry</u>, **24**(8), 1789 (1985).

VILLENOL

Source: *Sideritis chamaedryfolia* Cav.

Mol. formula: $C_{20}H_{34}O_2$

Mol. wt.: 306

Solvent: $CDCl_3$

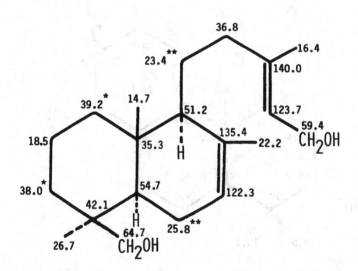

B. Rodriguez, <u>Phytochemistry</u>, **17**(2), 281 (1978).

DITERPENE BV-IIIa

Source: *Brickellia vernicosa* B.L. Robins.
Mol. formula: $C_{20}H_{36}O_2$
Mol. wt.: 308
Solvent: $CDCl_3$

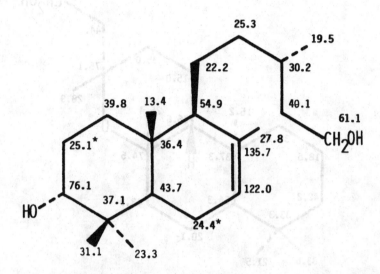

A.A. Ahmed, D.A. Gage, J.S. Calderon and T.J. Mabry, <u>Phytochemistry</u>, 25(6), 1385 (1986).

GOMEROL

Source: *Sideritis nutans* Svent.
Mol. formula: $C_{20}H_{36}O_2$
Mol. wt.: 308
Solvent: $CDCl_3$

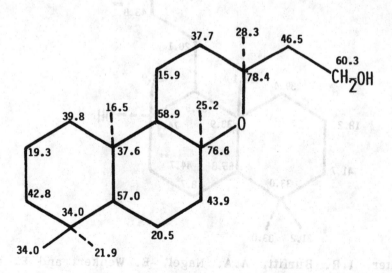

C. Fernandez, B.M. Fraga and M.G. Hernandez, <u>Phytochemistry</u>, 25 (12), 2825 (1986).

13-EPI-GOMEROL

Source: *Sideritis nutans* Svent.
Mol. formula: $C_{20}H_{36}O_2$
Mol. wt.: 308
Solvent: $CDCl_3$

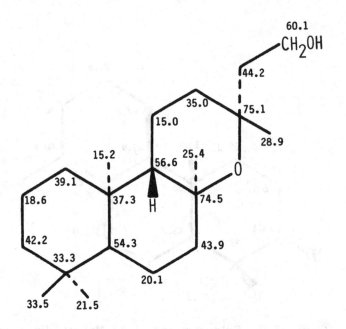

C. Fernandez, B.M. Fraga and M.G. Hernandez, <u>Phytochemistry</u>, 25 (12), 2825 (1986).

SCLAREOL

Source: *Salvia sclarea* L.
Mol. formula: $C_{20}H_{36}O_2$
Mol. wt.: 308
Solvent: $CDCl_3$

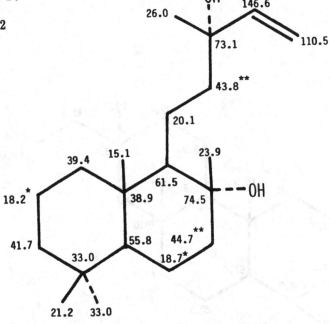

1. B.L. Buckwalter, I.R. Burfitt, A.A. Nagel, E. Wenkert and F. Naf, <u>Helv. Chim. Acta.</u>, 58(6), 1567 (1975).

2. L. Ruzicka and M.M. Janot, <u>Helv. Chim. Acta.</u>, 14, 645 (1931).

STEREBIN A

Source: *Stevia rebaudiana* Bertoni
Mol. formula: $C_{18}H_{30}O_4$
Mol. wt.: 310
Solvent: C_5D_5N

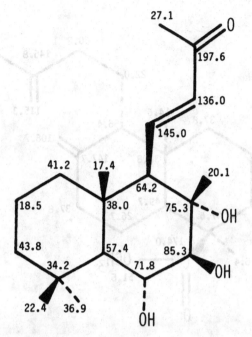

Y. Oshima, J.-I. Saito and H. Hikino, <u>Tetrahedron</u>, 42 (23), 6443 (1986).

||||||||||||||||||||||||||||||||||||||||||||||||||||||||||||||||||||||||||||||||||||||||||||||||||||

POTAMOGETONIN

Source: *Potamogeton ferrugineus* Hagstr.
Mol. formula: $C_{20}H_{26}O_3$
Mol. wt.: 314
Solvent: $CDCl_3$

C.R. Smith, Jr., R.V. Madrigal, D. Weisleder, K.L. Mikolajczak and R.J. Highet, <u>J. Org. Chem.</u>, 41(4), 593 (1976).

DITERPENE XA-I

Source: *Xylopia aromatica* Lam. (Mart.)
Mol. formula: $C_{21}H_{32}O_2$
Mol. wt.: 316
Solvent: $CDCl_3$

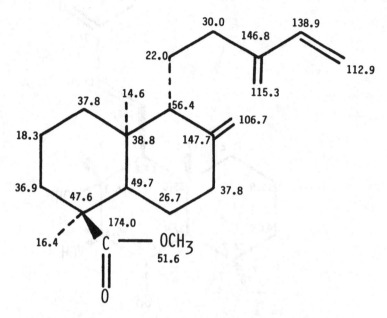

M.P.L. Moraes and N.F. Roque, Phytochemistry, 27(10), 3205 (1988).

||||||||||||||||||||||||||||||||||||||||||||||||||||||||||||||||||||||||||||||||||||||||||||||||||

METHYL COMMUNATE

Source: *Metasequoia glyptostroboides* Hu et Cheng
Mol. formula: $C_{21}H_{32}O_2$
Mol. wt.: 316
Solvent: $CDCl_3$

S. Braun and H. Breitenbach, Tetrahedron, 33(1), 145 (1977).

ANDROGRAPANIN

Source: *Andrographis paniculata* Nees

Mol. formula: $C_{20}H_{30}O_3$

Mol. wt.: 318

Solvent: $CDCl_3$

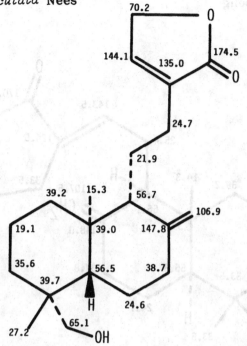

T. Fujita, R. Fujitani, Y. Takeda, Y. Takaishi, T. Yamada, M. Kido and I. Miura, Chem. Pharm. Bull , (Tokyo), 32(6), 2117 (1984).

|||||||||||||||||||||||||||||||||||||||||||||||||||||||||||||||||||||||||||||||||||||||||||||||||||||||||||||||||

CORONARIN C

Source: *Hedychium coronarium* Koeng

Mol. formula: $C_{20}H_{30}O_3$

Mol. wt.: 318

Solvent: $CDCl_3$

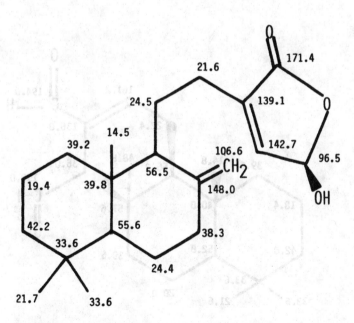

H. Itokawa, H. Morita, I. Katou, K. Takeya, A.J. Cavalheiro, R.C.B. De Oliveira, M. Ishige and M. Motidome, Planta Med., 54(4), 311 (1988).

Source: *Hedychium coronarium* Koeng

Mol. formula: $C_{20}H_{30}O_3$

Mol. wt.: 318

Solvent: $CDCl_3$

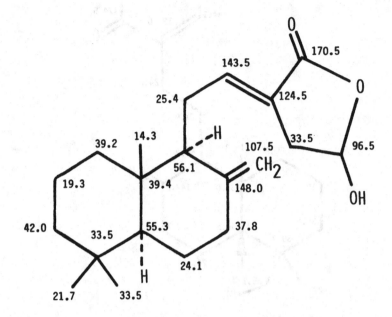

H. Itokawa, H. Morita, I. Katou, K. Takeya, A. J. Cavalheiro, R.C.B. De Oliveira, M. Ishige and M. Motidome, <u>Planta Med.</u>, **54**(4), 311 (1988).

‖‖‖

(<u>E</u>)-8β,17-EPOXYLABD-12-ENE-15,16-DIAL

Source: *Afromomum daniellii*

Mol. formula: $C_{20}H_{30}O_3$

Mol. wt.: 318

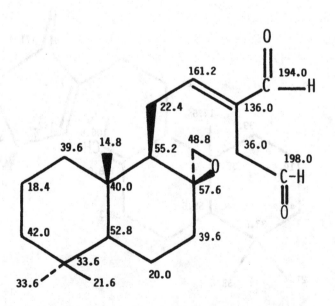

S.F. Kimbu, T.K. Njimi, B.L. Sondengam, J.A. Akinniyi and J.D. Connolly, <u>J. Chem. Soc. Perkin Trans, I</u>, (5), 1303 (1979).

GALANOLACTONE

Source: *Alpinia galanga* (L.) Willd.

Mol. formula: $C_{20}H_{30}O_3$

Mol. wt.: 318

Solvent: $CDCl_3$

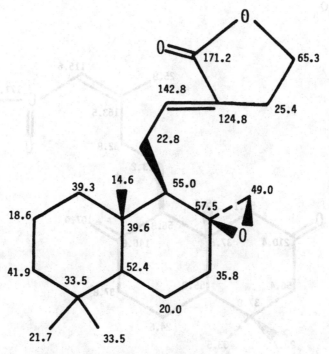

H. Morita and H.Itokawa, <u>Planta Med.</u>, 54(2), 117 (1988).

HISPANOLONE

Source: *Galeopsis angustifolia* Hoffm.

Mol. formula: $C_{20}H_{30}O_3$

Mol. wt.: 318

Solvent: $CDCl_3$

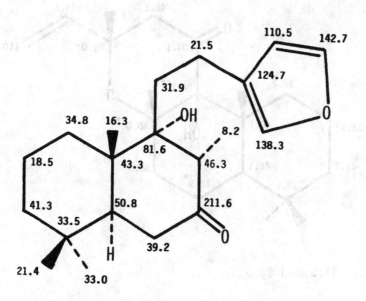

B. Rodriguez and G. Savona, <u>Phytochemistry</u>, 19(8), 1805 (1980).

61

Source: *Nolana rostrata* (Lindley) Mier
Mol. formula: $C_{20}H_{30}O_3$
Mol. wt.: 318
Solvent: $CDCl_3$

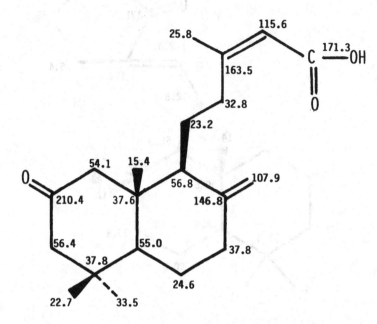

J.A. Garbarino, M.C. Chamy and V. Gambaro, Phytochemistry, 25 (12), 2833 (1986).

||||||||||||||||||||||||||||||||||||||||||||||||||||||||||||||||||||||||||||||||||||||

COLEOL

Source: *Coleus forskohlii* Briq.
Mol. formula: $C_{20}H_{32}O_3$
Mol. wt.: 320
Solvent: $CDCl_3$

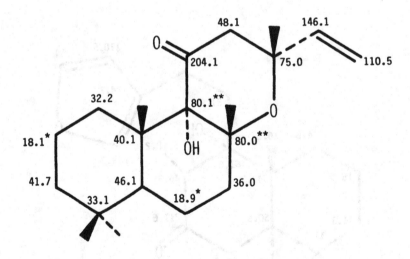

CH_3: 16.6, 21.6, 25.9, 31.2 and 33.5

B. Gabetta, G. Zini and B. Danieli, Phytochemistry, 28 (3), 859 (1989).

HAVARDIC ACID F (METHYL ESTER)

Source: *Grindelia havardii* Steyerm.

Mol. formula: $C_{20}H_{32}O_3$

Mol. wt.: 320

Solvent: $CDCl_3$

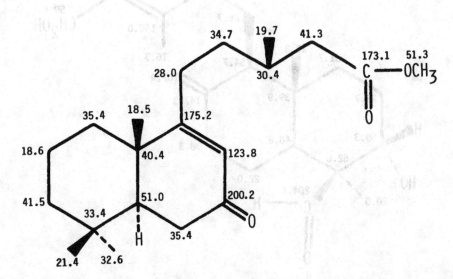

S.D. Jolad, B.N. Timmermann, J.J. Hoffmann, R.B. Bates and T.J. Siahaan, <u>Phytochemistry</u>, 26(2), 483 (1987).

||||||||||||||||||||||||||||||||||||||||||||||||||||||||||||||||||||||||||||||||||||||||||||||||||||||||||||||||||||

6β-HYDROXY-8,13-EPOXY-LABD-14-EN-11-ONE

Source: *Coleus forskohlii* Briq.

Mol. formula: $C_{20}H_{32}O_3$

Mol. wt.: 320

Solvent: $CDCl_3$

CH_3: 16.8, 23.7, 23.9, 31.4 and 33.2

B. Gabetta, G. Zini and B. Danieli, <u>Phytochemistry</u>, 28 (3), 859 (1989).

3α-HYDROXY-ISOAGATHOLAL

Source: *Juniperus thurifera*

Mol. formula: $C_{20}H_{32}O_3$

Mol. wt.: 320

Solvent: $CDCl_3$

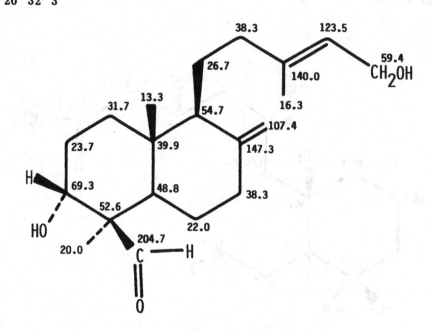

A. San Feliciano, M. Medarde, J.L. Lopez, J.M.M. Del Corral, P. Puebla and A.F. Barrero, <u>Phytochemistry</u>, 27(7), 2241 (1988).

<u>ENT-8α-HYDROXYLABDA-13(16),14-DIEN-18-OIC ACID</u>

Source: *Sideritis varoi* Soc.

Mol. formula: $C_{20}H_{32}O_3$

Mol. wt.: 320

Solvent: $CDCl_3$

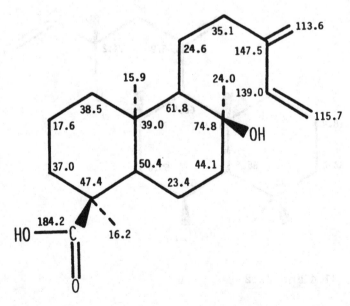

A. Garcia-Granados, A. Martinez, A. Molina, M.E. Onorato, M. Rico, A.S.De Buruaga and J.M.S.De Buruaga, <u>Phytochemistry</u>, 24(8), 1789 (1985).

18-HYDROXY-7-OXO-9β,13β-EPOXY-ENT-LABD-14-ENE

Source: *Austrobrickellia patens* (D. Don. ex. H. et A.) K. et R.

Mol. formula: $C_{20}H_{32}O_3$

Mol. wt.: 320

Solvent: $CDCl_3$

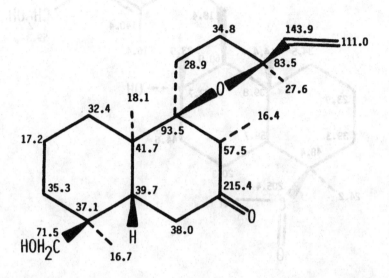

J. Jakupovic, E. Ellmauerer, F. Bohlmann, R.M. King and H. Robinson, Phytochemistry, 25 (8) 1927 (1986).

|||||||||||||||||||||||||||||||||||||||||||||||||||||||||||||||||||||||||||||||||||||||||||||||||||||||||||||

AUSTROINULIN

Source: *Stevia rebaudiana* Bertoni

Mol. formula: $C_{20}H_{34}O_3$

Mol. wt.: 322

Solvent: $CDCl_3$

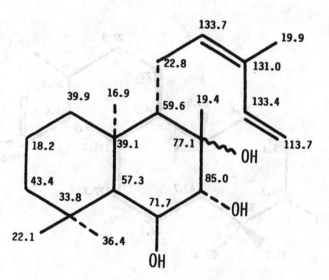

M. Sholichin, K. Yamasaki, R. Miyama, S. Yahara and O. Tanaka, Phytochemistry, 19(2), 326 (1980).

Source: *Juniperus thurifera*

Mol. formula: $C_{20}H_{34}O_3$

Mol. wt.: 322

Solvent: $CDCl_3$

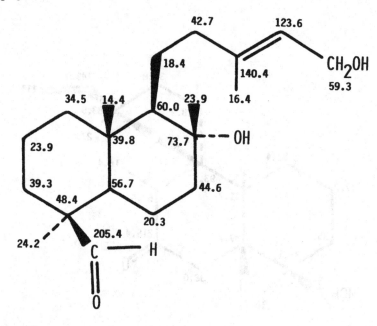

A. San Feliciano, M. Medarde, J.L. Lopez, J.M.M. Del Corral, P. Puebla and A.F. Barrero, <u>Phytochemistry</u>, 27(7), 2241 (1988).

3α-HYDROXYCATIVIC ACID

Source: *Brickellia varnicosa* B.L. Robins.

Mol. formula: $C_{20}H_{34}O_3$

Mol. wt.: 322

Solvent: $CDCl_3$

A.A. Ahmed, D.A. Gage, J.S. Calderon and T.J. Mabry, <u>Phytochemistry</u>, 25(6), 1385 (1986).

15-HYDROXY-7-LABDEN-17-OIC ACID

Source: *Halimium viscosum*
Mol. formula: $C_{20}H_{34}O_3$
Mol. wt.: 322
Solvent: $CDCl_3$

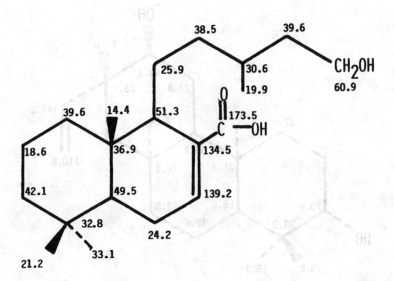

J.G. Urones, I.S. Marcos, D.D. Martin, F.M.S.B. Palma and J.M. Rodilla, <u>Phytochemistry</u>, **26**(11), 3037 (1987).

ENT-LABDA-13-EN-8β-OIC ACID

Source: *Hymeneae coubaril* L.
Mol. formula: $C_{20}H_{34}O_3$
Mol. wt.: 322
Solvent: $CDCl_3$

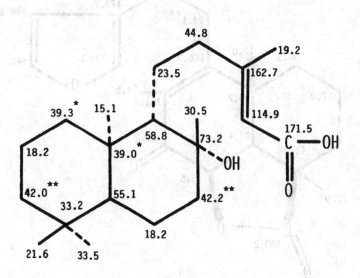

P.M. Imamura, A.J. Marsaioli, L.E.S. Barata and E.A. Ruveda, <u>Phytochemistry</u>, **16**(11), 1842 (1977).

VARODIOL

Source: *Sideritis varoi* Soc.

Mol. formula: $C_{20}H_{34}O_3$

Mol. wt.: 322

Solvent: $CDCl_3$

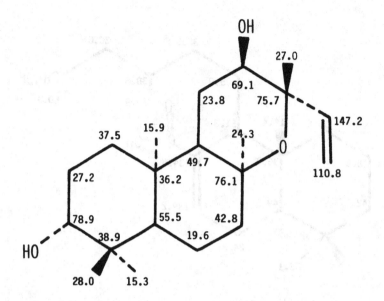

A. Garcia-Granados, A. Martinez, A. Molina, M.E. Onorato, M. Rico, A.S.De Buruaga and J.M.De Buruaga, <u>Phytochemistry</u>, **24**(8), 1789 (1985).

||||||||||||||||||||||||||||||||||||||||||||||||||||||||||||||||||||||||||||||||||||||||||||||||||||||||||||||||||||||||||||||||||||||||||

BALLONIGRIN

Source: *Ballota rupestris* Vis.

Mol. formula: $C_{20}H_{24}O_4$

Mol. wt.: 328

Solvent: $CDCl_3$

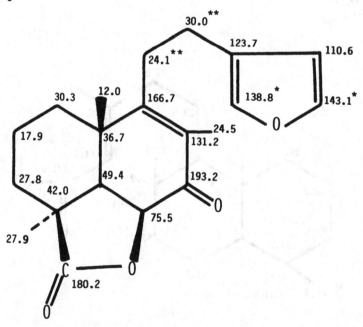

G. Savona, F. Piozzi, J.R. Hanson and M. Siverns, <u>J. Chem. Soc. Perkin Trans. I</u>, (3), 322 (1977).

SCIADIN

Source: *Sciadopitys verticillata* Sieb. et Zucc.

Mol. formula: $C_{20}H_{24}O_4$

Mol. wt.: 328

Solvent: $CDCl_3$

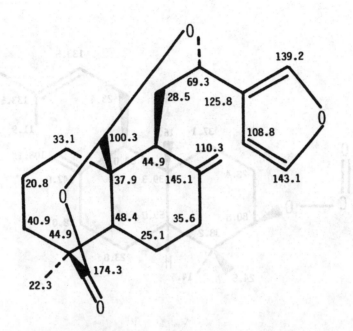

S. Hasegawa and Y. Hirose, <u>Phytochemistry</u>, **24(9)**, 2041 (1985).

SCIADINONE

Source: *Sciadopitys verticillata* Sieb. et Zucc.

Mol. formula: $C_{20}H_{24}O_4$

Mol. wt.: 328

Solvent: $CDCl_3$

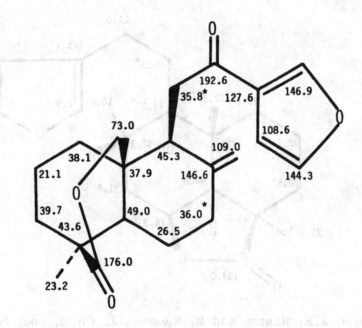

S. Hasegawa and Y. Hirose, <u>Phytochemistry</u>, **24(9)**, 2041 (1985).

Source: *Palafoxia rosea* (Bush.) Cory

Mol. formula: $C_{22}H_{34}O_2$

Mol. wt.: 330

Solvent: $CDCl_3$

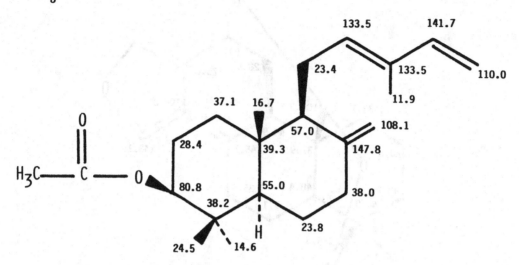

F. Bohlmann and H. Czerson, <u>Phytochemistry</u>, 18 (1), 115 (1979).

|||||||||||||||||||||||||||||||||||||||||||||||||||||||||||||||||||||||||||||||||||||||||||||||||||||||||||||||||||||

MARRUBIIN

Source: *Ballota nigra*

Mol. formula: $C_{20}H_{28}O_4$

Mol. wt.: 332

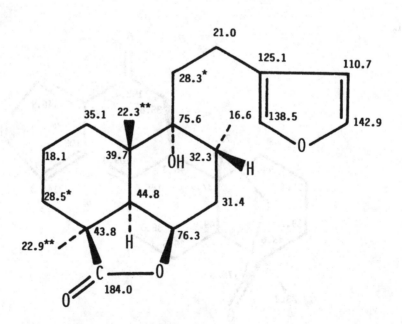

G. Savona, F. Piozzi, J.R. Hanson and M. Siverns, <u>J. Chem. Soc. Perkin Trans.I</u>, (15), 1607 (1976).

DITERPENE GC-XIb (METHYL ESTER)

Source: *Grindelia camporum* Greene and *Chrysothamnus paniculatus* (Gray) Hall

Mol. formula: $C_{21}H_{32}O_3$

Mol. wt.: 332

Solvent: $CDCl_3$

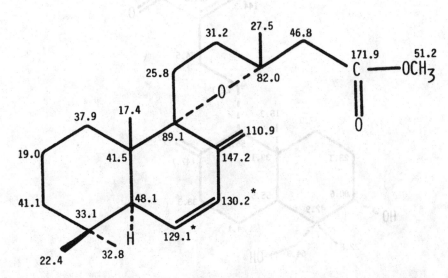

B.N. Timmermann, D.J. Luzbetak, J.J. Hoffmann, S.D. Jolad, K.H. Schram, R.B. Bates and R.E. Klenck, Phytochemistry, 22(2), 523 (1983).

CORONARIN B

Source: *Hedychium coronarium* Koeng

Mol. formula: $C_{20}H_{30}O_4$

Mol. wt.: 334

Solvent: $CDCl_3$

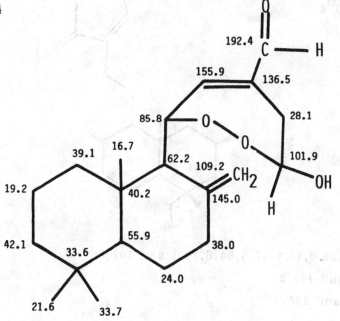

H. Itokawa, H. Morita, I. Katou, K. Takeya, A.J. Cavalheiro, R.C.B. De Oliveira, M. Ishige and M. Motidome, Planta Med., 54(4), 311 (1988).

14-DEOXYANDROGRAPHOLIDE

Source: *Andrographis paniculata* Nees

Mol. formula: $C_{20}H_{30}O_4$

Mol. wt.: 334

Solvent: $CDCl_3$

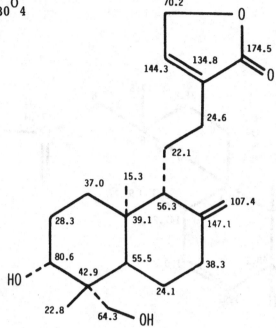

T. Fujita, R. Fujitani, Y. Takeda, Y. Takaishi, T. Yamada, M. Kido and I. Miura, Chem. Pharm. Bull. (Tokyo), 32 (6), 2117 (1984).

PHLOGANTHOLIDE A

Source: *Phlogacanthus thyrsiflorus* Nees

Mol. formula: $C_{20}H_{30}O_4$

Mol. wt.: 334

Solvent: $CDCl_3$

CH_3: 16.1 and 27.2

CH_2: 21.7, 23.6, 24.3, 38.0, 43.8, 47.5, 64.6, 70.1 and 107.5

CH: 55.3, 56.2, 64.5 and 144.3

C: 40.3, 40.6, 134.3 and 146.5

C=O: 174.4

A.K. Barua, M.K. Chowdhury, S. Biswas, C.D. Gupta, S.K. Banerjee, S.K. Saha, A. Patra and A.K. Mitra, Phytochemistry, 24(9), 2037 (1985).

1,9-DIDEOXY-7-DEACETYLFORSKOLIN

Source: *Coleus forskohlii* Briq.

Mol. formula: $C_{20}H_{32}O_4$

Mol. wt.: 336

Solvent: $CDCl_3$

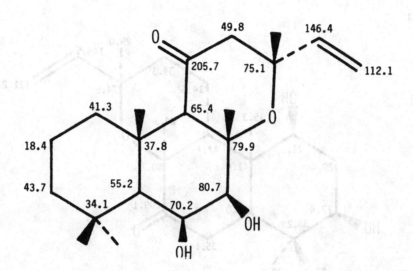

CH_3: 16.9, 23.7, 23.9, 31.4 and 33.2

B. Gabetta, G. Zini and B. Danieli, <u>Phytochemistry</u>, 28 (3), 859 (1989).

2β,3β-DIHYDROXYLABDA-8(17),13Z-DIEN-15-OIC ACID

Source: *Nolana rostrata* (Lindley) Mier

Mol. formula: $C_{20}H_{32}O_4$

Mol. wt.: 336

Solvent: $CDCl_3$

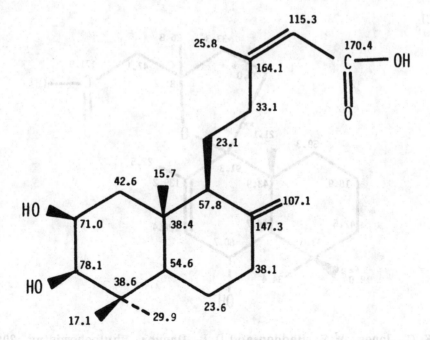

J.A. Garbarino, M.C. Chamy and V. Gambaro, <u>Phytochemistry</u>, 25 (12), 2833 (1986).

Source: *Helichrysum ambiguum* Turcz.

Mol. formula: $C_{20}H_{32}O_4$

Mol. wt.: 336

Solvent: $CDCl_3$

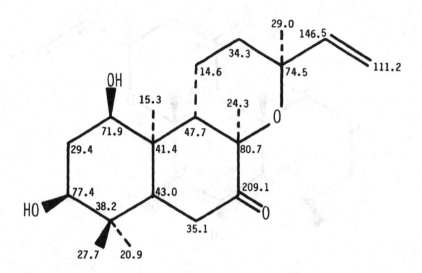

J. Jakupovic, A. Schuster, F. Bohlmann, U. Ganzer, R.M. King and H. Robinson, <u>Phytochemistry</u>, 28(2), 543 (1989).

6α-HYDROXYGRINDELIC ACID

Source: *Grindelia humilis* (Hook. and Arn.)

Mol. formula: $C_{20}H_{32}O_4$

Mol. wt.: 336

Solvent: $CDCl_3$

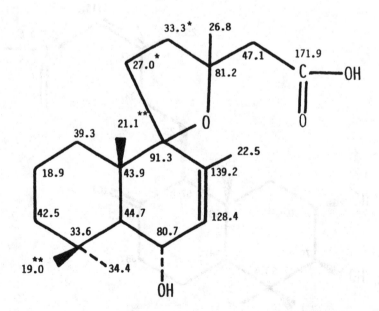

A.F. Rose, K.C. Jones, W.F. Haddon and D.L. Dreyer, <u>Phytochemistry</u>, 20(9), 2249 (1981).

PINIFOLIC ACID

Source: *Pinus sylvestris* L.

Mol. formula: $C_{20}H_{32}O_4$

Mol. wt.: 336

Solvent: CD_3OD

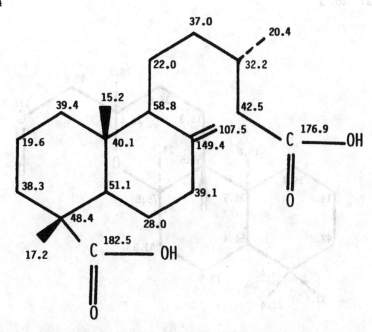

I.I. Bardyshev, A.S. Degtyarenko, T.I. Pekhk and S.A. Makhnach, <u>chem. Nat. Compds.</u>, **17** (5), 408 (1981); <u>Khim. Prir. Soedin.</u>, **17** (5), 568 (1981).

‖‖

DISCOIDIC ACID (METHYL ESTER)

Source: *Grindelia discoidea* Hook and Arn

Mol. formula: $C_{21}H_{36}O_3$

Mol. wt.: 336

Solvent: $CDCl_3$

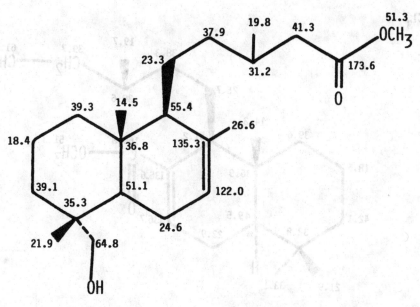

B.N. Timmermann, J.J. Hoffmann, S.D. Jolad, R.B. Bates and T.J. Siahaan, <u>Phytochemistry</u>, **25** (3), 723 (1986).

GOMERIC ACID (METHYL ESTER)

Source: *Sideritis nutans* Svent.

Mol. formula: $C_{21}H_{36}O_3$

Mol. wt.: 336

Solvent: $CDCl_3$

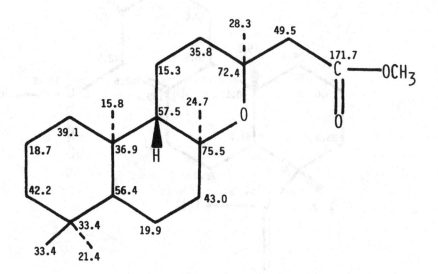

C. Fernandez, B.M. Fraga and M.G. Hernandez, <u>Phytochemistry</u>, 25 (12), 2825 (1986).

HAVARDIC ACID B (METHYL ESTER)

Source: *Grindelia havardii* Steyerm.

Mol. formula: $C_{21}H_{36}O_3$

Mol. wt.: 336

Solvent: $CDCl_3$

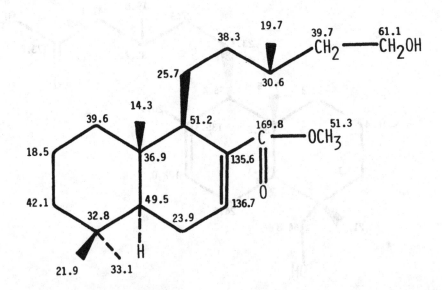

S.D. Jolad, B.N. Timmermann, J.J. Hoffmann, R.B. Bates and T.J. Siahaan, <u>Phytochemistry</u>, 26(2), 483 (1987).

3β-HYDROXYLABD-7-EN-15-OIC ACID (METHYL ESTER)

Source: *Grindelia discoidea* Hook and Arn.

Mol. formula: $C_{21}H_{36}O_3$

Mol. wt.: 336

Solvent: $CDCl_3$

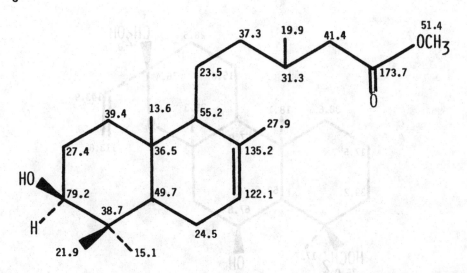

B.N. Timmermann, J.J. Hoffmann, S.D. Jolad, R.B. Bates and T.J. Siahaan, <u>Phyto-</u>chemistry, **25** (3), 723 (1986).

||||||||||||||||||||||||||||||||||||||||||||||||||||||||||||||||||||||||||||||||||||||||||||||||||||||||

4β-HYDROXYMETHYLLABD-7-EN-15-OIC ACID (METHYL ESTER)

Source: *Grindelia discoidea* Hook and Arn

Mol. formula: $C_{21}H_{36}O_3$

Mol. wt.: 336

Solvent: $CDCl_3$

B.M. Timmermann, J.J. Hoffmann, S.D. Jolad, R.B. Bates and T.J. Siahaan, <u>Phytochemistry</u>, **25** (3), 723 (1986).

Source: *Sideritis arborescens* Salzm. ex Bentham

Mol. formula: $C_{20}H_{34}O_4$

Mol. wt.: 338

Solvent: $CDCl_3$

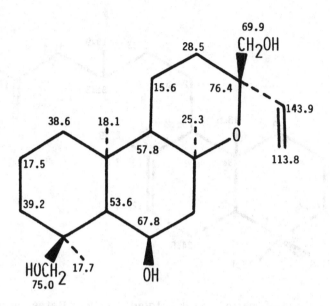

A. Garcia-Granados, A. Martinez and M.E. Onorato, <u>Phytochemistry</u>, 24(3), 517 (1985).

STEREBIN E

Source: *Stevia rebaudiana* Bertoni

Mol. formula: $C_{20}H_{34}O_4$

Mol. wt.: 338

Solvent: C_5D_5N

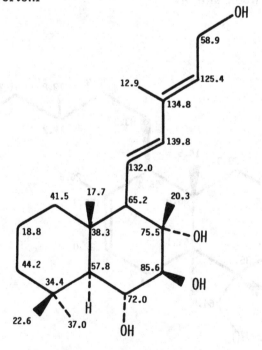

Y. Oshima, J.-I. Saito and H. Hikino, <u>Phytochemistry</u>, 27(2), 624 (1988).

STEREBIN F

Source: *Stevia rebaudiana* Bertoni

Mol. formula: $C_{20}H_{34}O_4$

Mol. wt.: 338

Solvent: C_5D_5N

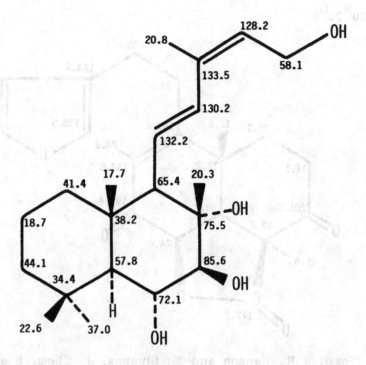

Y. Oshima, J.-I. Saito and H. Hikino, <u>Phytochemistry</u>, 27(2), 624 (1988).

14,15,19-TRIHYDROXY-13-<u>EPI-ENT</u>-MANOYLOXIDE

Source: *Florestina tripteris* DC.

Mol. formula: $C_{20}H_{36}O_4$

Mol. wt.: 340

Solvent: $CDCl_3$

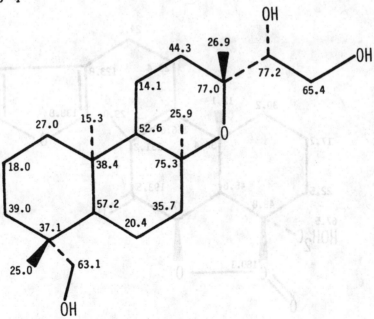

X.A. Dominguez, H. Sanchez, J. Slim, J. Jakupovic, T.V. Chau-Thi and F. Bohlmann, <u>Phytochemistry</u>, **27** (2), 613 (1988).

BALLONIGRINONE

Source: *Ballota rupestris* Vis.

Mol. formula: $C_{20}H_{22}O_5$

Mol. wt.: 342

Solvent: $CDCl_3$

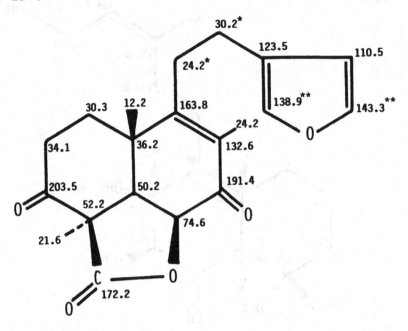

G. Savona, F. Piozzi, J.R. Hanson and M. Siverns, <u>J. Chem. Soc. Perkin Trans. I</u>, (3), 322 (1977).

18-HYDROXYBALLONIGRIN

Source: *Ballota acetobulosa*

Mol. formula: $C_{20}H_{24}O_5$

Mol. wt.: 344

Solvent: $CDCl_3$

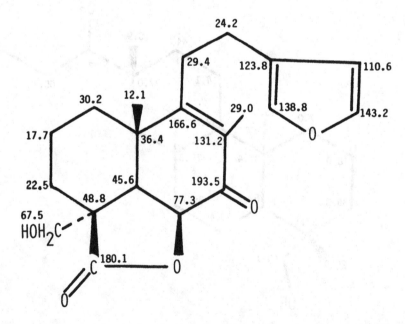

G. Savona, F. Piozzi, J.R. Hanson and M. Siverns, <u>J. Chem. Soc. Perkin Trans. I</u>, (10), 1271 (1978).

METHYL-12-OXO-LAMBERTIANATE

Source: *Sciadopitys verticillata* Sieb. et. Zucc.

Mol. formula: $C_{21}H_{28}O_4$

Mol. wt.: 344

Solvent: $CDCl_3$

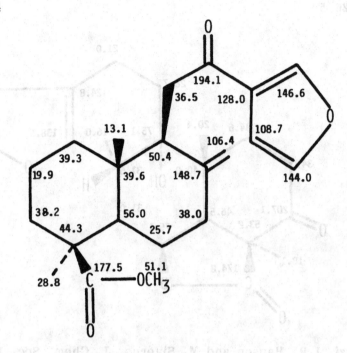

S. Hasegawa and Y. Hirose, Phytochemistry, 24(9), 2041 (1985).

||||||||||||||||||||||||||||||||||||||||||||||||||||||||||||||||||||||||||||||||||||||||||||||||||||||||||||||||||||||||||||||||||||||||||||

BALLOTINONE

Source: *Ballota nigra*

Mol. formula: $C_{20}H_{26}O_5$

Mol. wt.: 346

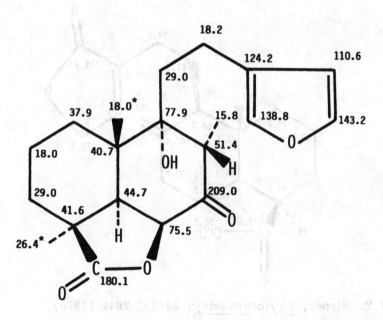

G. Savona, F. Piozzi, J.R. Hanson and M. Siverns, J. Chem. Soc. Perkin Trans. I, (15), 1607 (1976).

81

PEREGRINONE

Source: *Ballota nigra*

Mol. formula: $C_{20}H_{26}O_5$

Mol. wt.: 346

Solvent: $CDCl_3$

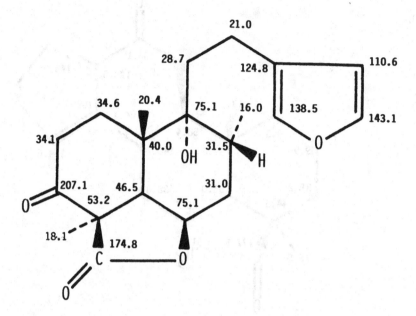

G. Savona, F. Piozzi, J.R. Hanson and M. Siverns, J. Chem. Soc. Perkin Trans. I, (15), 1607 (1976).

||||||||||||||||||||||||||||||||||||||||||||||||||||||||||||||||||||||||||||||||||||||||||||||||||||||||||||

PINUSOLIDE

Source: *Pinus koraensis* Sieb. et. Zucc. and *P. sibirica* R. Mayr.

Mol. formula: $C_{21}H_{30}O_4$

Mol. wt.: 346

Solvent: $CDCl_3$

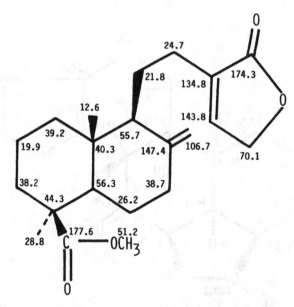

S. Hasegawa and Y. Hirose, Phytochemistry, 24(9), 2041 (1985).

V.A. Raldugin, A.I. Lisina, N.K. Kashtanova and V.A. Pentegova, Chem. Nat. Compds., 6(5), 559 (1970); Khim. Prir. Soedin, 6(5), 541 (1970).

ENT-6β-ACETOXY-17-HYDROXY-LABDA-7,12E,14-TRIENE

Source: *Rutidosis murchisonii* F. Muell.

Mol. formula: $C_{22}H_{34}O_3$

Mol. wt.: 346

Solvent: $CDCl_3$

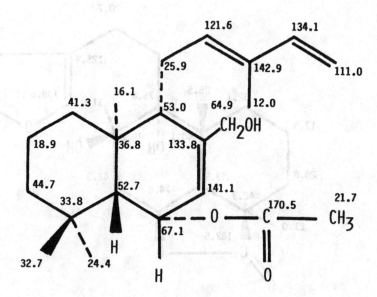

C. Zdero, F. Bohlmann, R.M. King and H. Robinson, <u>Phytochemistry</u>, **26** (6), 1759 (1987).

||||||||||||||||||||||||||||||||||||||||||||||||||||||||||||||||||||||||||||||||||||||||||||||||||||||||||||||||||||||

BALLOTINOL

Source: *Ballota nigra*

Mol. formula: $C_{20}H_{28}O_5$

Mol. wt.: 348

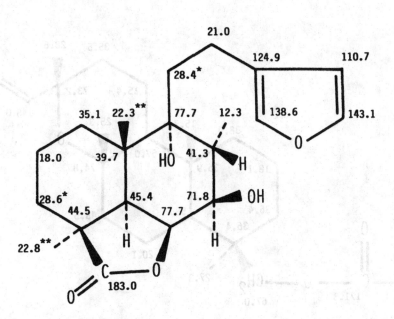

G. Savona, F. Piozzi, J.R. Hanson and M. Siverns, <u>J. Chem. Soc. Perkin Trans. I</u>,
(15), 1607 (1976).

LEONOTIN

Source: *Ballota nigra*
Mol. formula: $C_{20}H_{28}O_5$
Mol. wt.: 348
Solvent: $CDCl_3$

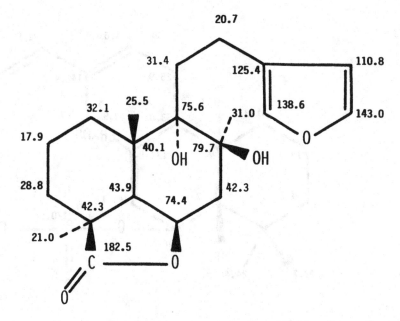

G. Savona, F. Piozzi, J.R. Hanson and M. Siverns, J. Chem. Soc. Perkin Trans. I, (15), 1607 (1976).

||||||||||||||||||||||||||||||||||||||||||||||||||||||||||||||||||||||||||||||||||||||||||||||||||||||||||

19-ACETOXYMANOYL OXIDE

Source: *Polemonium viscosum*
Mol. formula: $C_{22}H_{36}O_3$
Mol. wt.: 348
Solvent: $CDCl_3$

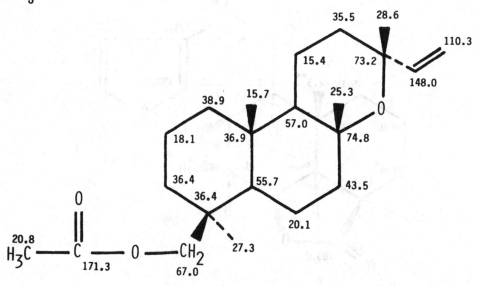

D.B. Stierle, A.A. Stierle and R.D. Larsen, Phytochemistry, 27(2), 517 (1988).

JHANOL ACETATE

Source: *Eupatorium jhanii* Robinson

Mol. formula: $C_{22}H_{36}O_3$

Mol. wt.: 348

Solvent: $CDCl_3$

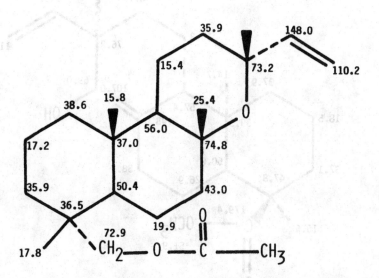

A.G. Gonzalez, J.M. Arteaga, J.L. Breton and B.M. Fraga, Phytochemistry, 16(1), 107 (1977).

||||||||||||||||||||||||||||||||||||||||||||||||||||||||||||||||||||||||||||||||||||||||||||||||||||||||||||||||||||||||||||||||||||||||||||||||||||||||||||||

ANDROGRAPHOLIDE

Source: *Andrographis paniculata* Nees

Mol. formula: $C_{20}H_{30}O_5$

Mol. wt.: 350

Solvent: C_5D_5N

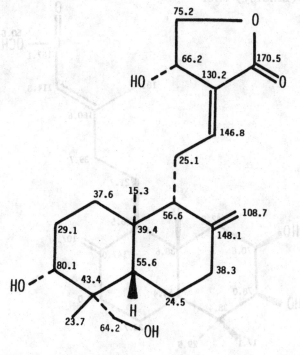

T.Fujita, R.Fujitani, Y.Takeda, Y.Takaishi, T.Yamada, M.Kido and I.Miura, Chem.Pharm. Bull. (Tokyo),32(6), 2117 (1984).

ENT-LAB-8(17),14-DIEN-13,16-DIOL-18-OIC ACID METHYL ESTER

Source: *Xylopia aromatica* Lam. (Mart.)
Mol. formula: $C_{21}H_{34}O_4$
Mol. wt.: 350
Solvent: $CDCl_3$

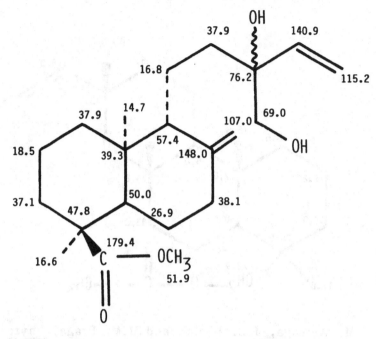

M.P.L. Moraes and N.F. Roque, <u>Phytochemistry</u>, 27(10), 3205 (1988).

METHYL-2β,3β-DIHYDROXYLABDA-8(17),13E-DIEN-15-OATE

Source: *Nolana rostrata* (Lindley) Mier
Mol. formula: $C_{21}H_{34}O_4$
Mol. wt.: 350
Solvent: $CDCl_3$

J.A. Garbarino, M.C. Chamy and V. Gambaro, <u>Phytochemistry</u>, 25 (12), 2833 (1986).

METHYL 15-METHOXY-7-LABDEN-17-OATE

Source: *Halimium verticillatum*

Mol. formula: $C_{22}H_{38}O_3$

Mol. wt.: 350

Solvent: $CDCl_3$

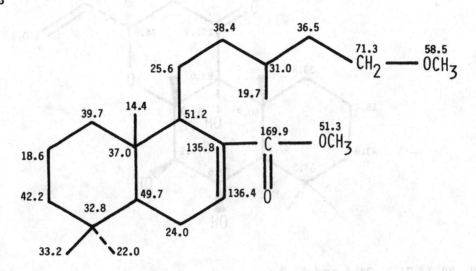

J.G. Urones, I.S. Marcos, D.D. Martin, M.C.A. Alonso, F.M.S.B. Palma and J.M.L. Rodilla, <u>Phytochemistry</u>, **28** (2), 557 (1989).

||||||||||||||||||||||||||||||||||||||||||||||||||||||||||||||||||||||||||||||||||||||||||||||||||||||||||||

STEREBIN B

Source: *Stevia rebaudiana* Bertoni

Mol. formula: $C_{20}H_{32}O_5$

Mol. wt.: 352

Solvent: $CDCl_3$

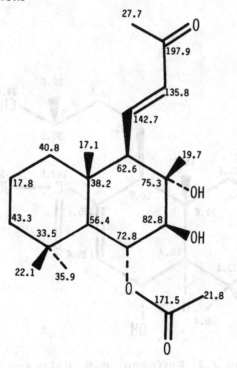

Y. Oshima, J.-I. Saito and H. Hikino, <u>Tetrahedron</u>, **42** (23), 6443 (1986).

Source: *Coleus forskohlii* Briq.

Mol. formula: $C_{20}H_{32}O_5$

Mol. wt.: 352

Solvent: $CDCl_3$

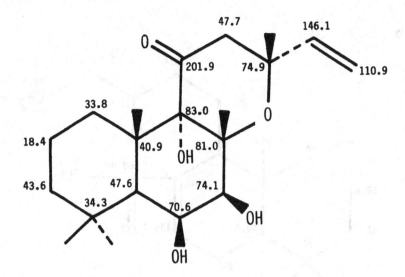

CH_3: 18.2, 22.2, 24.3, 31.5 and 33.3

B. Gabetta, G. Zini and B. Danieli, <u>Phytochemistry</u>, 28 (3), 859 (1989).

||||||||||||||||||||||||||||||||||||||||||||||||||||||||||||||||||||||||||||||||||||||||||||||||||||||||||||||||||

HAVARDIC ACID C (METHYL ESTER)

Source: *Grindelia havardii* Steyerm.

Mol. formula: $C_{21}H_{36}O_4$

Mol. wt.: 352

Solvent: $CDCl_3$

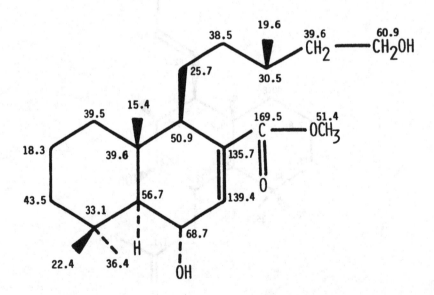

S.D. Jolad, B.N. Timmermann, J.J. Hoffmann, R.B. Bates and T.J. Siahaan, <u>Phytochemistry</u>, 26(2), 483 (1987).

3α-HYDROXYGOMERIC ACID (METHYL ESTER)

Source: *Sideritis nutans* Svent.
Mol. formula: $C_{21}H_{36}O_4$
Mol. wt.: 352
Solvent: $CDCl_3$

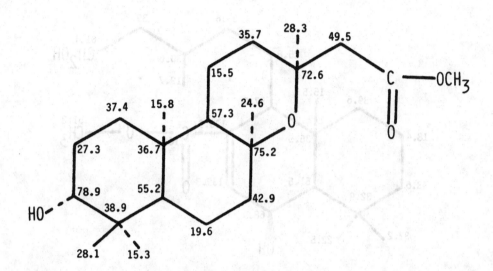

C. Fernandez, B.M. Fraga and M.G. Hernandez, <u>Phytochemistry</u>, 25 (12), 2825 (1986).

LAURIFOLIC ACID (METHYL ESTER)

Source: *Cistus laurifolius*
Mol. formula: $C_{21}H_{36}O_4$
Mol. wt.: 352
Solvent: $CDCl_3$

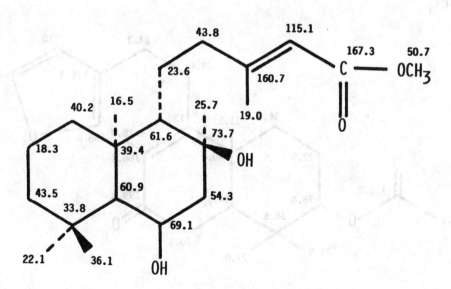

J. De P. Teresa, J.G. Urones, I.S. Marcos, P.B. Barcala and N.M. Garrido, <u>Phytochemistry</u>, 25 (5), 1185 (1986).

METHYL-6β,15-DIHYDROXY-7-LABDEN-17-OATE

Source: *Halimium verticillatum*

Mol. formula: $C_{21}H_{36}O_4$

Mol. wt.: 352

Solvent: $CDCl_3$

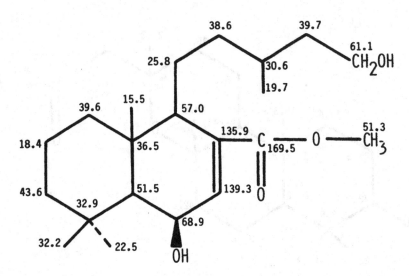

J.G. Urones, I.S. Marcos, D.D. Martin, M.C.A. Alonso, F.M.S.B. Palma and J.M.L. Rodilla, Phytochemistry, **28** (2), 557 (1989).

||||||||||||||||||||||||||||||||||||||||||||||||||||||||||||||||||||||||||||||||||||

CALYENONE

Source: *Roylea calycina* (Roxb.) Briq.

Mol. formula: $C_{22}H_{30}O_4$

Mol. wt.: 358

Solvent: $CDCl_3$

Om Prakash, D.S. Bhakuni, R.S. Kapil, G.S.R. Subba Rao and B. Ravindranath, J. Chem. Soc. Perkin Trans. I, (5), 1305 (1979).

RUPESTRALIC ACID

Source: *Ballota rupestris*

Mol. formula: $C_{20}H_{24}O_6$

Mol. wt.: 360

Solvent: C_5D_5N

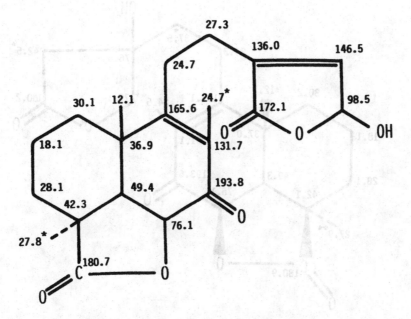

G. Savona, F. Piozzi and J.R. Hanson, <u>Phytochemistry</u>, 17(12), 2132 (1978).

20-OXO-PINUSOLIDE

Source: *Sciadopitys verticillata* Sieb. et. Zucc.

Mol. formula: $C_{21}H_{28}O_5$

Mol. wt.: 360

Solvent: $CDCl_3$

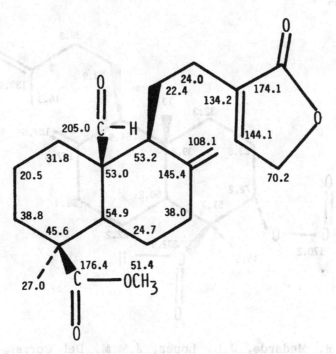

S. Hasegawa and Y. Hirose, <u>Phytochemistry</u>, 24(9), 2041 (1985).

13-HYDROXYBALLONGRINOLIDE

Source: *Ballota lanata*

Mol. formula: $C_{20}H_{26}O_6$

Mol. wt.: 362

Solvent: C_5D_5N

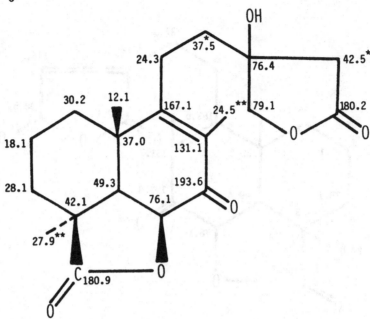

G. Savona, F. Piozzi and J.R. Hanson, <u>Phytochemistry</u>, 17 (12), 2132 (1978).

3α-ACETOXY-ISOAGATHOLAL

Source: *Juniperus thurifera*

Mol. formula: $C_{22}H_{34}O_4$

Mol.wt.: 362

Solvent: $CDCl_3$

A. San Feliciano, M. Medarde, J.L. Lopez, J.M.M. Del Corral, P. Puebla and A.F. Barrero, <u>Phytochemistry</u>, 23(7), 2241 (1988).

3-ACETOXYLABDA-8(20),13-DIENE-15-OIC ACID

Source: *Metaesequoia glyptostroboides* Hu et Cheng

Mol. formula: $C_{22}H_{34}O_4$

Mol. wt.: 362

Solvent: $CDCl_3$

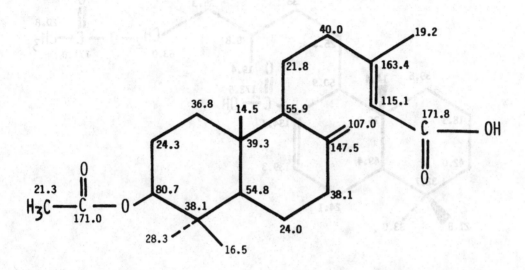

S. Braun and H. Breitenbach, <u>Tetrahedron</u>, 33(1), 145 (1977).

SEMPERVIRENIC ACID

Source: *Solidago sempervirens* L.

Mol. formula: $C_{22}H_{34}O_4$

Mol. wt.: 362

K.K. Purushothaman, A. Sarada, A. Saraswathy and J.D. Connolly, <u>Phytochemistry</u>, 22(4), 1042 (1983).

15-ACETOXY-7-LABDEN-17-OIC ACID

Source: *Halimium viscosum*

Mol. formula: $C_{22}H_{36}O_4$

Mol. wt.: 364

Solvent: $CDCl_3$

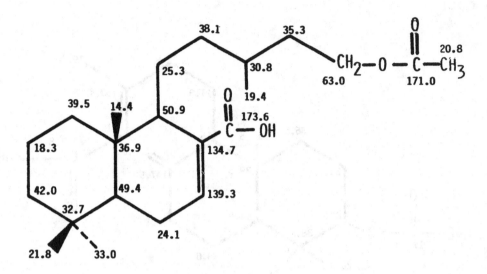

J.G. Urones, I.S. Marcos, D.D. Martin, F.M.S.B. Palma and J.M. Rodilla, Phytochemistry, 26(11), 3037 (1987).

|||||||||||||||||||||||||||||||||||||||||||||||||||||||||||||||||||||||||||||||||||||||||

6-ACETYL-ISO-ANDALUSOL

Source: *Sideritis foetens* Clemen.

Mol. formula: $C_{22}H_{36}O_4$

Mol. wt.: 364

Solvent: $CDCl_3$

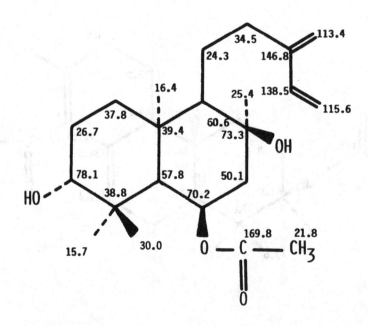

M.C. Garcia-Alvarez and B.Rodriguez, Phytochemistry, 19(11), 2405 (1980).

3-ACETYLVARODIOL

Source: *Sideritis varoi* Soc.

Mol. formula: $C_{22}H_{36}O_4$

Mol. wt.: 364

Solvent: $CDCl_3$

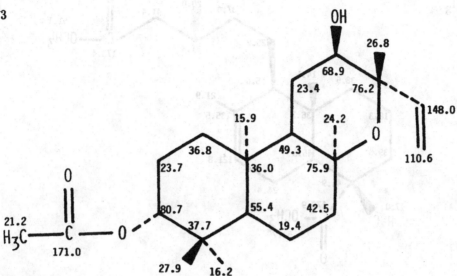

A. Garcia-Granados, A. Martinez, A. Molina, M.E. Onorato, M. Rico, A.S.De Buruaga and J.M.S.De Buruaga, Phytochemistry, 24(8), 1789 (1985).

|||||||||||||||||||||||||||||||||||||||||||||||||||||||||||||||||||||||||||||||||||||||||||||||||||||||||||

12-ACETYLVARODIOL

Source: *Sideritis varoi* Soc.

Mol. formula: $C_{22}H_{36}O_4$

Mol. wt.: 364

Solvent: $CDCl_3$

A. Garcia-Granados, A. Martinez, A. Molina, M.E. Onorato, M. Rico, A.S.De Buruaga and J.M.S.De Buruaga, Phytochemistry, 24(8), 1789 (1985).

8,17H-7,8-DEHYDROPINIFOLIC ACID (METHYL ESTER)

Source: *Ericameria linearifolia* (D.C.) Urb. and Wuss.

Mol. formula: $C_{22}H_{36}O_4$

Mol. wt.: 364

Solvent: $CDCl_3$

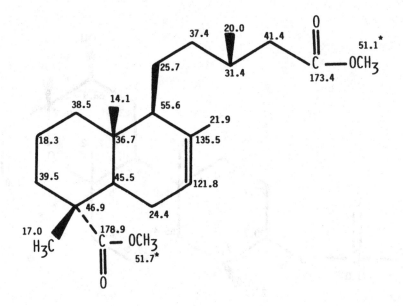

S.J. Dentali, J.J. Hoffmann, S.D. Jolad and B.N. Timmermann, <u>Phytochemistry</u>, **26** (11), 3025 (1987).

||||||||||||||||||||||||||||||||||||||||||||||||||||||||||||||||||||||||||||||||||||||||||||||

HAMACHILOBENE E

Source: *Frullania hamachiloba* Steph.

Mol. formula: $C_{22}H_{36}O_4$

Mol. wt.: 364

Solvent: $CDCl_3$

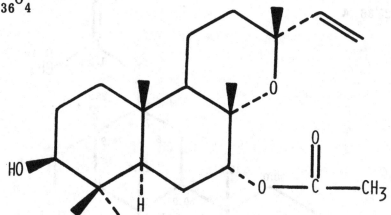

CH_3: 15.2, 15.2, 21.6, 25.2, 27.8 and 28.8

CH_2: 14.9, 24.3, 27.2, 35.2, 37.1 and 109.7

CH: 47.8, 49.4, 74.6, 78.7 and 147.5

C: 36.5, 38.3, 73.5 and 84.1

C=O: 170.4

M. Toyota, F. Nagashima and Y. Asakawa, <u>Phytochemistry</u>, 27(6), 1789 (1988).

HAVARDIC ACID A (METHYL ESTER)

Source: *Grindelia havardii* Steyerm.

Mol. formula: $C_{22}H_{36}O_4$

Mol. wt.: 364

Solvent: $CDCl_3$

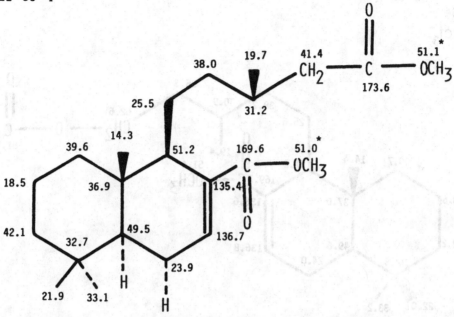

S.D. Jolad, B.N. Timmermann, J.J. Hoffmann, R.B. Bates and T.J. Siahaan, <u>Phytochemistry</u>, 26(2), 483 (1987).

||||||||||||||||||||||||||||||||||||||||||||||||||||||||||||||||||||||||||||||||||||||||||||||||||||||||||||||

JHANIDIOL-18-MONOACETATE

Source: *Eupatorium jhanii* Robinson

Mol. formula: $C_{22}H_{36}O_4$

Mol. wt.: 364

Solvent: $CDCl_3$

A.G. Gonzalez, J.M. Arteaga, J.L. Breton and B.M. Fraga, <u>Phytochemistry</u>, 16 (1), 107 (1977).

METHYL-15-FORMYLOXY-7-LABDEN-17-OATE

Source: *Halimium viscosum* and *H. verticilatum*

Mol. formula: $C_{22}H_{36}O_4$

Mol. wt.: 364

Solvent: $CDCl_3$

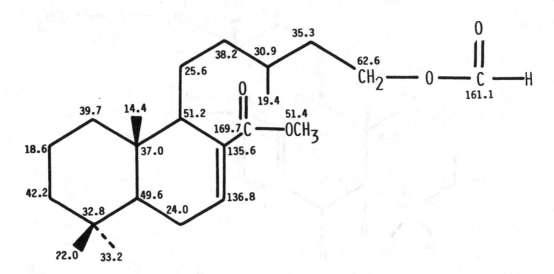

J.G. Urones, I.S. Marcos, D.D. Martin, F.M.S.B. Palma and J.M. Rodilla, <u>Phytochemistry</u>, **26** (11), 3037 (1987).

||||||||||||||||||||||||||||||||||||||||||||||||||||||||||||||||||||||||||||||||||||||||||||

METHYL-7-LABDEN-15,17-DIOATE

Source: *Halimium verticillatum*

Mol. formula: $C_{22}H_{36}O_4$

Mol. wt.: 364

Solvent: $CDCl_3$

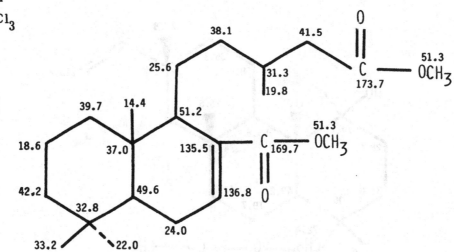

J.G. Urones, I.S. Marcos, D.D. Martin, M.C.A. Alonso, F.M.S.B. Palma and J.M.L. Rodilla, <u>Phytochemistry</u>, 28 (2), 557 (1989).

6-O-ACETYL-AUSTROINULIN

Source: *Stevia rebaudiana*

Mol. formula: $C_{22}H_{36}O_4$

Mol. wt.: 364

Solvent: $CDCl_3$

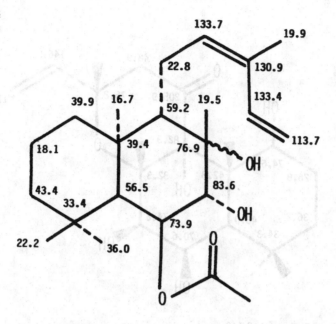

M. Sholichin, K. Yamasaki, R. Miyama, S. Yahara and O. Tanaka, <u>Phytochemistry</u>, 19(2), 326 (1980).

METHYL-15-HYDROXY-7α-METHOXY-8-LABDEN-17-OATE

Source: *Halimium verticillatum*

Mol. formula: $C_{22}H_{38}O_4$

Mol. wt.: 366

Solvent: $CDCl_3$

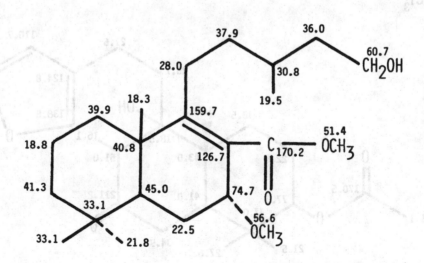

J.G. Urones, I.S. Marcos, D.D. Martin, M.C.A. Alonso, F.M.S.B. Palma and J.M.L. Rodilla, <u>Phytochemistry</u>, 28 (2), 557 (1989).

DEACETYLFORSKOLIN

Source: *Coleus forskohlii* Briq.

Mol. formula: $C_{20}H_{32}O_6$

Mol. wt: 368

Solvent: $CDCl_3$

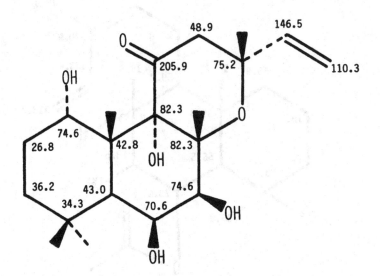

CH_3: 20.0, 23.4, 24.2, 30.7 and 33.0

B. Gabetta, G. Zini and B. Danieli, <u>Phytochemistry</u>, 28 (3), 859 (1989).

CALYONE

Source: *Roylea calycina* (Roxb.) Briq.

Mol. formula: $C_{22}H_{32}O_5$

Mol. wt.: 376

Solvent: $CDCl_3$

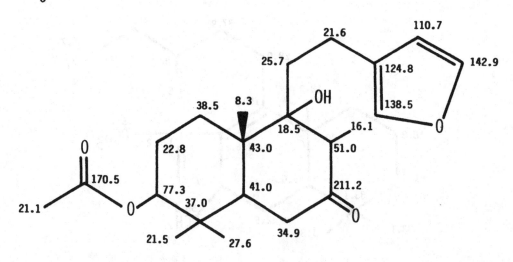

Om Prakash, D.S. Bhakuni, R.S. Kapil, G.S.R. Subba Rao and B. Ravindranath, <u>J. Chem. Soc. Perkin Trans. I</u>, (5), 1305 (1979).

GALEOPSIN

Source: *Galeopsis angustifolia* Hoffm.

Mol. formula: $C_{22}H_{32}O_5$

Mol. wt.: 376

Solvent: $CDCl_3$

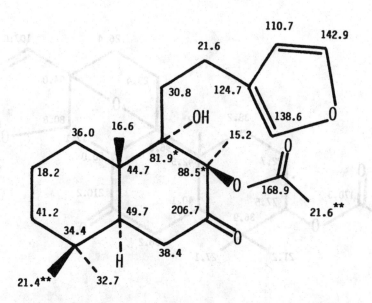

B. Rodriguez and G. Savona, <u>Phytochemistry</u>, 19(8), 1805 (1980).

||||||||||||||||||||||||||||||||||||||||||||||||||||||||||||||||||||||||||||||||||||||||||

3-OXO-18-ACETOXYLABDA-8(17),13Z-DIEN-15-OIC ACID

Source: *Nolana rostrata* (Lindley) Mier

Mol. formula: $C_{22}H_{32}O_5$

Mol. wt.: 376

Solvent: $CDCl_3$

25.8
115.5
163.4
171.0
OH
32.9
22.9
35.8
13.9
55.7
107.8
35.2
39.1
146.4
213.3
47.8
37.6
50.4
24.9
67.0
170.5
20.9
17.8
CH_2—O—C—CH_3
O

J.A. Garbarino, M.C. Chamy and V. Gambaro, <u>Phytochemistry</u>, 25 (12), 2833 (1986).

PRECALYONE

Source: *Roylea calycina* (Roxb.) Briq.

Mol. formula: $C_{22}H_{32}O_5$

Mol. wt.: 376

Solvent: $CDCl_3$

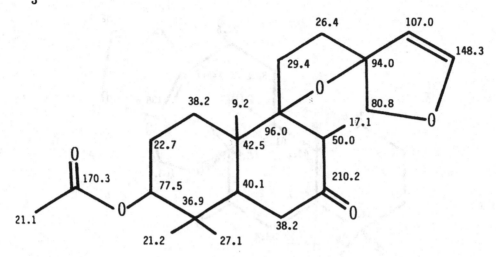

Om Prakash, D.S. Bhakuni, R.S. Kapil, G.S.R. Subba Rao and B. Ravindranath, <u>J. Chem. Soc. Perkin Trans.I</u>, (5), 1305 (1979).

PREGALEOPSIN

Source: *Galeopsis angustifolia* Hoffm.

Mol. formula: $C_{22}H_{32}O_5$

Mol. wt.: 376

Solvent: $CDCl_3$

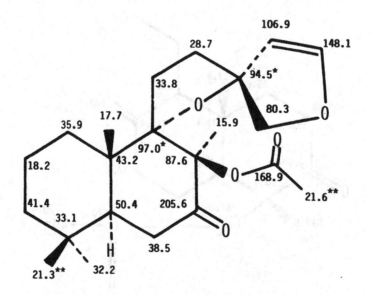

B. Rodriguez and G. Savona, <u>Phytochemistry</u>, 19(8), 1805 (1980).

2β-ACETOXYLABDA-8(17), 13E-DIEN-15-OIC ACID (METHYL ESTER)

Source: *Nolana filifolia* (Hook et Arn) Johnston
Mol. formula: $C_{23}H_{36}O_4$
Mol. wt.: 376
Solvent: $CDCl_3$

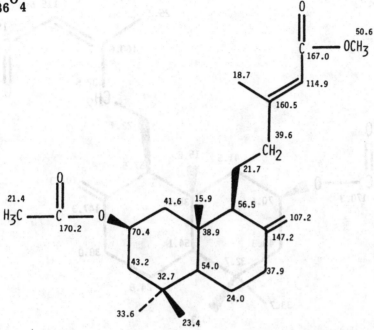

J.A. Garbarino, M.C. Chamy, M. Piovano and V. Gambaro, <u>Phytochemistry</u>, 27(6), 1795 (1988).

DITERPENE NT-I (METHYL ESTER)

Source: *Nepeta tuberosa*
Mol. formula: $C_{23}H_{36}O_4$
Mol. wt.: 376

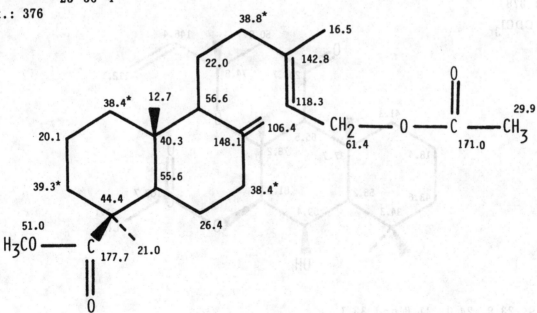

J.G. Urones, I.S. Marcos, J.F. Ferreras and P.B. Barcala, <u>Phytochemistry</u>, 27(2), 523 (1988).

METHYL-2β-ACETOXYLABDA-8(17), 13Z-DIEN-15-OATE

Source: *Nolana filifolia* (Hook et Arn) Johnston

Mol. formula: $C_{23}H_{36}O_4$

Mol. wt.: 376

Solvent: $CDCl_3$

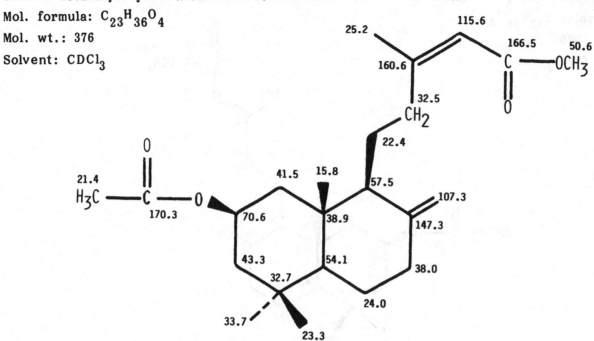

J.A. Garbarino, M.C. Chamy, M. Piovano and V.Gambaro, <u>Phytochemistry</u>, 27(6), 1795 (1988).

||||||||||||||||||||||||||||||||||||||||||||||||||||||||||||||||||||||||||||||||||||||||||||||||||||

1,9-DIDEOXYFORSKOLIN

Source: *Coleus forskohlii* Briq.

Mol. formula: $C_{22}H_{34}O_5$

Mol. wt.: 378

Solvent: $CDCl_3$

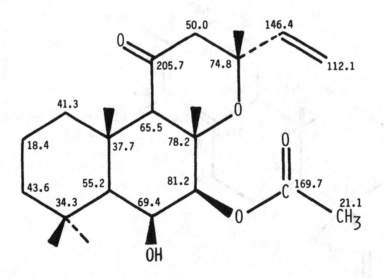

CH_3: 16.9, 23.8, 24.0, 31.6 and 33.1

B. Gabetta, G. Zini annd B. Danieli, <u>Phytochemistry</u>, 28 (3), 859 (1989).

DITERPENE HA-II

Source: *Helichrysum ambiguum* Turcz.

Mol. formula: $C_{22}H_{34}O_5$

Mol. wt.: 378

Solvent: $CDCl_3$

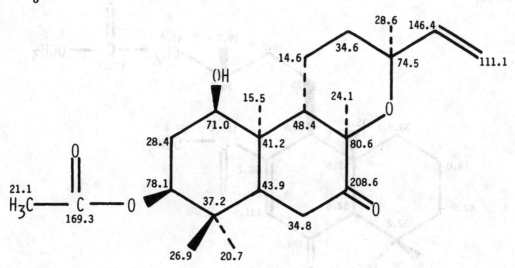

J. Jakupovic, A. Schuster, F. Bohlmann, U. Ganzer, R.M. King and H. Robinson, <u>Phytochemistry</u>, 28(2), 543 (1989).

EVILLOSIN

Source: *Eupatorium villosum* Sw.

Mol. formula: $C_{22}H_{34}O_5$

Mol. wt.: 378

Solvent: $CDCl_3$

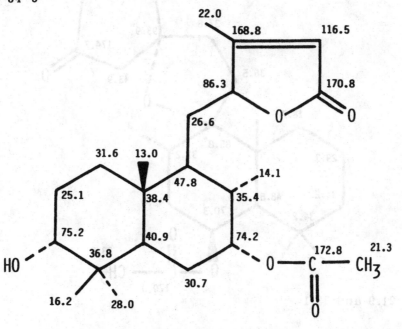

P.S. Manchand, J.F. Blount T. McCabe and J. Clardy, <u>J. Org. Chem.</u>, 44(8), 1322 (1979).

HAVARDIC ACID D (METHYL ESTER)

Source: *Grindelia havardii* Steyerm.

Mol. formula: $C_{22}H_{34}O_5$

Mol. wt.: 378

Solvent: $CDCl_3$

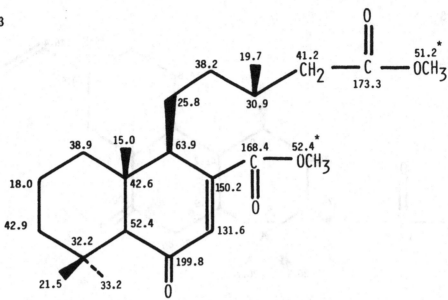

S.D. Jolad, B.N. Timmermann, J.J. Hoffmann, R.B. Bates and T.J. Siahaan, <u>Phytochemistry</u>, **26** (2), 483 (1987).

||||||||||||||||||||||||||||||||||||||||||||||||||||||||||||||||||||||||||||||||||||||||||||||||||||||||||||||||

PREVITEXILACTONE

Source: *Vitex rotundifolia* L.Fil.

Mol. formula: $C_{22}H_{34}O_5$

Mol. wt.: 378

Solvent: $CDCl_3$

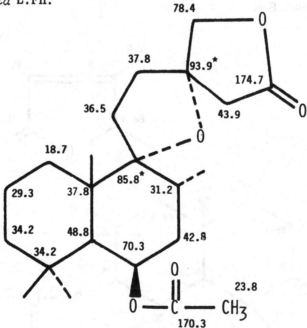

CH_3: 17.2, 19.8, 21.9 and 33.1

Y. Kondo, K. Sugiyama and S. Nozoe, <u>Chem. Pharm. Bull.</u> (Tokyo), 34(11), 4829 (1986).

VITEXILACTONE

Source: *Vitex rotundifolia* L.Fil.

Mol. formula: $C_{22}H_{34}O_5$

Mol. wt.: 378

Solvent: $CDCl_3$

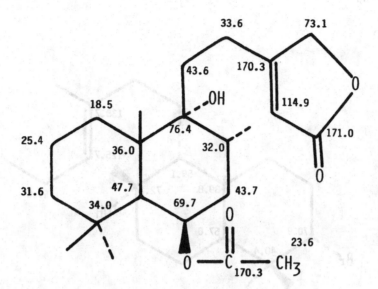

CH_3: 16.1, 19.0, 21.9 and 33.6

Y. Kondo, K. Sugiyama and S. Nozoe, Chem. Pharm. Bull. (Tokyo), 34(11), 4829 (1986).

METHYL-15-ACETOXY-7-LABDEN-17-OATE

Source: *Halimium viscosum*

Mol. formula: $C_{23}H_{38}O_4$

Mol. wt.: 378

Solvent: $CDCl_3$

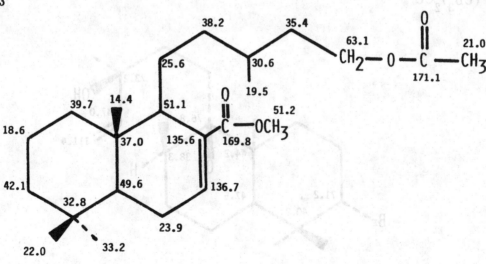

J.G. Urones, I.S. Marcos, D.D. Martin, F.M.S.B. Palma and J.M. Radilla, Phytochemistry, 26(11), 3037 (1987).

Source: *Laurencia snyderae* Dawson

Mol. formula: $C_{20}H_{35}BrO_2$

Mol. wt.: 387

Solvent: $(CD_3)_2CO$

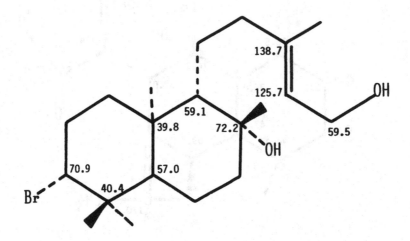

CH_3 and CH_2: 15.7, 18.8, 20.7, 24.7, 30.7, 30.7, 31.6, 31.7, 41.3, 43.3 and 44.0

B.M. Howard and W. Fenical, Phytochemistry, 19(12), 2774 (1980).

CONCINNDIOL

Source: *Laurencia snyderae* Dawson

Mol. formula: $C_{20}H_{35}BrO_2$

Mol. wt.: 387

Solvent: $(CD_3)_2CO$

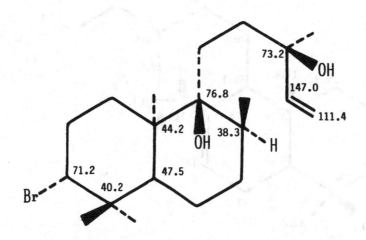

CH_3 and CH_2: 16.7, 16.8, 18.7, 24.0, 32.1, 32.7, 34.5 and 37.1

B.M. Howard and W. Fenical, Phytochemistry, 19 (12), 2774 (1980).

ISOCONCINNDIOL

Source: *Laurencia snyderae* Dawson
Mol. formula: $C_{20}H_{35}BrO_2$
Mol. wt.: 387
Solvent: $(DC_3)_2CO$

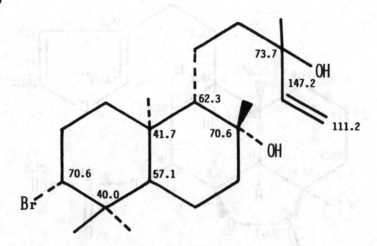

CH_3 and CH_2: 15.8, 18.5, 20.0, 22.7, 24.4, 29.3, 30.6, 30.8, 32.7, 45.0 and 45.9.

B.M. Howard and W. Fenical, <u>Phytochemistry</u>, 19(12), 2774 (1980).

||||||||||||||||||||||||||||||||||||||||||||||||||||||||||||||||||||||||||||||||||||||||||||||||||||||||||||||||||||

METHYL-12R-ACETOXY-LAMBERTIANATE

Source: *Sciadopitys verticillata* Sieb. et. Zucc.
Mol. formula: $C_{23}H_{32}O_5$
Mol. wt.: 388
Solvent: $CDCl_3$

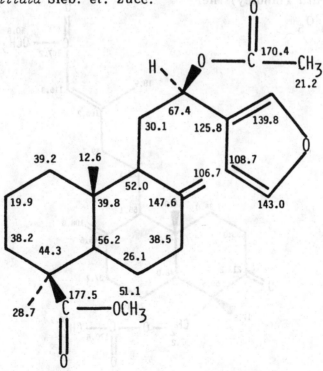

S. Hasegawa and Y. Hirose, <u>Phytochemistry</u>, 24(9), 2041 (1985).

Source: *Ballota nigra* L.

Mol. formula: $C_{22}H_{30}O_6$

Mol. wt.: 390

Solvent: $CDCl_3$

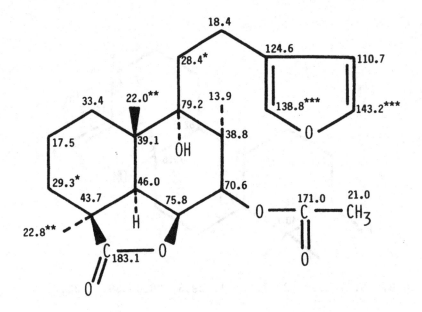

G. Savona, F. Piozzi, J.R. Hanson and M. Siverns, <u>J. Chem. Soc. Perkin Trans. I</u>, (3), 322 (1977).

||||||||||||||||||||||||||||||||||||||||||||||||||||||||||||||||||||||||||||||||||||||||||

METHYL-3-OXO-18-HYDROXYLABDA-8(17),13<u>E</u>-DIEN-15-OATE (ACETATE)

Source: *Nolana rostrata* (Lindley) Mier

Mol. formula: $C_{23}H_{34}O_5$

Mol. wt.: 390

Solvent: $CDCl_3$

J.A. Garbarino, M.C. Chamy and V. Gambaro, <u>Phytochemistry</u>, 25 (12), 2833 (1986).

2α,15-DIHYDROXY-<u>ENT</u>-LABDA-7,13E-DIENE (DIACETATE)

Source: *Baccharis sternbergiana* Steud.

Mol. formula: $C_{24}H_{38}O_4$

Mol. wt.: 390

Solvent: $CDCl_3$

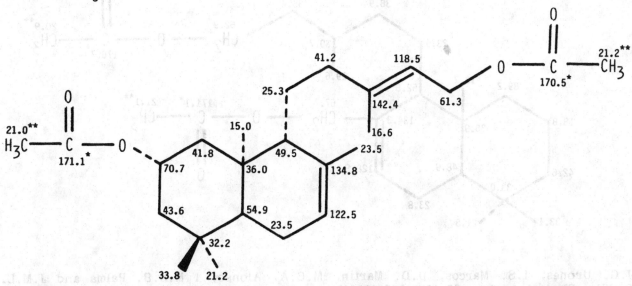

C. Zdero, F. Bohlmann, J.C. Solomon, R.M. King and H. Robinson, <u>Phytochemistry</u>, 28(2), 531 (1989).

|||||||||||||||||||||||||||||||||||||||||||||||||||||||||||||||||||||||||||||||||||||||||||||||||||||||||||||||||||||||||||||||||||||||||||

AUSTROEUPATORION

Source: *Austroeupatorium inulaefolium*

Mol. formula: $C_{21}H_{28}O_7$

Mol. wt.: 392

Solvent: $CDCl_3$

F. Bohlmann, G. Schmeda-Hirschmann and J. Jakupovic, <u>Planta Med.</u>, 50 (2), 199 (1984).

15,17-DIACETOXY-7-LABDENE

Source: *Halimium verticillatum*

Mol. formula: $C_{24}H_{40}O_4$

Mol. wt.: 392

Solvent: $CDCl_3$

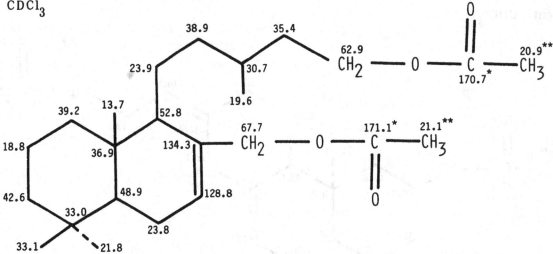

J.G. Urones, I.S. Marcos, D.D. Martin, M.C.A. Alonso, F.M.S.B. Palma and J.M.L. Rodilla, <u>Phytochemistry</u>, **28** (2), 557 (1989).

||||||||||||||||||||||||||||||||||||||||||||||||||||||||||||||||||||||||||||||||||||||||||||||||

7β-ACETOXY-6β,9α-DIHYDROXY-8,13-EPOXY-LABD-14-EN-11-ONE

Source: *Coleus forskohlii* Briq.

Mol. formula: $C_{22}H_{34}O_6$

Mol. wt.: 394

Solvent: $CDCl_3$

CH_3: 18.5, 22.5, 24.2, 31.8 and 33.2

B. Gabetta, G. Zini and B. Danieli, <u>Phytochemistry</u>, **28** (3), 859 (1989).

9-DEOXYFORSKOLIN

Source: *Coleus forskohlii* Briq.

Mol. formula: $C_{22}H_{34}O_6$

Mol. wt.: 394

Solvent: $CDCl_3$

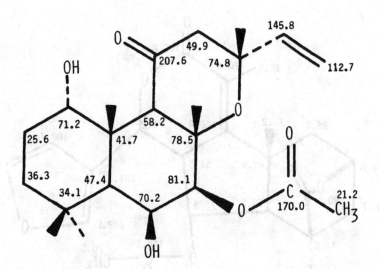

CH$_3$: 18.2, 23.6, 24.5, 31.5 and 32.8

B. Gabetta, G. Zini and B. Danieli, <u>Phytochemistry</u>, 28 (3), 859 (1989).

||||||||||||||||||||||||||||||||||||||||||||||||||||||||||||||||||||||||||||||||||||||||||||||||||||

CORDOBIC ACID-18-ACETATE (METHYL ESTER)

Source: *Grindelia discoidea* Hook and Arn

Mol. formula: $C_{23}H_{38}O_5$

Mol. wt.: 394

Solvent: $CDCl_3$

B.N. Timmermann, J.J. Hoffmann, S.D. Jolad, R.B. Bates and T.J. Siahaan, <u>Phytochemistry</u>, 25 (6), 1389 (1986).

RICHARDIANIDIN-I

Source: *Cluytia richardiana* L.

Mol. formula: $C_{22}H_{24}O_7$

Mol. wt.: 400

Solvent: $CDCl_3$

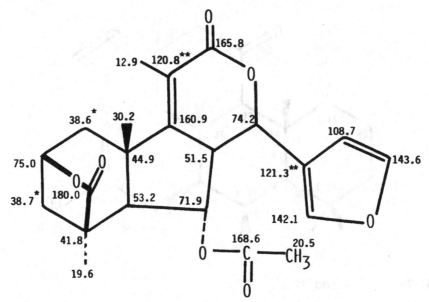

J.S. Mossa, J.M. Cassady, J.F. Kozlowski, T.M. Zennie, M.D. Antoun, M.G. Pellechia, A.T. McKenzie and S.R. Byrn, Tetrahedron Lett., 29(30), 3627 (1988).

||||||||||||||||||||||||||||||||||||||||||||||||||||||||||||||||||||||||||||||||||||||||||||||||||||||||||||

3α-ANGELOYLOXY-18-HYDROXY-13-FURYL-ENT-LABDA-8(17)-ENE

Source: *Gutierrezia grandis* S.F. Blake

Mol. formula: $C_{25}H_{36}O_4$

Mol. wt.: 400

Solvent: $CDCl_3$

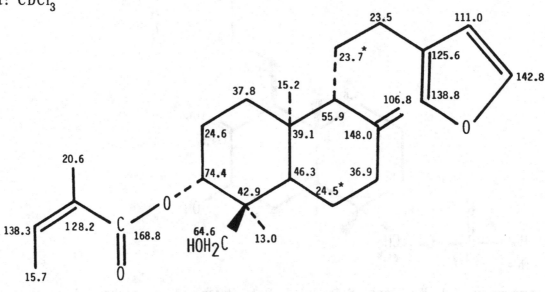

F. Gao, M. Leidig and T.J. Mabry, Phytochemistry, 24(7), 1541 (1985).

3α-HYDROXY-18-ANGELOYLOXY-13-FURYL-<u>ENT</u>-LABDA-8(17)-ENE

Source: *Gutierrezia grandis* S.F.Blake

Mol. formula: $C_{25}H_{36}O_4$

Mol. wt.: 400

Solvent: $CDCl_3$

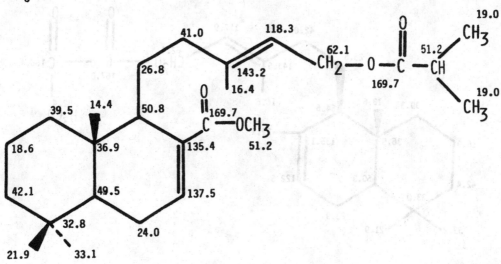

F. Gao, M. Leidig and T.J. Mabry, <u>Phytochemistry</u>, 24(7), 1541 (1985).

15-ISOBUTYLOXY-7,13E-LABDADIEN-17-OIC-ACID(METHYL ESTER)

Source: *Halimium viscosum*

Mol. formula: $C_{25}H_{40}O_4$

Mol. wt.: 404

Solvent: $CDCl_3$

J.G. Urones, I.S. Marcos, D.D. Martin, F.M.S.B. Palma and J.M. Rodilla, <u>Phytochemistry</u>, 26(11), 3037 (1987).

3,12-DIACETYLVARODIOL

Source: *Sideritis varoi* Soc.
Mol. formula: $C_{24}H_{38}O_5$
Mol. wt.: 406
Solvent: $CDCl_3$

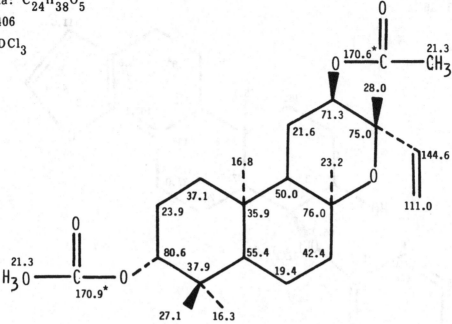

A. Garcia-Granados, A. Martinez, A. Molina, M.E. Onorato, M. Rico, A.S.De Buruaga and J.M.S.De Buruaga, Phytochemistry, 24(8), 1789 (1985).

|||||||||||||||||||||||||||||||||||||||||||||||||||||||||||||||||||||||||||||||||||||||||||||||||||||||||||||||

7,13E-LABDADIEN-15-YL-METHYL MALONIC ACID DIESTER

Source: *Parentucellia latifolia* (L) Caruel
Mol. formula: $C_{24}H_{38}O_4$
Mol. wt.: 390
Solvent: $CDCl_3$

J.G. Urones, I.S. Marcos, L. Cubillo, V.A. Monje, J.M. Hernandez and P. Basabe, Phytochemistry, 28 (2), 651 (1989).

6β-ISOVALEROXYLABDA-8'13-DIEN-7α,15-DIOL

Source: *Trimusculus reticulatus*

Mol. formula: $C_{25}H_{42}O_4$

Mol. wt.: 406

Solvent: $CDCl_3$

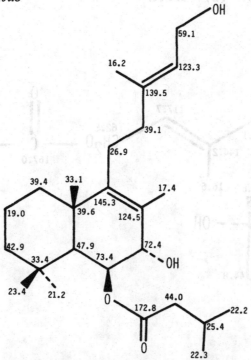

D.C. Manker and D.J. Faulkner, Tetrahedron, **43** (16), 3677 (1987).

6α-HYDROXY-17-ACETOXYGRINDELIC ACID (METHYL ESTER)

Source: *Grindelia acutifolia* Steyerm.

Mol. formula: $C_{23}H_{36}O_6$

Mol. wt.: 408

Solvent: $CDCl_3$

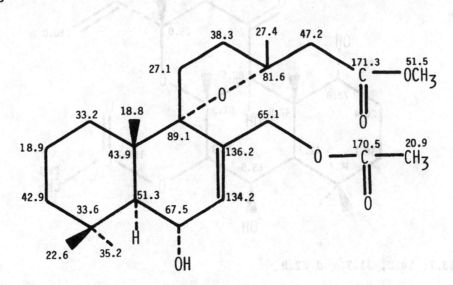

B.N. Timmermann, J.J. Hoffmann, S.D. Jolad, R.B. Bates and T.J. Siahaan, Phytochemistry, **26**(2), 467 (1987).

8-HYDROXY-13E-LABDEN-15-YL-METHYL MALONIC ACID (DIESTER)

Source: *Parentucellia latifolia* (L) Caruel.

Mol. formula: $C_{24}H_{40}O_5$

Mol. wt.: 408

Solvent: $CDCl_3$

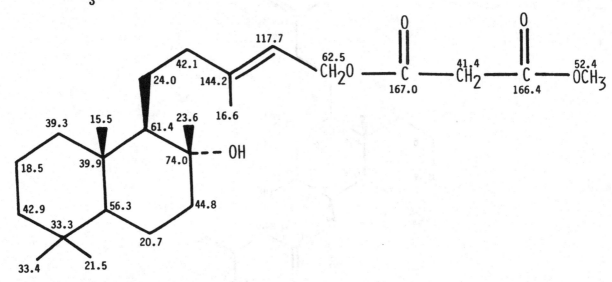

J.G. Urones, I.S. Marcos, L. Cubillo, V.A. Monje, J.M. Hernandez and P. Basabe, <u>Phytochemistry</u>, 28(2), 651 (1989).

FORSKOLIN

Source: *Coleus forskohlii* Briq.

Mol. formula: $C_{22}H_{34}O_7$

Mol. wt.: 410

Solvent: $CDCl_3$

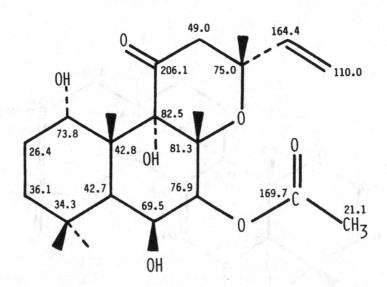

CH_3: 19.7, 23.7, 24.2, 31.7 and 32.9

B. Gabetta, G. Zini and B. Danieli, <u>Phytochemistry</u>, 28 (3), 859 (1989).

118

HAMACHILOBENE B

Source: *Frullania hamachiloba* Steph.

Mol. formula: $C_{24}H_{36}O_6$

Mol. wt.: 420

Solvent: $CDCl_3$

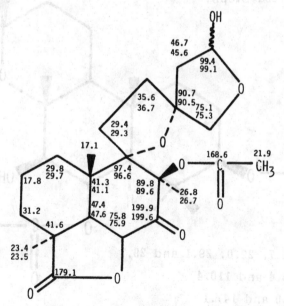

CH_3: 15.9, 16.3, 21.0, 21.1, 24.5, 29.8 and 30.1

CH_2: 15.5, 22.8, 33.2, 36.4 and 111.8

CH: 52.7, 57.4, 74.0, 79.7 and 146.3

C: 37.2, 38.4, 74.7 and 79.9

C=O: 169.9, 170.8 and 202.2

M. Toyota, F. Nagashima and Y. Asakawa, <u>Phytochemistry</u>, 27(6), 1789 (1988).

||||||||||||||||||||||||||||||||||||||||||||||||||||||||||||||||||||||||||||||||||||||||||||||||||||||||||

LEOCARDIN

Source: *Leonurus cardiaca*

Mol. formula: $C_{22}H_{30}O_8$

Mol. wt.: 422

Solvent: $CDCl_3$

Note: The double values of some carbons is due to the presence of two compounds epimeric at C-15.

P. Malakov, G. Papanov, J. Jakupovic, M. Grenz and F. Bohlmann, <u>Phytochemistry</u>, 24 (10), 2341 (1985).

HAMACHILOBENE A

Source: *Frullania hamachiloba* Steph.

Mol. formula: $C_{24}H_{38}O_6$

Mol. wt.: 422

Solvent: $CDCl_3$

CH_3: 15.9, 16.8, 21.2, 21.9, 25.2, 28.9 and 29.9

CH_2: 14.8, 23.0, 34.5, 36.4 and 110.8

CH: 47.4, 49.2, 71.3, 75.5, 80.3 and 146.4

C: 37.1, 37.3, 74.0 and 76.2

C=O: 170.4 and 170.8

M. Toyota, F. Nagashima and Y. Asakawa, <u>Phytochemistry</u>, **27** (6), 1789 (1988).

HAMACHILOBENE C

Source: *Frullania hamachiloba* Steph.

Mol. formula: $C_{24}H_{38}O_6$

Mol. wt.: 422

Solvent: $CDCl_3$

CH_3: 16.6, 16.9, 21.2, 21.7, 22.0, 29.1 and 29.7

CH_2: 15.2, 23.0, 34.4, 36.4 and 110.4

CH: 51.6, 56.3, 80.0, 84.0 and 147.1

C: 37.5, 37.8, 73.6 and 77.3

C=O: 170.9 and 171.0

M. Toyota, F. Nagashima and Y. Asakawa, <u>Phytochemistry</u>, 27(6), 1789 (1988).

HAMACHILOBENE D

Source: *Frullania hamachiloba* Steph.

Mol. formula: $C_{24}H_{38}O_6$

Mol. wt.: 422

Solvent: $CDCl_3$

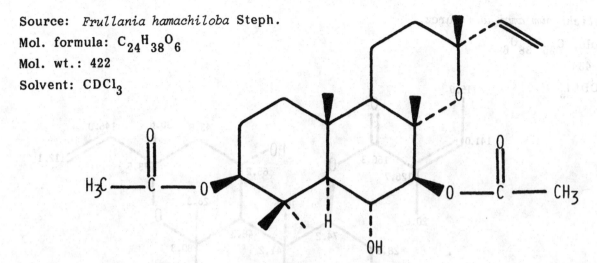

CH$_3$: 16.4, 16.6, 21.2, 21.2, 22.5, 29.5 and 30.4

CH$_2$: 15.2, 23.1, 34.6, 36.5 and 110.6

CH: 52.4, 58.6, 71.2, 80.4, 86.4 and 147.1

C: 37.1, 38.4, 73.3 and 76.0

C=O: 170.9 and 172.2

M. Toyota, F. Nagashima and Y. Asakawa, <u>Phytochemistry</u>, 27(6), 1789 (1988).

||||||||||||||||||||||||||||||||||||||||||||||||||||||||||||||||||||||||||||||||||||||||||||||||||||||||||||||||||||||||||||||||

DITERPENE H.A–XVIII

Source: *Helichrysum ambiguum* Turcz.

Mol. formula: $C_{25}H_{36}O_6$

Mol. wt.: 432

Solvent: $CDCl_3$

J. Jakupovic, A. Schuster, F. Bohlmann, U. Ganzer, R.M. King and H. Robinson, <u>Phyto-chemistry</u>, **28** (2), 543 (1989).

121

DITERPENE H.A–XII

Source: *Helichrysum ambiguum* Turcz.

Mol. formula: $C_{25}H_{38}O_6$

Mol. wt.: 434

Solvent: $CDCl_3$

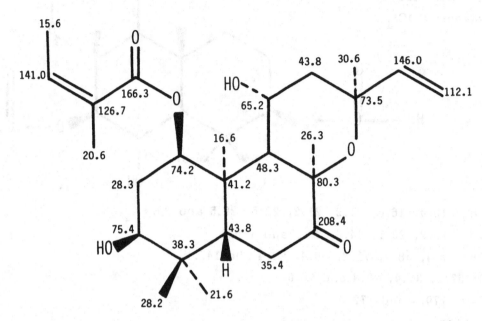

J. Jakupovic, A. Schuster, F. Bohlmann, U. Ganzer, R.M. King and H. Robinson, <u>Phytochemistry</u>, **28** (2), 543 (1989)

||||||||||||||||||||||||||||||||||||||||||||||||||||||||||||||||||||||||||||||||||||||||||||||||||||||||||||

6-OXO-17-ISOBUTYROXYGRINDELIC ACID (METHYL ESTER)

Source: *Grindelia acutifolia* Steyerm.

Mol. formula: $C_{25}H_{38}O_6$

Mol. wt.: 434

Solvent: $CDCl_3$

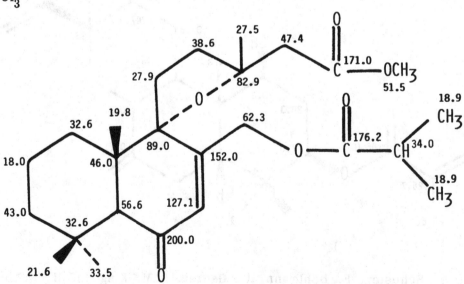

B.N. Timmermann, J.J. Hoffmann, S.D. Jolad, R.B. Bates and T.J. Siahaan, <u>Phytochemistry</u>, 26(2), 467 (1987).

ISOAGATHOLAL-15-O-β-D-XYLOPYRANOSIDE

Source: *Thujopsis dolabrata* Sieb et Zucc.

Mol. formula: $C_{25}H_{40}O_6$

Mol. wt.: 436

Solvent: $CDCl_3$

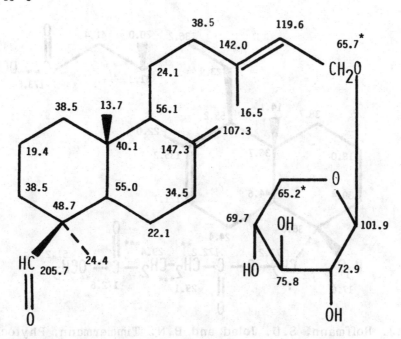

S. Hasegawa and Y. Hirose, <u>Phytochemistry</u>, 19(11), 2479 (1980).

METHYL-7β,15-DIACETOXY-8-LABDEN-17-OATE

Source: *Halimium verticillatum*

Mol. formula: $C_{25}H_{40}O_6$

Mol. wt.: 436

Solvent: $CDCl_3$

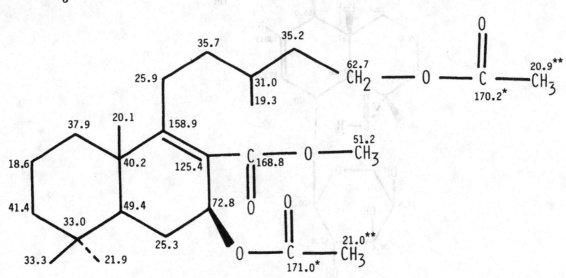

J.G. Urones, I.S. Marcos, D.D. Martin, M.C.A. Alonso, F.M.S.B. Palma and J.M.L. Rodilla, <u>Phytochemistry</u>, 28 (2), 557 (1989).

18α-SUCCINYLOXY-LABD-7-EN-15-OIC ACID (DIMETHYL ESTER)

Source: *Ericameria linearifolia* (D.C.) Urb. and Wuss.

Mol. formula: $C_{26}H_{42}O_6$

Mol. wt.: 450

Solvent: $CDCl_3$

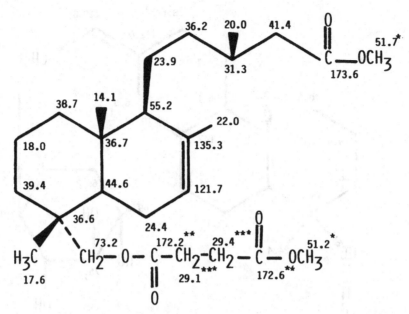

S.J. Dentali, J.J. Hoffmann, S.D. Jolad and B.N. Timmermann, <u>Phytochemistry</u>, **26** (11), 3025 (1987).

DITERPENE GS-II

Source: *Gutierrezia sphaerocephala* Gray

Mol. formula: $C_{25}H_{40}O_7$

Mol. wt.: 452

Solvent: $CDCl_3$

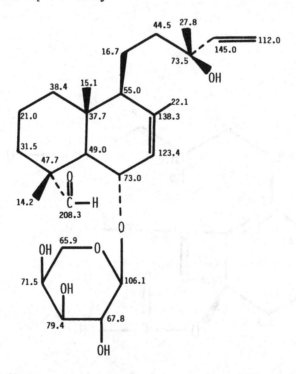

F. Gao, M. Leidig and T.J. Mabry, <u>Phytochemistry</u>, 25(6), 1371 (1986).

DITERPENE AS-III

Source: *Aster spathulifolius* Maxim.

Mol. formula: $C_{26}H_{46}O_6$

Mol. wt.: 454

Solvent: C_5D_5N

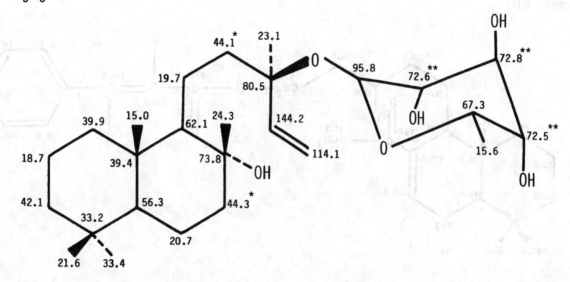

Y. Uchio, M. Nagasaki, S. Eguchi, A. Matsuo, M. Nakayama and S. Hayashi, <u>Tetrahedron Lett.</u>, 21(39), 3775 (1980).

||||||||||||||||||||||||||||||||||||||||||||||||||||||||||||||||||||||||||||||||||||||||||||||||||||||||

CORONARIN F

Source: *Hedychium coronarium*

Mol. formula: $C_{30}H_{46}O_3$

Mol. wt.: 454

Solvent: $CDCl_3$

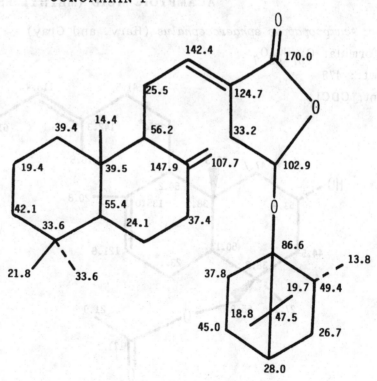

H. Itokawa, H. Morita, K. Takeya and M. Motidome, <u>Chem. Pharm. Bull.</u> (Tokyo), 36(7), 2682 (1988).

METHYL-15-CINNAMOYLOXY-7-LABDEN-17-OATE

Source: *Halimium viscosum*

Mol. formula: $C_{30}H_{42}O_4$

Mol. wt.: 466

Solvent: $CDCl_3$

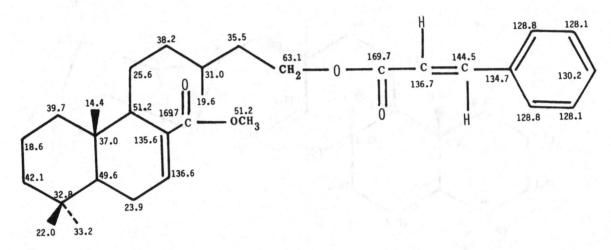

J.G. Urones, I.S. Marcos, D.D. Martin, F.M.S.B. Palma and J.M. Rodilla, <u>Phytochemistry</u>, 26 (11), 3037 (1987).

||||||||||||||||||||||||||||||||||||||||||||||||||||||||||||||||||||||||||||||||||||||||||||||||||||||||||||

ACAMPTOIC ACID (METHYL ESTER)

Source: *Acamptopappus sphaerocephalus* (Harv. and Gray)

Mol. formula: $C_{27}H_{42}O_7$

Mol. wt.: 478

Solvent: $CDCl_3$

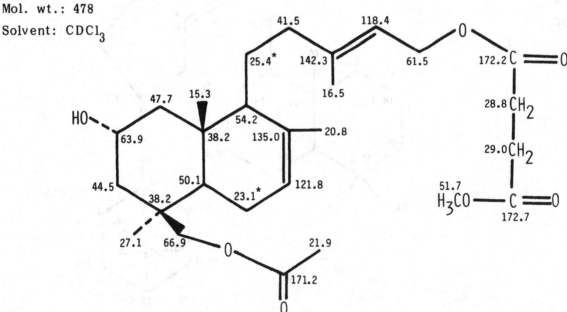

S.D. Jolad, J.J. Hoffmann, B.N. Timmermann, S.P. Mclaughlin, R.B. Bates, F.A. Camou and J.R. Cole, <u>Phytochemistry</u>, 27 (10), 3197 (1988).

DITERPENE AS-II

Source: *Aster spathulifolius* Maxim.

Mol. formula: $C_{28}H_{46}O_6$

Mol. wt.: 478

Solvent: C_5D_5N

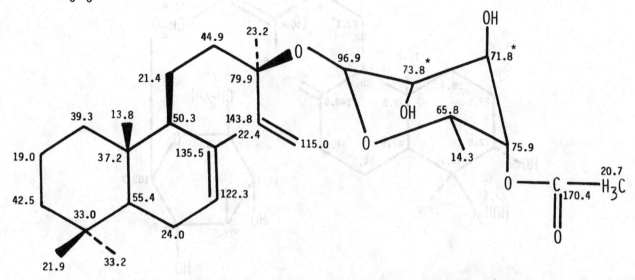

Y. Uchio, M. Nagasaki, S. Eguchi, A. Matsuo, M. Nakayama and S. Hayashi, <u>Tetrahedron</u> <u>Lett.</u>, 21(39), 3775 (1980).

||||||||||||||||||||||||||||||||||||||||||||||||||||||||||||||||||||||||||||||||||||||||||||||||||||||||||||||||||

NEOANDROGRAPHOLIDE

Source: *Andrographis paniculata* Nees

Mol. formula: $C_{26}H_{40}O_8$

Mol. wt.: 480

Solvent: C_5D_5N

T. Fujita, R. Fujitani, Y. Takeda, Y. Takaishi, T. Yamada, M. Kido and I. Miura, <u>Chem. Pharm.</u> <u>Bull.</u> (Tokyo), 32(6), 2117 (1984).

GOSHONOSIDE-F1

Source: *Rubus chingii* Hu

Mol. formula: $C_{26}H_{44}O_8$

Mol. wt.: 484

Solvent: C_5D_5N

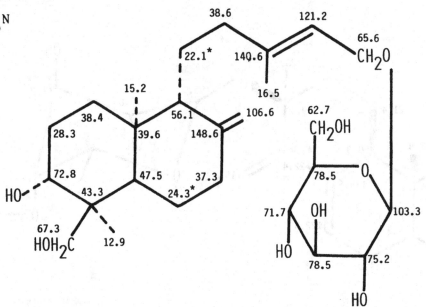

T. Tanaka, K. Kawamura, T. Kitahara, H. Kohda and O. Tanaka, <u>Phytochemistry</u>, 23 (3), 615 (1984).

GOSHONOSIDE-F2

Source: *Rubus chingii* Hu

Mol. formula: $C_{26}H_{44}O_8$

Mol. wt.: 484

Solvent: C_5D_5N

T. Tanaka, K. Kawamura, T. Kitahara, H. Kohda and O. Tanaka, <u>Phytochemistry</u>, 23 (3), 615 (1984).

VENUSTANOL (DIACETATE)

Source: *Laurencia venusta* Yamada
Mol. formula: $C_{24}H_{39}O_5Br$
Mol. wt.: 487
Solvent: $CDCl_3$

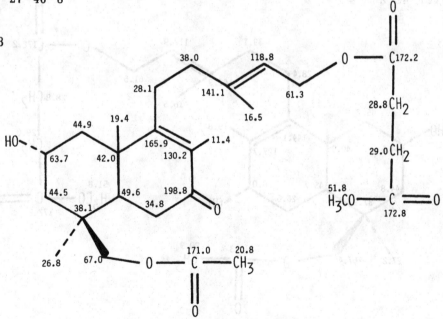

M. Suzuki, E. Kurosawa and K. Kurata, Phytochemistry, **27**(4), 1209 (1988).

7-OXO-8-EN-ACAMPTOIC ACID (METHYL ESTER)

Source: *Acamptopappus sphaerocephalus* (Harv. and Gray)
Mol. formula: $C_{27}H_{40}O_8$
Mol. wt.: 492
Solvent: $CDCl_3$

S.D. Jolad, J.J. Hoffmann, B.N. Timmermann, S.P. Mclaughlin, R.B. Bates, F.A. Camou and J.R. Cole, Phytochemistry, **27** (10), 3197 (1988).

Source: *Gutierrezia sphaerocephala* Gray

Mol. formula: $C_{27}H_{42}O_8$

Mol. wt.: 494

Solvent: $CDCl_3$

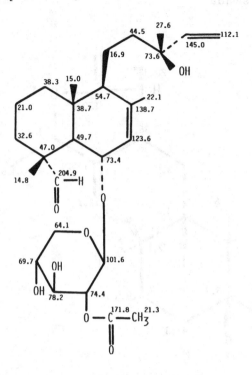

F. Gao, M. Leidig and T.J. Mabry, Phytochemistry, 25(6), 1371 (1986).

7α-HYDROXY-8-EN-ACAMPTOIC ACID (METHYL ESTER)

Source: *Acamptopappus sphaerocephalus* (Harv. and Gray)

Mol. formula: $C_{27}H_{42}O_8$

Mol. wt.: 494

Solvent: $CDCl_3$

S.D. Jolad, J.J. Hoffmann, B.N. Timmermann, S.P. Mclaughlin, R.B. Bates, F.A. Camou and J.R. Cole, Phytochemistry, 27 (10), 3197 (1988).

ANDROPANOSIDE

Source: *Andrographis paniculata* Nees

Mol. formula: $C_{26}H_{40}O_9$

Mol. wt.: 496

Solvent: C_5D_5N

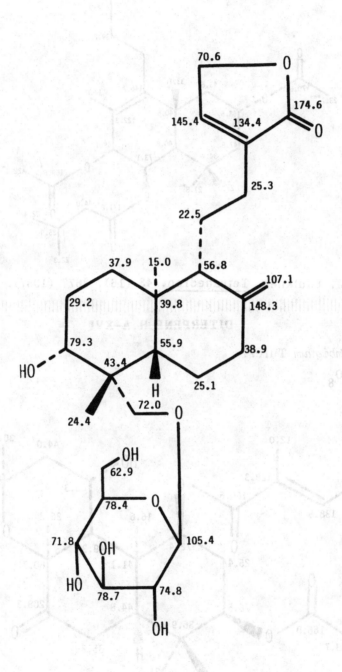

T. Fujita, R. Fujitani, Y. Takeda, Y. Takaishi, T. Yamada, M. Kido and I. Miura, Chem. Pharm. Bull. (Tokyo), 32(6), 2117 (1984).

2α,7α-DIACETOXY-6β-ISOVALEROXYLABDA-8,13-DIEN-15-OL

Source: *Trimusculus reticulatus*

Mol. formula: $C_{29}H_{46}O_7$

Mol. wt.: 506

Solvent: $CDCl_3$

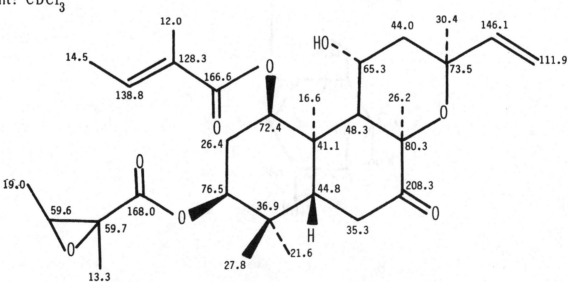

D.C. Manker and D.J. Faulkner, <u>Tetrahedron</u>, **43** (16), 3677 (1987).

DITERPENE H.A-XVI

Source: *Helichrysum ambiguum* Turcz.

Mol. formula: $C_{30}H_{44}O_8$

Mol. wt.: 532

Solvent: $CDCl_3$

J. Jakupovic, A. Schuster, F. Bohlmann, U. Ganzer, R.M. King and H. Robinson, <u>Phyto-chemistry</u>, **28** (2), 543 (1989).

DITERPENE GS-I

Source: *Gutierrezia sphaerocephala* Gray

Mol. formula: $C_{30}H_{46}O_8$

Mol. wt.: 534

Solvent: $CDCl_3$

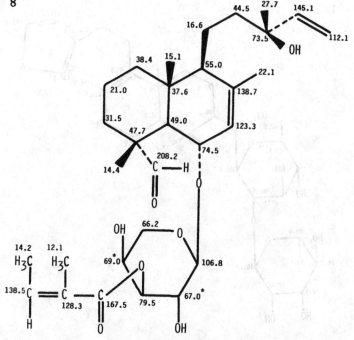

F. Gao, M. Leidig and T.J. Mabry, <u>Phytochemistry</u>, 25(6), 1371 (1986).

||||||||||||||||||||||||||||||||||||||||||||||||||||||||||||||||||||||||||||||||||||||||||||||||||||||||||||

BAIYUNOSIDE

Source: *Phlomis betonicoides* Diels

Mol. formula: $C_{31}H_{48}O_{11}$

Mol. wt.: 596

Solvent: C_5D_5N

T. Tanaka, O. Tanaka, Z.-W. Lin and J. Zhou, <u>Chem. Pharm. Bull.</u> (Tokyo), 33 (10), 4275 (1985).

133

PHLOMISOSIDE-I

Source: *Phlomis betonicoides* Diels

Mol. formula: $C_{32}H_{50}O_{11}$

Mol. wt.: 610

Solvent: C_5D_5N

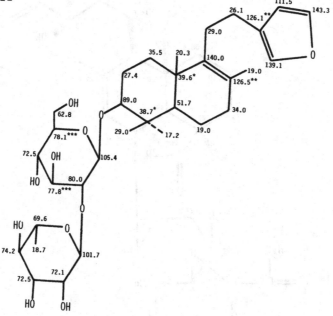

T.Tanaka, O.Tanaka, Z.-W.Lin and J.Zhou, <u>Chem.Pharm.Bull.</u> (Tokyo), 33(10), 4275 (1985).

||||||||||||||||||||||||||||||||||||||||||||||||||||||||||||||||||||||||||||||||||||||||||||||||||||

PHLOMISOSIDE-II

Source: *Phlomis betonicoides* Diels

Mol. formula: $C_{32}O_{50}O_{12}$

Mol. wt.: 626

Solvent: C_5D_5N

T. Tanaka, O. Tanaka, Z.-W. Lin and J. Zhou, <u>Chem. Pharm. Bull.</u> (Tokyo), 33(10), 4275 (1985).

GOSHONOSIDE-F4

Source: *Rubus chingii* Hu

Mol. formula: $C_{32}H_{54}O_{12}$

Mol. wt.: 630

Solvent: C_5D_5N

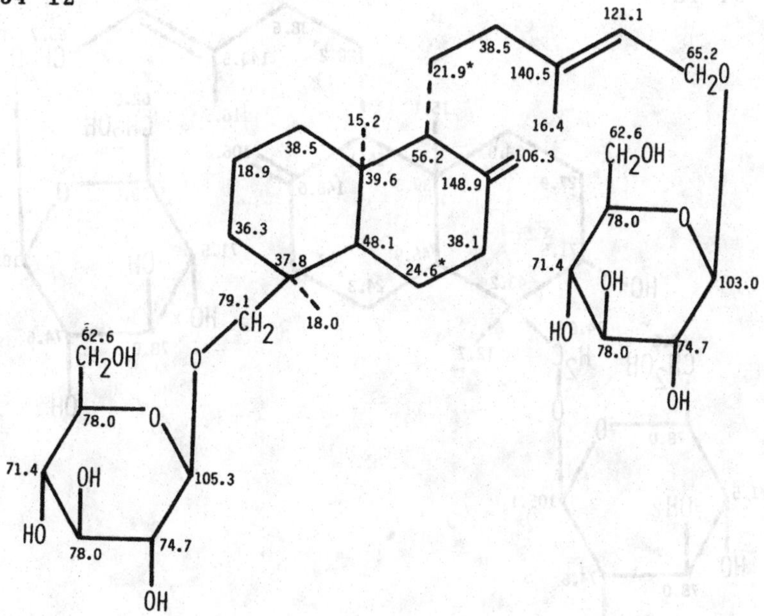

T. Tanaka, K. Kawamura, T. Kitahara, H. Kohda and O. Tanaka, <u>Phytochemistry</u>, 23 (3), 615 (1984).

GOSHONOSIDE-F3

Source: *Rubus chingii* Hu

Mol. formula: $C_{32}H_{52}O_{13}$

Mol. wt.: 644

Solvent: C_5D_5N

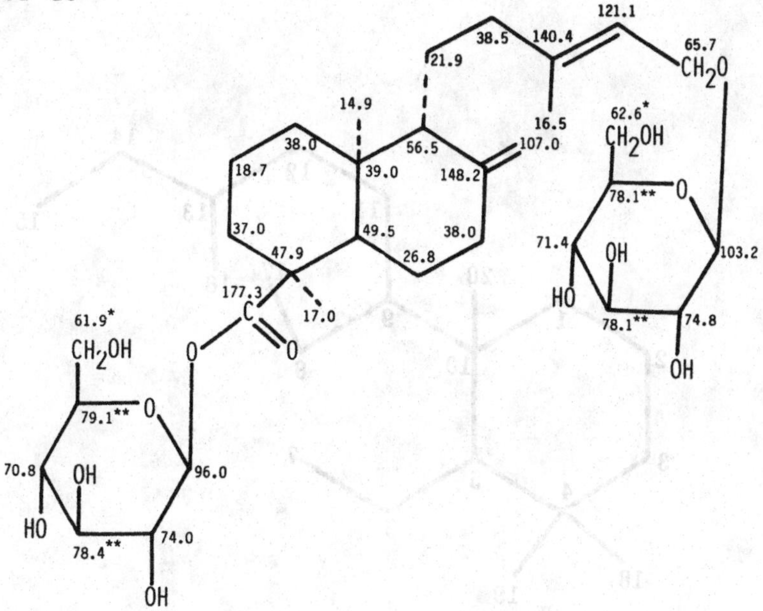

T. Tanaka, K. Kawamura, T. Kitahara, H. Kohda and O. Tanaka, <u>Phytochemistry</u>, 23 (3), 615 (1984).

Source: *Rubus chingii* Hu

Mol. formula: $C_{32}H_{54}O_{13}$

Mol. wt.: 646

Solvent: C_5D_5N

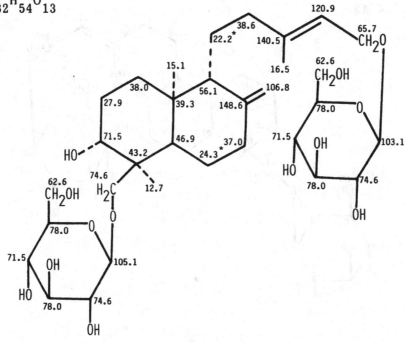

T. Tanaka, K. Kawamura, T. Kitahara, H. Kohda and O. Tanaka, <u>Phytochemistry</u>, **23** (3), 615 (1984).

Seco-LABDANE

Basic skeleton:

GALEOPSITRIONE

Source: *Galeopsis angustifolia* Hoffm.
Mol. formula: $C_{20}H_{28}O_4$
Mol. wt.: 332
Solvent: $CDCl_3$

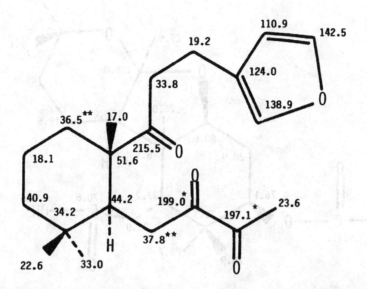

L.Perez-Sirvent, B.Rodriguez, G.Savona and O.Servettaz, <u>Phytochemistry</u>, 22(2), 527 (1983).

||||||||||||||||||||||||||||||||||||||||||||||||||||||||||||||||||||||||||||||||||||||||||||||||||||||||||||||||||||||||||||||||

ROTALIN B

Source: *Mycale rotalis*
Mol. formula: $C_{20}H_{33}BrO_2$
Mol. wt.: 385
Solvent: C_6D_6

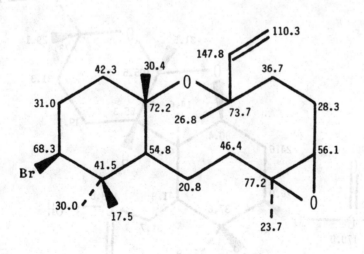

G. Corriero, A. Madaio, L. Mayol, V. Piccialli and D. Sica, <u>Tetrahedron</u>, 45 (1), 277 (1989).

DITERPENE HA-XX

Source: *Helichrysum ambiguum* Turcz.

Mol. formula: $C_{22}H_{34}O_6$

Mol. wt.: 394

Solvent: $CDCl_3$

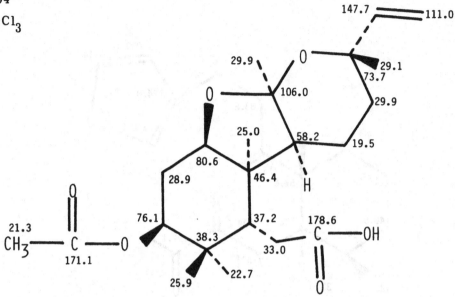

J. Jakupovic, A. Schuster, F. Bohlmann, U. Ganzer, R.M. King and H. Robinson, <u>Phyto-chemistry</u>, **28** (2), 543 (1989).

DITERPENE HA-XXII

Source: *Helichrysum ambiguum* Turcz.

Mol. formula: $C_{22}H_{34}O_6$

Mol. wt.: 394

Solvent: $CDCl_3$

J. Jakupovic, A. Schuster, F. Bohlmann, U. Ganzer, R.M. King and H. Robinson, <u>Phyto-chemistry</u>, **28** (2), 543 (1989).

Source: *Cluytia richardiana* L.

Mol. formula: $C_{22}H_{24}O_7$

Mol. wt.: 400

Solvent: $CDCl_3$

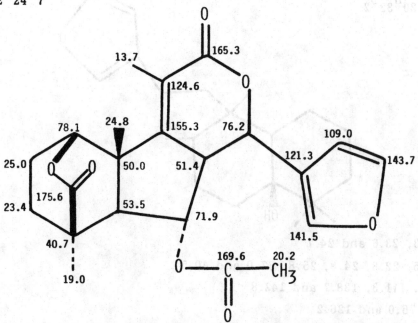

J.S. Mossa, J.M. Cassady, J.F. Kozlowski, T.M. Zennie, M.D. Antoun, M.G. Pellechia, A.T. McKenzie and S.R. Byrn, Tetrahedron Lett., **29**(30), 3627 (1988).

||||||||||||||||||||||||||||||||||||||||||||||||||||||||||||||||||||||||||||||||||||||||||||||||||||||||||||

CHETTAPHANANE

Basic skeleton:

AMBLIOL C

Source: *Dysidea amblia* (de Laubenfels)

Mol. formula: $C_{20}H_{32}O_2$

Mol. wt.: 304

Solvent: C_6D_6

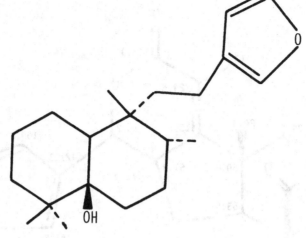

CH_3: 14.3, 21.3, 23.6 and 24.7

CH_2: 18.4, 21.5, 22.8, 24.8, 25.8, 37.0 and 40.5

CH: 35.5, 40.7, 111.3, 138.7 and 142.8

C: 37.5, 39.1, 76.0 and 126.2

R.P. Walker, R.M. Rosser, D.J. Faulkner, L.S. Bass, He Cun-Heng and J. Clardy, <u>J. Org. Chem.</u>, **49** (26), 5160 (1984).

||||||||||||||||||||||||||||||||||||||||||||||||||||||||||||||||||||||||||||||||||||||||

3α-HYDROXY-5,6-DEHYDROCHILIOLIDE

Source: *Chiliotrichium rosmarinifolium* Less. and *Nardophyllum lanatum* (Meyen) Cabrera

Mol. formula: $C_{20}H_{26}O_4$

Mol. wt.: 330

Solvent: $CDCl_3$

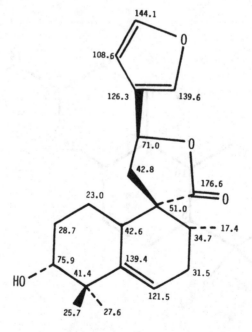

J. Jakupovic, S. Banerjee, F. Bohlmann, R.M. King and H. Robinson, <u>Tetrahedron</u>, **42**(5), 1305 (1986).

3α-HYDROXY-5β,10β-EPOXYCHILIOLIDE

Source: *Chiliotrichium rosmarinifolium* Less.

Mol. formula: $C_{20}H_{26}O_5$

Mol. wt.: 346

Solvent: $CDCl_3$

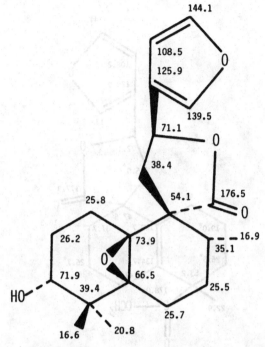

J. Jakupovic, S. Banerjee, F. Bohlmann, R.M. King and H. Robinson, <u>Tetrahedron</u>, 42(5), 1305 (1986).

3α,5α-DIHYDROXYCHILIOLIDE

Source: *Chiliotrichium rosmarinifolium* Less.

Mol. formula: $C_{20}H_{28}O_5$

Mol. wt.: 348

Solvent: $CDCl_3$

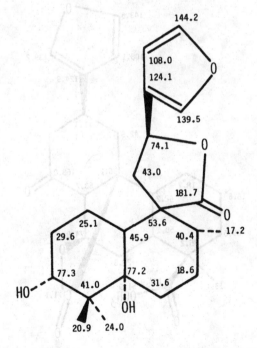

J. Jakupovic, S. Banerjee, F. Bohlmann, R.M. King and H. Robinson, <u>Tetrahedron</u>, 42(5), 1305 (1986).

PENDULIFLAWOROSIN

Source: *Croton penduliflorus* Hutch.

Mol. formula: $C_{21}H_{26}O_5$

Mol. wt.: 358

Solvent: $CDCl_3$

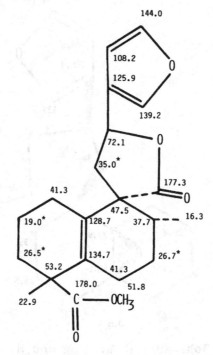

E.K. Adesogan, <u>J. Chem. Soc. Perkin Trans. I</u>, (4), 1151 (1981).

DIASIN

Source: *Croton diasii* Pires

Mol. formula: $C_{21}H_{24}O_7$

Mol. wt.: 388

Solvent: $CDCl_3$

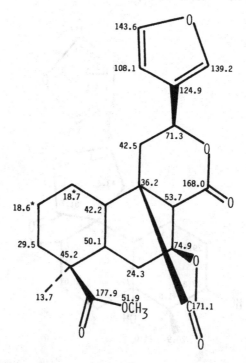

M.A. De Alvarenga, H.E. Gottlieb, O.R. Gottlieb, M.T. Magalhaes and V.O. Da Silva, <u>Phytochemistry</u>, 17 (10), 1773 (1978).

AGELASIMINE A

Source: *Agelas mauritiana*
Mol. formula: $C_{27}H_{43}N_5O$
Mol. wt.: 453

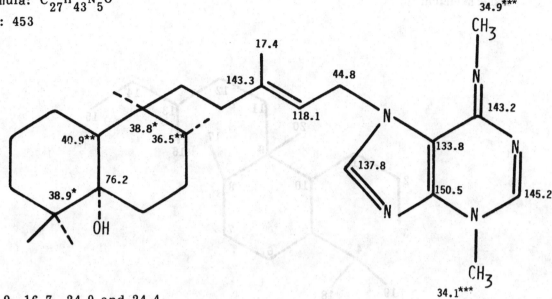

CH$_3$: 15.9, 16.7, 24.0 and 24.4
CH$_2$: 21.7, 22.1, 26.4, 32.0, 33.0, 36.1 and 36.9

R.Fathi-Afshar and T.M. Allen, <u>Can. J. Chem.</u>, 66(1), 45 (1988).

|||||||||||||||||||||||||||||||||||||||||||||||||||||||||||||||||||||||||||||||||||||||||||||||||||||||||||||||||||||

AGELASIMINE B

Source: *Agelas mauritiana*
Mol. formula: $C_{27}H_{45}N_5O$
Mol. wt.: 455

CH$_3$: 16.0 16.8, 24.1 and 24.5
CH$_2$: 21.8, 22.2, 26.5, 32.1, 33.1, 36.1 and 37.0

R. Fathi-Afshar and T.M. Allen, <u>Can. J. Chem.</u>, 66(1), 45 (1988).

Basic skeleton:

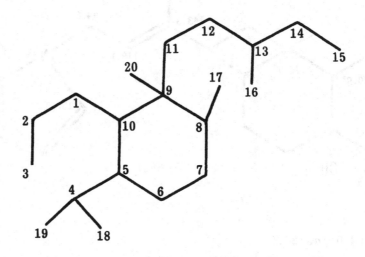

||||||||||||||||||||||||||||||||||||||||||||||||||||||||||||||||||||||||||||||||||||||||||||||||||||||||||||

SECOCHILIOTRIN

Source: *Chiliotrichium rosmarinifolium* Less.

Mol. formula: $C_{20}H_{30}O_3$

Mol. wt.: 318

Solvent: $CDCl_3$

J. Jakupovic, S. Banerjee, F. Bohlmann, R.M. King and H. Robinson, <u>Tetrahedron</u>, 42 (5), 1305 (1986).

ISOCHILIOLIDE LACTONE

Source: *Chiliotrichium rosmarinifolium* Less.

Mol. formula: $C_{20}H_{24}O_5$

Mol. wt.: 344

Solvent: $CDCl_3$

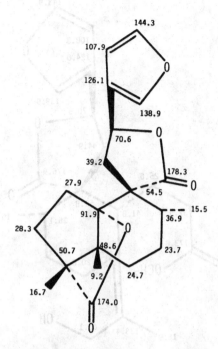

J. Jakupovic, S. Banerjee, F. Bohlmann, R.M. King and H. Robinson, <u>Tetrahedron</u>, 42(5), 1305 (1986).

SECOCHILIOLIDE LACTONE

Source: *Nardophyllum lanatum* (Meyen) Cabrera

Mol. formula: $C_{20}H_{24}O_5$

Mol. wt.: 344

Solvent: $CDCl_3$

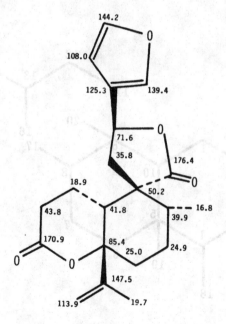

J. Jakupovic, S. Banerjee, F. Bohlmann, R.M. King and H. Robinson, <u>Tetrahedron</u>, 42 (5), 1305 (1986).

Source: *Chiliotrichium rosmarinifolium* Less.

Mol. formula: $C_{21}H_{28}O_6$

Mol. wt.: 376

Solvent: $CDCl_3$

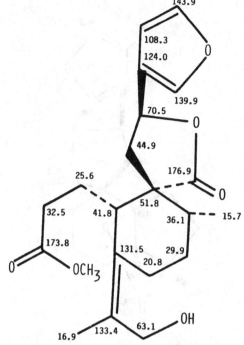

J. Jakupovic, S. Banerjee, F. Bohlmann, R.M. King and H. Robinson, <u>Tetrahedron</u>, 42(5), 1305 (1986).

||||||||||||||||||||||||||||||||||||||||||||||||||||||||||||||||||||||||||||||||||||||||||||

CLERODANE

Basic skeleton:

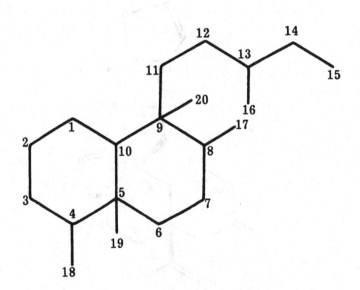

DITERPENE BG-IVA

Source: *Baccharis genistelloides*
Mol. formula: $C_{16}H_{22}O_5$
Mol. wt.: 294
Solvent: $CDCl_3+CD_3OD$

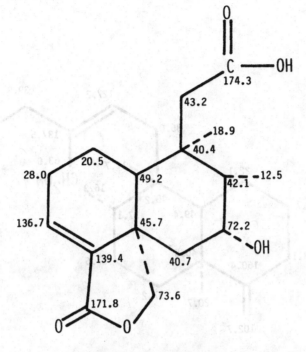

M. Kuroyanagi, K. Fujita, M. Kazaoka, S. Matsumoto, A. Ueno, S. Fukushima and M. Katsuoka, <u>Chem. Pharm. Bull.</u> (Tokyo), **33** (11), 5075 (1985).

CLEROD-3,13(16),14-TRIEN-17-OIC ACID

Source: *Jungermannia infusca*
Mol. formula: $C_{20}H_{30}O_2$
Mol. wt.: 302
Solvent: $CDCl_3$

M. Toyota, F. Nagashima and Y. Asakawa, <u>Phytochemistry</u>, 28 (9), 2507 (1989).

ISOLINARIDIOL

Source: *Linaria saxatilis* (L.) Chaz.
Mol. formula: $C_{20}H_{34}O_2$
Mol. wt.: 306
Solvent: $CDCl_3$

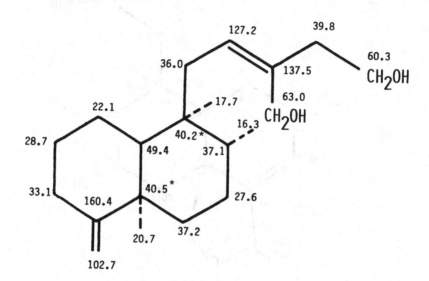

A.S. Feliciano, A.F. Barrero, J.M.M. Del Corral, M. Gordaliza and M. Medarde, <u>Tetrahedron</u>, **41** (4), 671 (1985).

DEHYDROCROTONIN

Source: *Croton cajucara* Benth
Mol. formula: $C_{19}H_{22}O_4$
Mol. wt.: 314
Solvent: $CDCl_3$

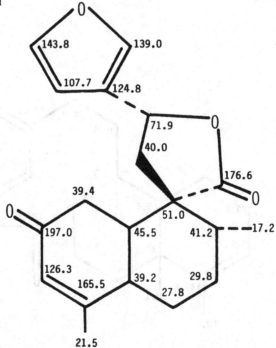

H. Itokawa, Y. Ichihara, H. Kojima, K. Watanabe and K. Takeya, <u>Phytochemistry</u>, **28** (6), 1667 (1989).

BACCHARODILYIC ACID

Source: Baccharis coridifolia (Wedd.) (?) left

Mol. formula: $C_{20}H_{28}O_3$

Mol. wt.: 316

Solvent: $CDCl_3$

C. Zdero, F. Bohlmann, J. C. Solomon, R. M. King, and H. Robinson, Phytochemistry, 28 (2), 531 (1989).

16-OXO-ENT-CLERODA-4(18),13-
DIEN-18,11-LACTONE

Source: Baccharis articulata (L.) Chez.

Mol. formula: $C_{20}H_{28}O_3$

Mol. wt.: 316

Solvent: $CDCl_3$

A. S. Feliciano, A. P. Barrero, M. M. Del Corral, M. Gordaliza, and M. Medarde, Tetrahedron, 41(4), 671 (1985).

BACCHABOLIVIC ACID

Source: *Baccharis boliviensis* (Wedd.) Cuatr.

Mol. formula: $C_{20}H_{28}O_3$

Mol. wt.: 316

Solvent: $CDCl_3$

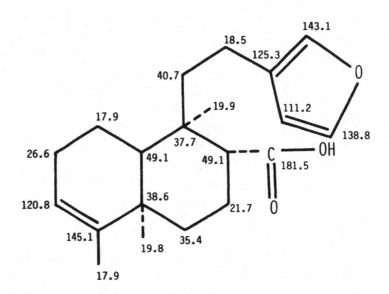

C. Zdero, F. Bohlmann, J.C. Solomon, R.M. King and H. Robinson, <u>Phytochemistry</u>, 28 (2), 531 (1989).

||||||||||||||||||||||||||||||||||||||||||||||||||||||||||||||||||||||||||||||||||||||||||||||||

16-OXO-<u>ENT</u>-CLERODA-4(18),12-DIEN-15, 11-LACTONE

Source: *Linaria saxatilis* (L.) Chaz.

Mol. formula: $C_{20}H_{28}O_3$

Mol. wt.: 316

Solvent: $CDCl_3$

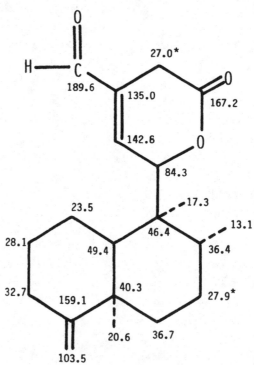

A.S. Feliciano, A.F. Barrero, J.M.M. Del Corral, M. Gordaliza and M. Medarde, <u>Tetrahedron</u>, 41(4), 671 (1985).

SOLIDAGOLACTONE V

Source: *Solidago altissima* L.
Mol. formula: $C_{20}H_{28}O_3$
Mol. wt.: 316
Solvent: $CDCl_3$

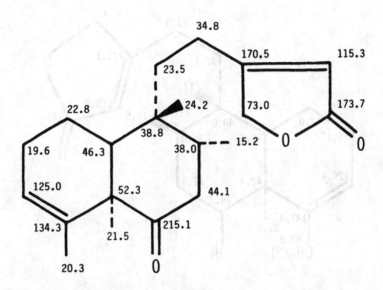

S. Manabe and C. Nishino, <u>Tetrahedron</u>, 42(13), 3461 (1986).

||||||||||||||||||||||||||||||||||||||||||||||||||||||||||||||||||||||||||||||||||||||||||||||||

BACCHOTRICUNEATIN D

Source: *Baccharis tricuneata* (L.f.) Pers.
Mol. formula: $C_{20}H_{30}O_3$
Mol. wt.: 318
Solvent: $CDCl_3$

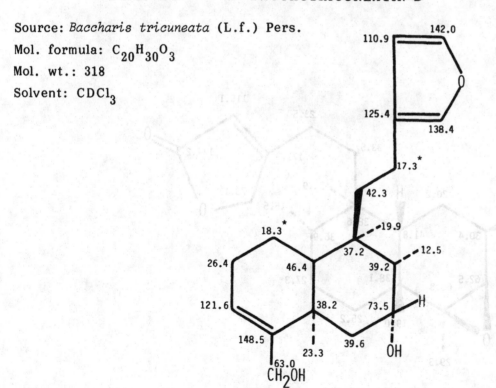

H. Wagner, R. Seitz, H. Lotter and W. Herz, <u>J. Org. Chem.</u>, 43(17), 3339 (1978).

6α,18-DIHYDROXY-<u>CIS</u>-CLERODA-3,13(14)-DIENE-15,16-OLIDE

Source: *Gutierrezia texana* (DC) T. and G.

Mol. formula: $C_{20}H_{30}O_4$

Mol. wt.: 334

Solvent: $CDCl_3$

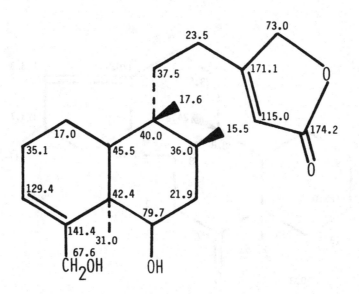

F. Gao and T.J. Mabry, <u>Phytochemistry</u>, 26 (1), 209 (1987).

||||||||||||||||||||||||||||||||||||||||||||||||||||||||||||||||||||||||||||||||||||||||||

3,4β-EPOXY-5β,10β-<u>CIS</u>-17α,20α-CLERODA-13(14)-EN-15,16-OLIDE

Source: *Solidago shortii*

Mol. formula: $C_{20}H_{30}O_3$

Mol. wt.: 318

Solvent: $CDCl_3$

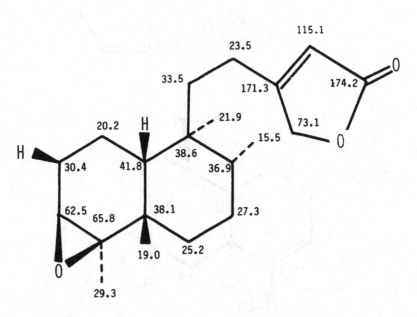

N. Fang, S.Yu, T.J. Mabry, K.A. Abboud and S.H. Simonsen, <u>Phytochemistry</u>, 27(10), 3187 (1988).

16α-HYDROXY-CLERODA-3,13(14)Z-DIEN-15,16-OLIDE

Source: *Polyalthia longifolia* Thw

Mol. formula: $C_{20}H_{30}O_3$

Mol. wt.: 318

Solvent: $CDCl_3$

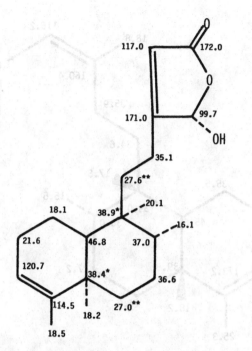

A.P. Phadnis, S.A. Patwardhan, N.N. Dhaneshwar, S.S. Tavale and T.N. Guru Row, Phytochemistry, 27(9), 2899 (1988).

||||||||||||||||||||||||||||||||||||||||||||||||||||||||||||||||||||||||||||||||||||||||||||||||||||||||||||||||||||||||||||||||||||||||||||||||||||||||||||||

2β-HYDROXY-5β,10β-CIS-17α,20α-CLERODA-3,13(14)-DIENE-15,16-OLIDE

Source: *Ageratina saltillensis* (B.L. Robo) King and Robinson

Mol. formula: $C_{20}H_{30}O_3$

Mol. wt.: 318

Solvent: $CDCl_3$

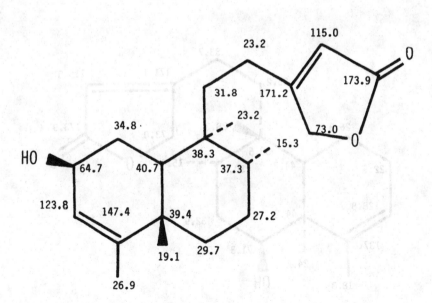

N. Fang, S.Yu, T.J. Mabry, K.A. Abboud and S.H. Simonsen, Phytochemistry, 27(10), 3187 (1988).

(13Z)-2-OXO-5β,10β-CIS-17α,20β-CLERODA-3,13(14)-DIENE-15-OIC ACID

Source: *Ageratina saltillensis* (B.L. Robo) King and Robinson

Mol. formula: $C_{20}H_{30}O_3$

Mol. wt.: 318

Solvent: $CDCl_3$

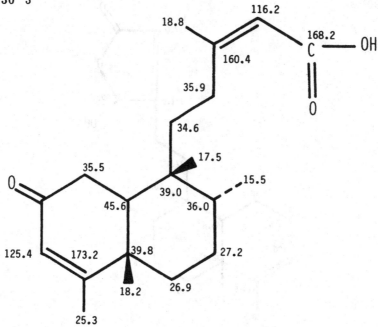

N. Fang, S. Yu, T.J. Mabry, K.A. Abboud and S.H. Simonsen, Phytochemistry, **27** (10), 3187 (1988).

||||||||||||||||||||||||||||||||||||||||||||||||||||||||||||||||||||||||||||||||||||||||||||||||||||||||||||||||

SOLIDAGOLACTONE-IV

Source: *Solidago altissima* L.

Mol. formula: $C_{20}H_{30}O_3$

Mol. wt.: 318

Solvent: $CDCl_3$

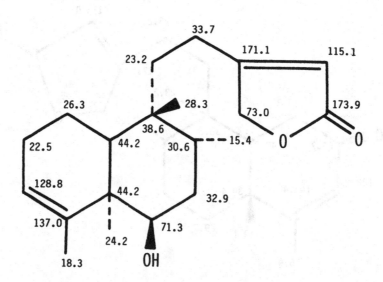

S. Manabe and C. Nishino, Tetrahedron, **42**(13), 3461 (1986).

ENT-CLERODA-4(18),13Z-DIEN-15-OIC ACID(METHYL ESTER)

Source: *Ageratina ixiocladon* (Benth. ex Oersted) King et Robins. (syn. *Eupatorium ixiocladon*)

Mol. formula: $C_{21}H_{34}O_2$

Mol. wt.: 318

Solvent: $CDCl_3$

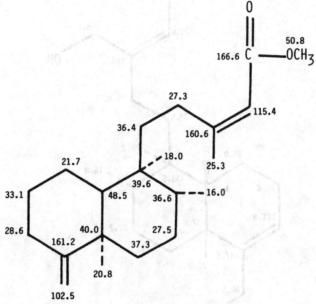

G. Tamayo-Castillo, J. Jakupovic, F. Bohlmann, V. Castro and R.M. King, Phytochemistry, 28 (1), 139 (1989).

||||||||||||||||||||||||||||||||||||||||||||||||||||||||||||||||||||||||||||||||||||||||||||||||

(13Z)-2β-HYDROXY-5β,10β-CIS-17α,20β-CLERODA-3,13(14)-DIENE-15-OIC ACID

Source: *Ageratina saltillensis* (B.L. Robo) King and Robinson

Mol. formula: $C_{20}H_{32}O_3$

Mol. wt.: 320

Solvent: $CDCl_3$

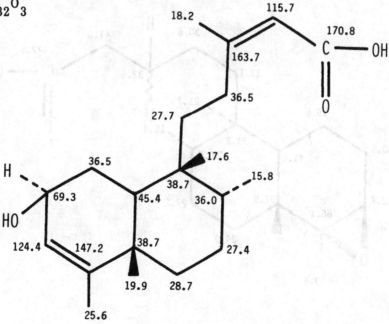

N. Fang, S. Yu, T.J. Mabry, K.A. Abboud and S.H. Simonsen, Phytochemistry, 27 (10), 3187 (1988).

BINCATRIOL

Source: *Baccharis incarum* Wedd

Mol. formula: $C_{20}H_{34}O_3$

Mol. wt.: 322

Solvent: $CDCl_3$

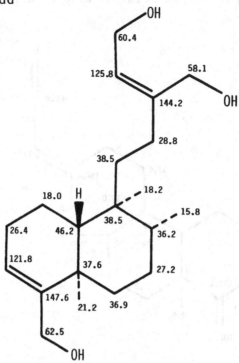

A. San-Martin, A. Givovich and M. Castillo, <u>Phytochemistry</u>, **25** (1), 264 (1986).

3,4β-EPOXY-5β,10β-<u>CIS</u>-17α,20α-CLERODA-15-OIC ACID

Source: *Ageratina saltillensis* (B.L. Robo) King and Robinson

Mol. formula: $C_{20}H_{34}O_3$

Mol. wt.: 322

Solvent: $CDCl_3$

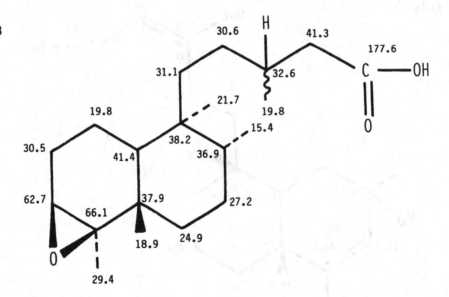

N. Fang, S.Yu, T.J. Mabry, K.A. Abboud and S.H. Simonsen, <u>Phytochemistry</u>, **27**(10), 3187 (1988).

16 HYDROXY-3,4β-EPOXY-5β,10β-<u>CIS</u>-17α,20α-CLERODA-15-OL

Source: *Ageratina saltillensis* (B.L. Robo) King and Robinson

Mol. formula: $C_{20}H_{36}O_3$

Mol. wt.: 324

Solvent: $CDCl_3$

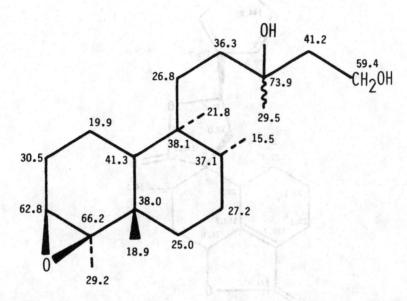

N. Fang, S.Yu, T.J. Mabry, K.A. Abboud and S.H. Simonsen, <u>Phytochemistry</u>, 27(10), 3187 (1988).

TEUSCOROLIDE

Source: *Teucrium scorodonia* L.

Mol. formula: $C_{19}H_{18}O_5$

Mol. wt.: 326

Solvent: C_5D_5N

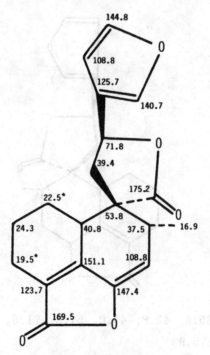

J.L. Marco, B. Rodriguez, C. Pascual, G. Savona and F. Piozzi, <u>Phytochemistry</u>, 22(3), 727 (1983).

TEUCVIDIN

Source: *Teucrium viscidium*

Mol. formula: $C_{19}H_{20}O_5$

Mol. wt.: 328

Solvent: $CDCl_3$

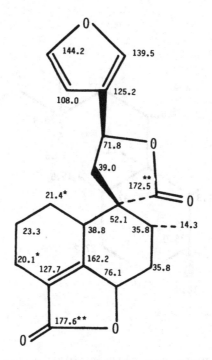

G. Savona, M.P. Paternostro, F. Piozzi, J.R. Hanson, P.B. Hitchcock and S.A. Thomas, J. Chem. Soc. Perkin Trans. I, (9), 1080 (1978).

||||||||||||||||||||||||||||||||||||||||||||||||||||||||||||||||||||||||||||||||||||||||||||||||||||||||||||||||||||||||||||

TEUFLIN (6-EPITEUCVIN)

Source: *Teucrium viscidum* Blume

Mol. formula: $C_{19}H_{20}O_5$

Mol. wt.: 328

17.6, 18.7, 23.3, 23.7, 31.8, 35.8, 42.8, 43.0, 51.0, 71.6, 76.6, 107.8, 123.7, 124.3, 139.8, 144.2, 166.2, 173.5 and 175.9

M. Node, M. Sai and E. Fujita, Phytochemistry, 20(4), 757 (1981).

BACCHASMACRANONE

Source: *Baccharis macraei* Hook et Arn

Mol. formula: $C_{20}H_{24}O_4$

Mol. wt.: 328

Solvent: $CDCl_3$

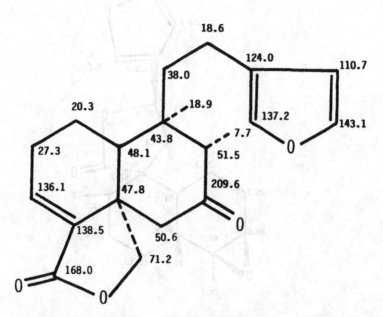

V. Gambaro, M.C.Chamy, J.A. Garbarino, A. San-Martin and M. Castillo, <u>Phytochemistry</u>, 25(9), 2175 (1986).

DEMETHYLMARRUBIAKETONE

Source: *Leonurus marrubiastrum* L.

Mol. formula: $C_{19}H_{22}O_5$

Mol. wt.: 330

Solvent: DMSO-d$_6$

R. Tschesche and B. Streuff, <u>Chem. Ber.</u>, 111 (6), 2130 (1978).

CHILIOMARIN

Source: *Chiliotrichium rosmarinifolium* Less.

Mol. formula: $C_{20}H_{26}O_4$

Mol. wt.: 330

Solvent: $CDCl_3$

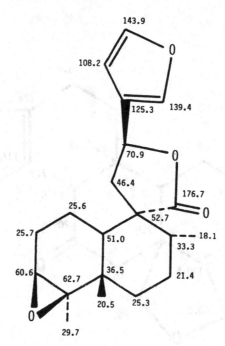

J. Jakupovic, S. Banerjee, F. Bohlmann, R.M. King and H. Robinson, <u>Tetrahedron</u>, 42(5), 1305 (1986).

||||||||||||||||||||||||||||||||||||||||||||||||||||||||||||||||||||||||||||||||||||||||||||||||||||||

CIS-CLERODA-3,13(14)-DIENE-15,16:18,19-DIOLIDE

Source: *Gutierrezia texana* (DC) T. and G.

Mol. formula: $C_{20}H_{26}O_4$

Mol. wt.: 330

Solvent: $CDCl_3$

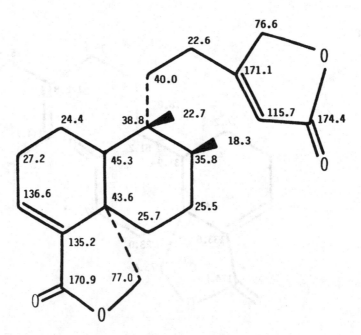

F. Gao and T.J. Mabry, <u>Phytochemistry</u>, 26(1), 209 (1987).

6-DEOXY-SOLIDAGOLACTONE IV-18,19-OLIDE

Source: *Solidago gigantea* Ait.

Mol. formula: $C_{20}H_{26}O_4$

Mol. wt.: 330

Solvent: $CDCl_3$

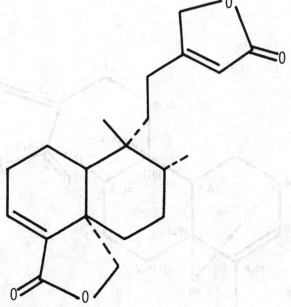

15.6, 19.0, 23.4, 25.5, 26.3, 27.0, 30.6, 31.0, 36.5, 37.9, 42.6, 43.0, 73.0, 76.1, 115.5, 135.7, 169.9, 170.1 and 173.6

J. Jurenitsch, J. Maurer, U. Rain and W. Robien, Phytochemistry, 27 (2), 626 (1988).

|||||||||||||||||||||||||||||||||||||||||||||||||||||||||||||||||||||||||||||||||||||||||||||||||||||||||||||||||||||||||||||||||||

DESOXYARTICULIN

Source: *Baccharis pedicellata* DC. and *B. marginalis* DC.

Mol. formula: $C_{20}H_{26}O_4$

Mol. wt.: 330

Solvent: $CDCl_3$

F. Faini, P. Rivera, M. Mahu and M. Castillo, Phytochemistry, 26 (12), 3281 (1987).

BREVIFLORALACTONE

Source: *Salvia breviflora* Moc and Sesse

Mol. formula: $C_{20}H_{28}O_4$

Mol. wt.: 332

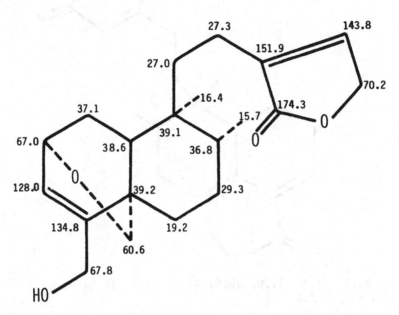

G. Cuevas, O. Collera, F. Garcia, J. Cardenas, E. Maldonado and A. Ortega, <u>Phytochemistry</u>, 26(7), 2019 (1987).

3,12-DIOXO-15,16-EPOXY-4-HYDROXYCLERODA-13(16),14-DIENE

Source: *Croton argyrophylloides*

Mol. formula: $C_{20}H_{28}O_4$

Mol. wt.: 332

Solvent: $CDCl_3$

F.J.Q. Monte, E.M.G. Dantas and R. Braz F., <u>Phytochemistry</u>, 27(10), 3209 (1988).

18,19-EPOXY-19α-HYDROXY-<u>CIS</u>-CLERODA-3,13(14)-DIENE-15,16-OLIDE

Source: *Gutierrezia texana* (DC) T. and G.

Mol. formula: $C_{20}H_{28}O_4$

Mol. wt.: 332

Solvent: $CDCl_3$

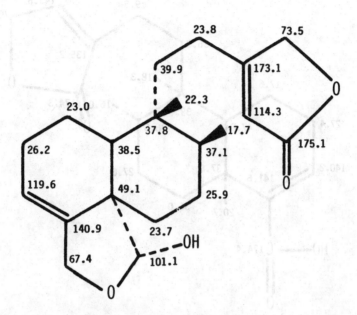

F. Gao and T.J. Mabry, <u>Phytochemistry</u>, 26(1), 209 (1987).

HAUTRIWAIC ACID

Source: *Baccharis sarothroides* A.Gray

Mol. formula: $C_{20}H_{28}O_4$

Mol. wt.: 332

Solvent: $CDCl_3$

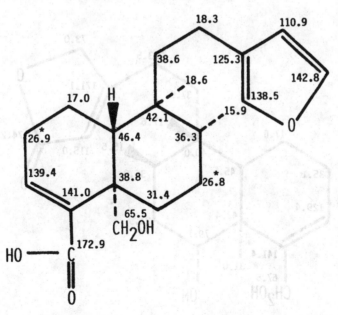

F.J. Arriaga-Giner, E. Wollenweber, I. Schober, P. Dostal and S. Braun, <u>Phytochemistry</u>, 25(3), 719 (1986).

Source: *Baccharis patagonica* Hook and Arn

Mol. formula: $C_{20}H_{28}O_4$

Mol. wt.: 332

Solvent: $CDCl_3$

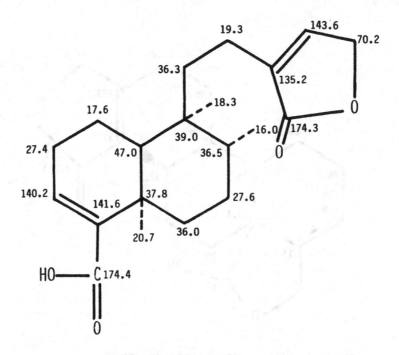

A.P. Rivera, F. Faini and M. Castillo, <u>J. Nat. Prod.</u>, (Lloydia), **51** (1), 155 (1988).

||||||||||||||||||||||||||||||||||||||||||||||||||||||||||||||||||||||||||||||||||||||||||||||||||||||||

6α,18-DIHYDROXY-<u>CIS</u>-CLERODA-3,13(14)-DIENE-15,16-OLIDE

Source: *Gutierrezia texana* (DC) T. and G.

Mol. formula: $C_{20}H_{30}O_4$

Mol. wt.: 334

Solvent: $CDCl_3$

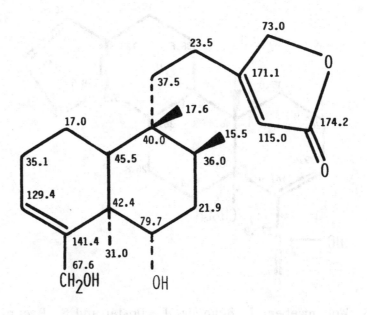

F. Gao and T.J. Mabry, <u>Phytochemistry</u>, **26** (1), 209 (1987).

18,19-DIHYDROXY-<u>CIS</u>-CLERODA-3,13(14)-DIENE-15,16-OLIDE

Source: *Gutierrezia texana* (DC) T. and G.

Mol. formula: $C_{20}H_{30}O_4$

Mol. wt.: 334

Solvent: $CDCl_3$

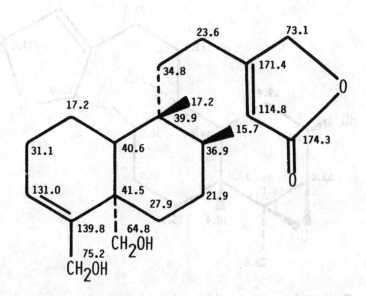

F. Gao and T.J. Mabry, <u>Phytochemistry</u>, **26**(1) 209 (1987).

||||||||||||||||||||||||||||||||||||||||||||||||||||||||||||||||||||||||||||||||||||||||||

4β,18-EPOXY-<u>ENT</u>-CLEROD-13<u>Z</u>-EN-15-OIC ACID (METHYL ESTER)

Source: *Ageratina ixiocladon* (Benth. ex Oersted) King et Robins. (syn. *Eupatorium ixiocladon*)

Mol. formula: $C_{21}H_{34}O_3$

Mol. wt.: 334

Solvent: $CDCl_3$

G. Tamayo-Castillo, J. Jakupovic, F. Bohlmann, V. Castro and R.M. King, <u>Phytochemistry</u>, **28** (1), 139 (1989).

2β-HYDROXY-3,4β-EPOXY-5β,10β-<u>CIS</u>-17α,20α-CLERODA-13(14)-EN-15,16-OLIDE

Source: *Ageratina saltillensis* (B.L. Robo) King and Robinson

Mol. formula: $C_{20}H_{30}O_4$

Mol. wt.: 334

Solvent: $CDCl_3$

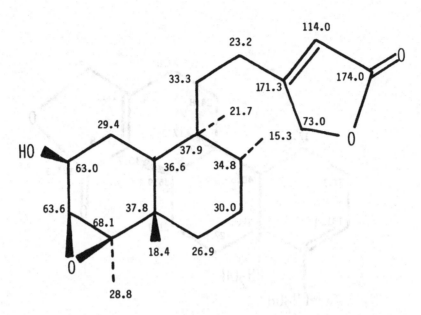

N. Fang, S. Yu, T.J. Mabry, K.A. Abboud and S.H. Simonsen, <u>Phytochemistry</u>, **27** (10), 3187 (1988).

16-HYDROXY-3,4β-EPOXY-5β,10β-<u>CIS</u>-17α,20α-CLERODA-13(14)-EN-15,16-<u>OLIDE</u>

Source: *Ageratina saltillensis* (B.L. Robo) King and Robinson

Mol. formula: $C_{20}H_{30}O_4$

Mol. wt.: 334

Solvent: $CDCl_3$

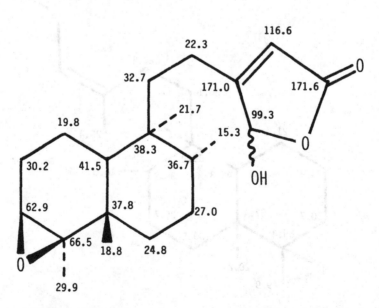

N. Fang, S. Yu, T.J. Mabry, K.A. Abboud and S.H. Simonsen, <u>Phytochemistry</u>, **27** (10), 3187 (1988).

3-OXO-4β-HYDROXY-5β,10β-CIS-17α,20α-CLERODA-13(14)-EN-15,$\overline{16}$-OLIDE

Source: *Ageratina saltillensis* (B.L. Robo) King and Robinson

Mol. formula: $C_{20}H_{30}O_4$

Mol. wt.: 334

Solvent: $CDCl_3$

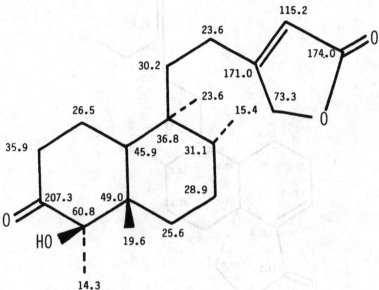

N. Fang, S. Yu, T.J. Mabry, K.A. Abboud and S.H. Simonsen, <u>Phytochemistry</u>, 27 (10), 3187 (1988).

‖‖

PANICULADIOL

Source: *Baccharis paniculata* DC.

Mol. formula: $C_{20}H_{30}O_4$

Mol. wt.: 334

Solvent: $CDCl_3$

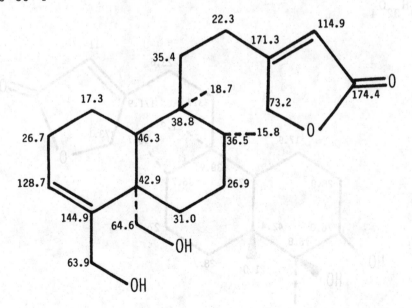

A.P. Rivera, L. Arancibia and M. Castillo, <u>J. Nat. Prod.</u>, (Lloydia), 52 (2), 433 (1989).

Source: *Portulaca* cv Jewel

Mol. formula: $C_{20}H_{30}O_4$

Mol. wt.: 334

Solvent: $CDCl_3$

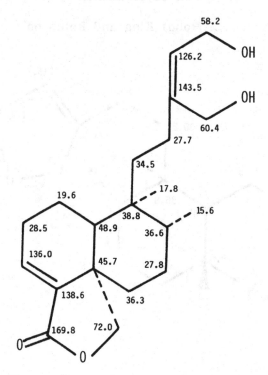

A. Ohsaki, N. Ohno, K. Shibata, T. Tokoroyama and T. Kubota, <u>Phytochemistry</u>, 25(10), 2414 (1986).

||||||||||||||||||||||||||||||||||||||||||||||||||||||||||||||||||||||||||||||||||||||||||||||||||||||||||||||||

3α,4β-DIHYDROXY-5β,10β-<u>CIS</u>-17α,20α-CLERODA-13 (14)-EN-15,$\overline{16}$-OLIDE

Source: *Ageratina saltillensis* (B.L. Robo) King and Robinson

Mol. formula: $C_{20}H_{32}O_4$

Mol. wt.: 336

Solvent: $CDCl_3$

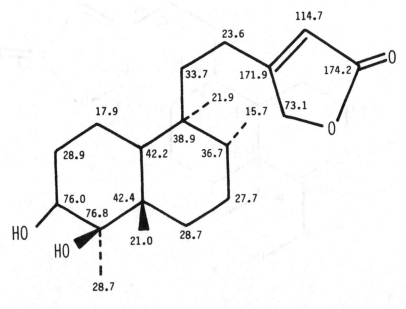

N. Fang, S. Yu, T.J. Mabry, K.A. Abboud and S.H. Simonsen, <u>Phytochemistry</u>, 27(10), 3187 (1988).

DITERPENE BR-I

Source: *Baccharis rhomboidalis* Remy
Mol. formula: $C_{21}H_{36}O_3$
Mol. wt.: 336
Solvent: $CDCl_3$

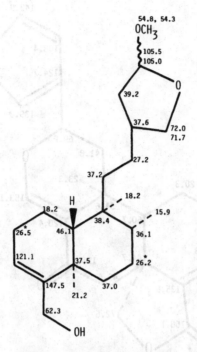

A. San-Martin, J. Rovirosa, C. Labbe, A. Givovich, M. Mahu and M. Castillo, <u>Phyto-</u>
<u>chemistry</u>, 25(6), 1393 (1986).

||||||||||||||||||||||||||||||||||||||||||||||||||||||||||||||||||||||||||||||||||||||||||||||||||||||||||||||||||

JEWENOL A

Source: *Portulaca* cv Jewel
Mol. formula: $C_{20}H_{34}O_4$
Mol. wt.: 338
Solvent: CD_3OD

A. Ohsaki, N. Ohno, K. Shibata, T. Tokoroyama, T. Kubota, K. Hirotsu and T. Higuchi,
<u>Phytochemistry</u>, 27(7), 2171 (1988).

CONYCEPHALOIDE

Source: *Conyza podocephala*
Mol. formula: $C_{20}H_{20}O_5$
Mol. wt.: 340
Solvent: $CDCl_3$

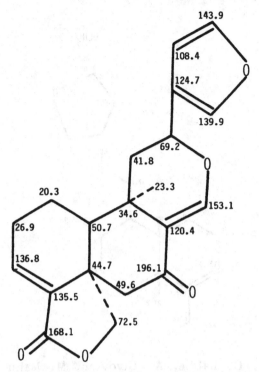

F. Bohlmann and P. Wegner, <u>Phytochemistry</u>, **21** (7), 1693 (1982).

1(10)-DEHYDROSALVIARIN

Source: *Salvia lineata* Benth.
Mol. formula: $C_{20}H_{20}O_5$
Mol. wt.: 340
Solvent: $CDCl_3$

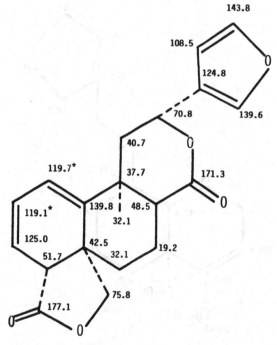

B. Esquivel, J. Cardenas, T.P. Ramamoorthy and L. Rodriguez-Hahn, <u>Phytochemistry</u>, **25**(10), 2381 (1986).

LINEARIFOLINE

Source: *Salvia lineata* Benth.

Mol. formula: $C_{20}H_{20}O_5$

Mol. wt.: 340

Solvent: $CDCl_3$

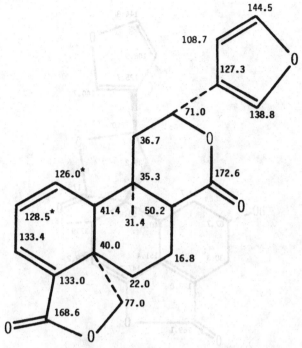

B. Esquivel, J. Cardenas, T.P. Ramamoorthy and L. Rodriguez-Hahn, <u>Phytochemistry</u>, 25(10), 2381 (1986).

SALVIFARICIN

Source: *Salvia farinacea* Benth.

Mol. formula: $C_{20}H_{20}O_5$

Mol. wt.: 340

Solvent: $CDCl_3$

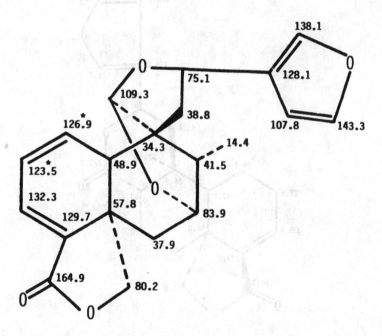

G. Savona, D. Raffa, M. Bruno and B. Rodriguez, <u>Phytochemistry</u>, 22(3), 784 (1983).

2-HYDROXYTEUSCOROLIDE

Source: *Teucrium scorodonia* L.

Mol. formula: $C_{19}H_{18}O_6$

Mol. wt.: 342

Solvent: C_5D_5N

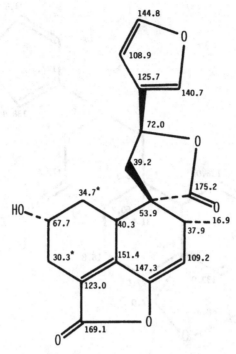

J.L. Marco, B. Rodriguez, C. Pascual, G. Savona and F. Piozzi, <u>Phytochemistry</u>, 22(3), 727 (1983).

BACCHOTRICUNEATIN A

Source: *Baccharis tricuneata* (L.f.) Pers.

Mol. formula: $C_{20}H_{22}O_5$

Mol. wt.: 342

Solvent: $CDCl_3$

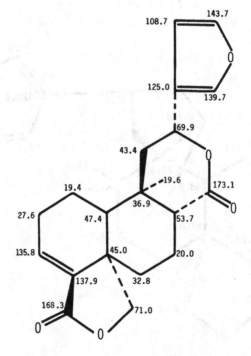

H. Wagner, R. Seitz, H. Lotter and W. Herz, <u>J. Org. Chem.</u>, 43(17), 3339 (1978).

BACCHOTRICUNEATIN B

Source: *Baccharis tricuneata* (L.f.) Pers.

Mol. formula: $C_{20}H_{22}O_5$

Mol. wt.: 342

Solvent: $CDCl_3$

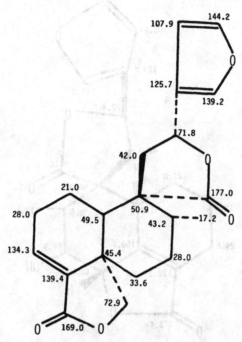

H. Wagner, R. Seitz, H. Lotter and W. Herz, J. Org. Chem., 43(17), 3339 (1978).

SALVIARIN

Source: *Tinospora cordifolia* Miers

Mol. formula: $C_{20}H_{22}O_5$

Mol. wt.: 342

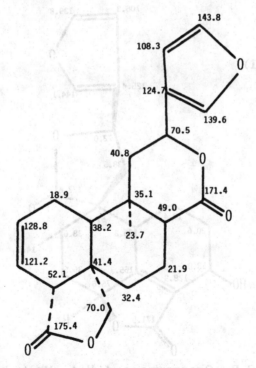

J.B. Hanuman, R.K. Bhatt and B.K. Sabata, Phytochemistry, 25(7), 1677 (1986).

TEUCRIN A

Source: *Teucrium chamaedrys* L.

Mol. formula: $C_{19}H_{20}O_6$

Mol. wt.: 344

Solvent: C_5D_5N

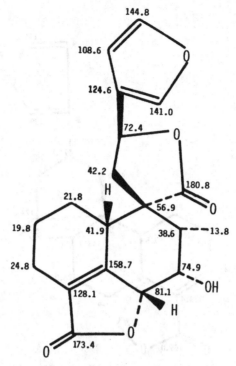

G. Savona, M.C. Garcia-Alvarez and B. Rodriguez, <u>Phytochemistry</u>, 21(3), 721 (1982).

TEUCRIN H1

Source: *Teucrium hyrcanicum*

Mol. formula: $C_{19}H_{20}O_6$

Mol. wt.: 344

Solvent: $CDCl_3$-DMSO-d_6 (4:1)

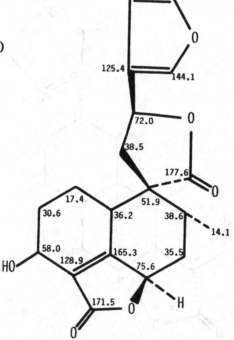

E. Gacs-Baitz, L. Radics, G.B. Oganessian and V.A. Mnatsakanian, <u>Phytochemistry</u>, 17 (11), 1967 (1978).

Source: *Teucrium hyrcanicum* L.

Mol. formula: $C_{19}H_{20}O_6$

Mol. wt.: 344

Solvent: $CDCl_3$ - DMSO-d_6 (4:1)

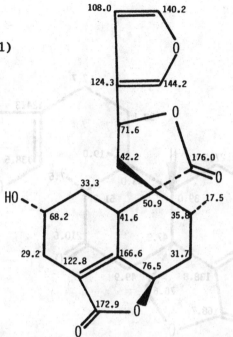

E. Gacs-Baitz, L. Radics, G.B. Oganessian and V.A. Mnatsakanian, <u>Phytochemistry</u>, 17(11), 1967 (1978).

||||||||||||||||||||||||||||||||||||||||||||||||||||||||||||||||||||||||||||||||||||||||||||||||||||

ALDEHYDOMARRUBIALACTONE

Source: *Leonurus marrubiastrum* L.

Mol. formula: $C_{20}H_{24}O_5$

Mol. wt.: 344

Solvent: $CDCl_3$

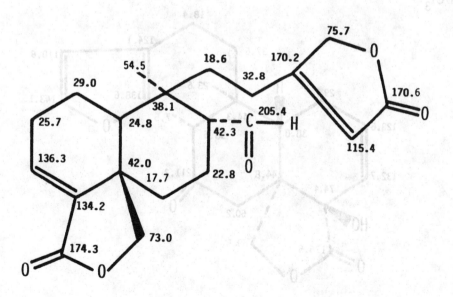

R. Tschesche and B. Streuff, <u>Chem. Ber.</u>, 111 (6), 2130 (1978).

2β-HYDROXYBACCHASMACRANONE

Source: *Baccharis macraei* Hook et Arn.

Mol. formula: $C_{20}H_{24}O_5$

Mol. wt.: 344

Solvent: $CDCl_3$

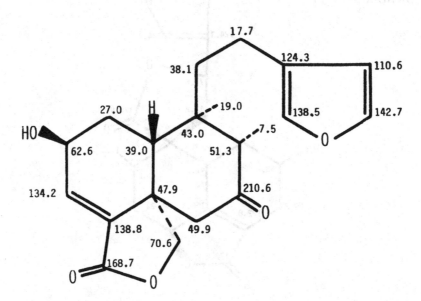

V. Gambaro, M.C. Chamy and J.A. Garbarino, Phytochemistry, 26(2), 475 (1987).

||||||||||||||||||||||||||||||||||||||||||||||||||||||||||||||||||||||||||||||||||||||||||||||||||||||||||||||||||||||||||||

4β-HYDROXYISOBACCHASMACRANONE

Source: *Baccharis macraei* Hook et Arn.

Mol. formula: $C_{20}H_{24}O_5$

Mol. wt.: 344

Solvent: $CDCl_3$

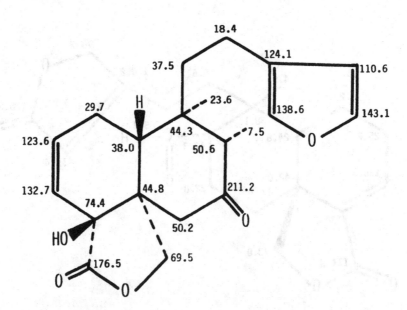

V. Gambaro, M.C. Chamy and J.A. Garbarino, Phytochemistry, 26(2), 475 (1987).

Source: *Salvia keerlii* Bentham

Mol. formula: $C_{20}H_{24}O_5$

Mol. wt.: 344

Solvent: DMSO-d$_6$

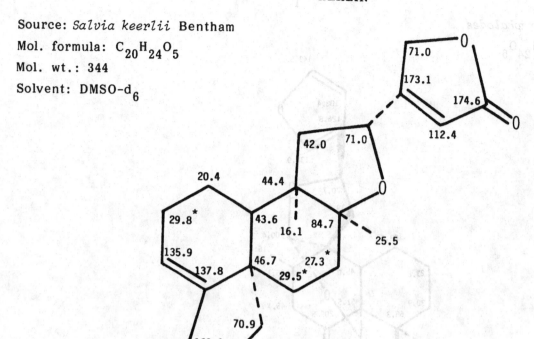

B. Esquivel, A. Mendez, A. Ortega, M. Soriano-Garcia, A. Toscano and L. Rodriguez-Hahn, <u>Phytochemistry</u>, 24(8), 1769 (1985).

7-KETO-<u>ENT</u>-CLERODAN-3,13-DIEN-18,19:16,15-DIOLIDE

Source: *Salvia melissodora* Lag.

Mol. formula: $C_{20}H_{24}O_5$

Mol. wt.: 344

Solvent: CDCl$_3$

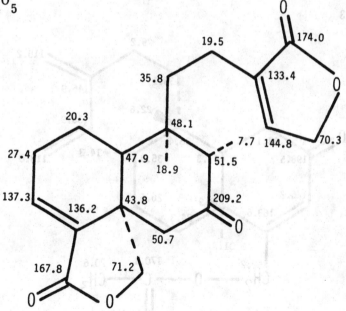

B. Esquivel, L.M. Hernandez, J. Cardenas, T. P. Ramamoorthy and L. Rodriguez-Hahn, <u>Phytochemistry</u>, 28 (2), 561 (1989).

TEUCRIN P₁

Source: *Teucrium gnaphalodes*
Mol. formula: $C_{20}H_{24}O_5$
Mol. wt.: 344
Solvent: $CDCl_3$

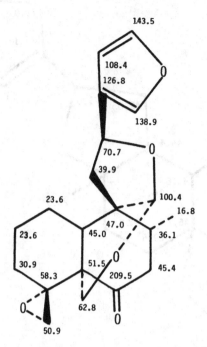

M. Martinez-Ripoll, J. Fayos, B. Rodriguez, M.C. Garcia-Alvarez, G. Savona, F. Piozzi, M. Paternostro and J.R. Hanson, <u>J. Chem. Soc. Perkin Trans. I</u>,(4), 1186 (1981).

||||||||||||||||||||||||||||||||||||||||||||||||||||||||||||||||||||||||||||||||||||||||||||

2-OXO-18-ACETOXY-10α,17α,19α,20β-(-)-CLERODA-3,13(16),14-TRIENE

Source: *Monodora brevipes* Benth.
Mol. formula: $C_{22}H_{32}O_3$
Mol. wt.: 344
Solvent: $CDCl_3$

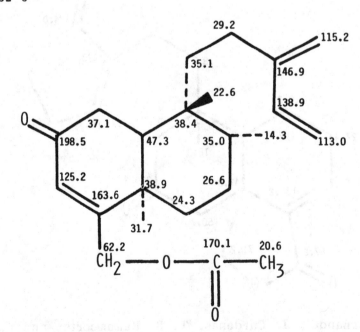

J.T. Etse, A.I. Gray, D.W. Thomas and P.G. Waterman, <u>Phytochemistry</u>, 28 (9), 2489 (1989).

DITERPENE BT-IIa

Source: *Baccharis trimera* (Less.) DC

Mol. formula: $C_{20}H_{26}O_5$

Mol. wt.: 346

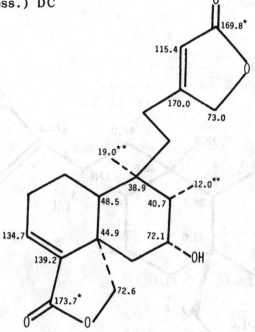

CH_2: 19.4, 22.4, 27.7, 35.7 and 40.6

W. Herz, A.-M. Pilotti, A.-C. Soderholm, I.K. Shuhama and W. Vichnewski, J. Org. Chem., **42** (24), 3913 (1977).

|||||||||||||||||||||||||||||||||||||||||||||||||||||||||||||||||||||||||||||||||||||||||||||||||||||||

7β-HYDROXY-ENT-CLERODAN-3,13-DIEN-18,19:16,15-DIOLIDE

Source: *Salvia melissodora* Lag.

Mol. formula: $C_{20}H_{26}O_5$

Mol. wt.: 346

Solvent: DMSO-d_6

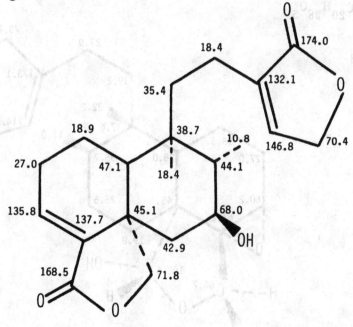

B. Esquivel, L.M. Hernandez, J. Cardenas, T.P. Ramamoorthy and L. Rodriguez-Hahn, Phytochemistry, **28** (2), 561 (1989).

Source: *Leonurus marrubiastrum* L.

Mol. formula: $C_{20}H_{26}O_5$

Mol. wt.: 346

Solvent: DMSO-d_6+$(CD_3)_2CO$

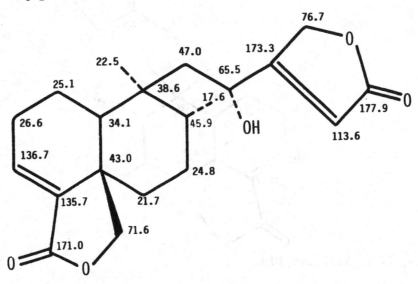

R. Tschesche and B. Streuff, <u>Chem. Ber.</u>, 111 (6), 2130 (1978).

|||||||||||||||||||||||||||||||||||||||||||||||||||||||||||||||||||||||||||||||||||||

3α,4:18,19-DIEPOXY-19-HYDROXY-<u>CIS</u>-CLERODA-13(14)-ENE-15,16-OLIDE

Source: *Gutierrezia texana* (DC) T. and G.

Mol. formula: $C_{20}H_{28}O_5$

Mol. wt.: 348

Solvent: $CDCl_3$

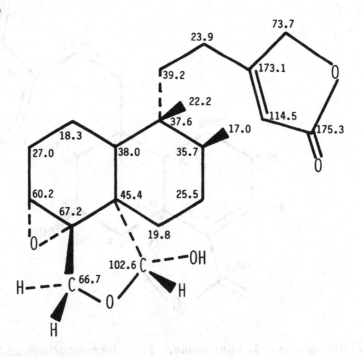

F. Gao and T.J. Mabry, <u>Phytochemistry</u>, 26 (1), 209 (1987).

DITERPENE BT-Ib

Source: *Baccharis trimera* (Less.) DC

Mol. formula: $C_{20}H_{28}O_5$

Mol. wt.: 348

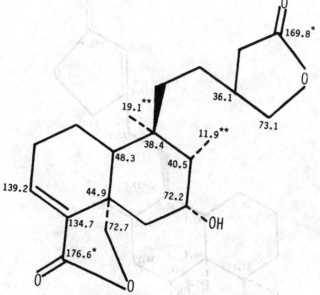

CH$_2$: 19.4, 26.5, 27.6, 34.6, 36.2 and 40.5

W. Herz, A.-M. Pilotti, A.-C. Soderholm, I.K. Shuhama and W. Vichnewski, <u>J. Org. Chem.</u>, **42** (24), 3913 (1977).

PORTULIDE D

Source: *Portulaca* cv Jewel

Mol. formula: $C_{20}H_{28}O_5$

Mol. wt.: 348

Solvent: $CDCl_3$

A. Ohsaki, N. Ohno, K. Shibata, T. Tokoroyama and T. Kubota, <u>Phytochemistry</u>, **25**(10), 2414 (1986).

DEACETYLAJUGARIN-II

Source: *Teucrium massiliense* L.

Mol. formula: $C_{20}H_{30}O_5$

Mol. wt.: 350

Solvent: $CDCl_3$

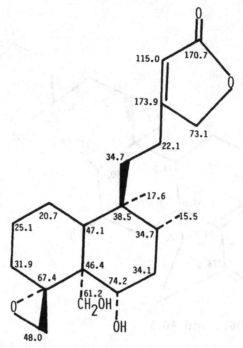

G. Savona, M. Bruno, F. Piozzi, O. Servettaz and B. Rodriguez, <u>Phytochemistry</u>, 23 (4), 849 (1984).

PORTULIDE A

Source: *Portulaca* sp.

Mol. formula: $C_{20}H_{30}O_5$

Mol. wt.: 350

Solvent: CD_3OD

A. Ohsaki, N. Ohno, K. Shibata, T. Tokoroyama and T. Kubota, <u>Phytochemistry</u>, 25 (10), 2414 (1986).

PORTULIDE C

Source: *Portulaca* cv Jewel

Mol. formula: $C_{20}H_{30}O_5$

Mol. wt.: 350

Solvent: CD_3OD

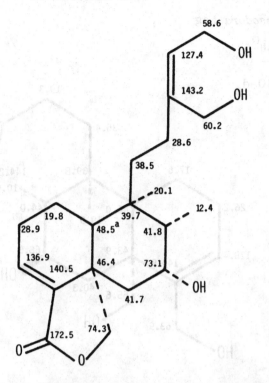

a: Value in C_5D_5N

A. Ohsaki, N. Ohno, K. Shibata, T. Tokoroyama and T. Kubota, <u>Phytochemistry</u>, **25**(10), 2414 (1986).

‖‖

TEUMASSILIN

Source: *Teucrium massiliense* L.

Mol. formula: $C_{20}H_{30}O_5$

Mol. wt.: 350

Solvent: $CDCl_3$

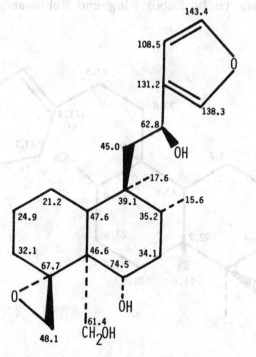

G. Savona, M. Bruno, F. Piozzi, O. Servettaz and B. Rodriguez, <u>Phytochemistry</u>, **23** (4), 849 (1984).

7β-18,19-TRIHYDROXY-<u>ENT</u>-CLERODAN-3,13-DIEN-16,15-DIOLIDE

Source: *Salvia melissodora* Lag.

Mol. formula: $C_{20}H_{30}O_5$

Mol. wt.: 350

Solvent: $CDCl_3$+DMSO-d_6

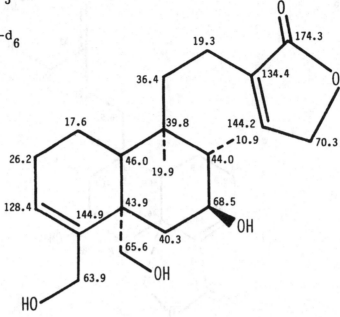

B. Esquivel, L.M. Hernandez, J. Cardenas, T.P. Ramamoorthy and L. Rodriguez-Hahn
<u>Phytochemistry</u>, 28 (2), 561 (1989).

|||||||||||||||||||||||||||||||||||||||||||||||||||||||||||||||||||||||||||||||||||||||||||||||||||||||||||||||||||||||||||||||||||||

3α-METHOXY-4β-HYDROXY-5β,10β-<u>CIS</u>-17α,20α-CLERODA-13 (14)-EN-15,16-<u>OLIDE</u>

Source: *Ageratina saltillensis* (B.L. Robo) King and Robinson

Mol. formula: $C_{21}H_{34}O_4$

Mol. wt.: 350

Solvent: $CDCl_3$

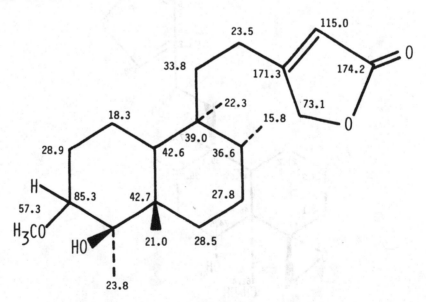

N. Fang, S. Yu, T.J. Mabry, K.A. Abboud and S.H. Simonsen, <u>Phytochemistry</u>, 27 (10), 3187 (1988).

3α,4β,16-TRIHYDROXY-5β,10β-<u>CIS</u>-17α,20α-CLERODA-13(14)-EN-15,16-OLIDE

Source: *Ageratina saltillensis* (B.L. Robo) King and Robinson
Mol. formula: $C_{20}H_{32}O_5$
Mol. wt.: 352
Solvent: $CDCl_3$+5% CD_3OD

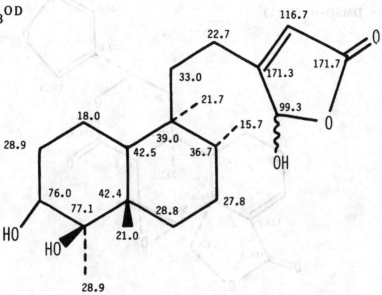

N. Fang, S. Yu, T.J. Mabry, K.A. Abboud and S.H. Simonsen, <u>Phytochemistry</u>, **27** (10)', 3187 (1988).

||||||||||||||||||||||||||||||||||||||||||||||||||||||||||||||||||||||||||||||||||||||||||||||||||||||||||||||||||||||||||||||||

DITERPENE BR-III

Source: *Baccharis rhomboidalis* Remy
Mol. formula: $C_{21}H_{36}O_4$
Mol. wt.: 352
Solvent: $CDCl_3$

54.7, 54.3
OCH₃

105.5, 105.0
O
42.8
37.3
72.3, 71.7
27.1
37.0
18.7
15.7
H
38.5
36.2
17.1
46.1
26.7
129.0
39.1
26.3
145.0
64.6
30.9
CH₂OH
64.0
OH

A. San-Martin, J. Rovirosa, C. Labbe, A. Givovich, M. Mahu and M. Castillo, <u>Phytochemistry</u>, 25(6), 1393 (1986).

1,2-DIHYDRO-6α,7α-EPOXYLINEARIFOLINE

Source: *Salvia sousae* Ramamoorthy

Mol. formula: $C_{20}H_{20}O_6$

Mol. wt.: 356

Solvent: $CDCl_3$ + DMSO-d_6 (1:1)

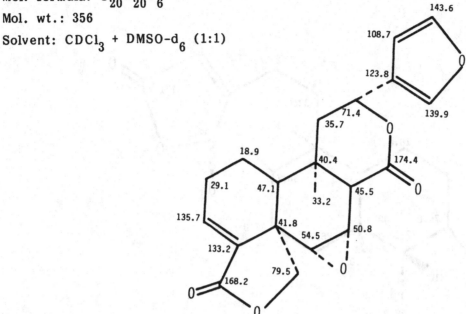

B. Esquivel, J. Ochoa, J. Cardenas, T.P. Ramamoorthy and L. Rodriguez-Hahn, <u>Phyto-chemistry</u>, **27** (2), 483 (1988).

||||||||||||||||||||||||||||||||||||||||||||||||||||||||||||||||||||||||||||||||||||||||||||||||||||||

1α,10α-EPOXYSALVIARIN

Source: *Salvia lineata* Benth.

Mol. formula: $C_{20}H_{20}O_6$

Mol. wt.: 356

Solvent: $CDCl_3$

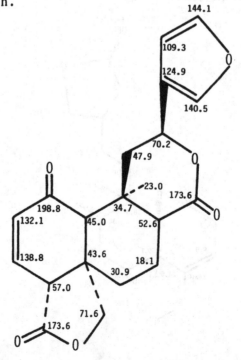

B. Esquivel, J. Cardenas, T.P. Ramamoorthy and L. Rodriguez-Hahn, <u>Phytochemistry</u>, 25(10), 2381 (1986).

RHYNCHOSPERIN A

Source: *Rhynchospermum verticillatum* Reinw.

Mol. formula: $C_{20}H_{20}O_6$

Mol. wt.: 356

Solvent: C_5D_5N

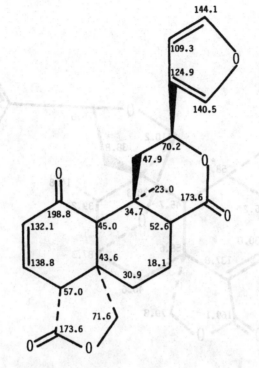

M. Seto, T. Miyase and A. Ueno, <u>Phytochemistry</u>, 26 (12), 3289 (1987).

||||||||||||||||||||||||||||||||||||||||||||||||||||||||||||||||||||||||||||||||||||||||||||||||||

SALVIACOCCIN

Source: *Salvia coccinea* Juss.

Mol. formula: $C_{20}H_{20}O_6$

Mol. wt.: 356

Solvent: C_5D_5N

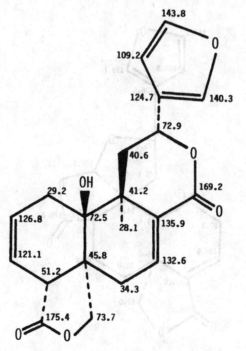

G. Savona, M. Bruno, M. Paternostro, J.L. Marco and B. Rodriguez, <u>Phytochemistry</u>, 21(10), 2563 (1982).

SALVIFARIN

Source: *Salvia farinacea* Benth.
Mol. formula: $C_{20}H_{20}O_6$
Mol. wt.: 356
Solvent: $CDCl_3$

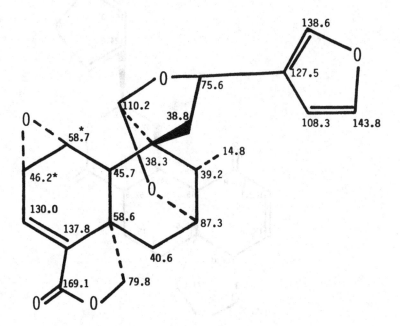

G. Savona, D. Raffa, M. Bruno and B. Rodriguez, <u>Phytochemistry</u>, 22(3), 784 (1983).

VITTAGRACILIOLIDE

Source: *Vittadinia gracilis* (Hook. f.) N. Burb.
Mol. formula: $C_{20}H_{20}O_6$
Mol. wt.: 356
Solvent: $CDCl_3$

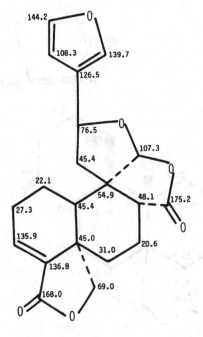

C. Zdero, F. Bohlmann, R.M. King and L. Haegi, <u>Phytochemistry</u>, 27 (7), 2251 (1988).

DIHYDROTUCUMANOIC ACID

Source: *Baccharis marginalis* DC. and *B.Pedicellata* DC.

Mol. formula: $C_{20}H_{36}O_5$

Mol. wt.: 356

Solvent: $CDCl_3$-DMSO-d_6

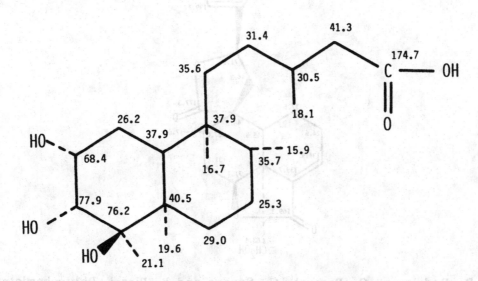

F. Faini, P. Rivera, M. Mahu and M. Castillo, <u>Phytochemistry</u>, 26(12), 3281 (1987).

BARTEMIDIOLIDE

Source: *Baccharis artemisioides* H. and A.

Mol. formula: $C_{20}H_{22}O_6$

Mol. wt.: 358

Solvent: $CDCl_3$

C.E. Tonn, O.S. Giordano, R. Bessalle, F. Frolow and D. Lavie, <u>Phytochemistry</u>, 27(2), 489 (1988).

TEUSCORODONIN

Source: *Teucrium scorodonia* L.

Mol. formula: $C_{20}H_{22}O_6$

Mol. wt.: 358

Solvent: C_5D_5N

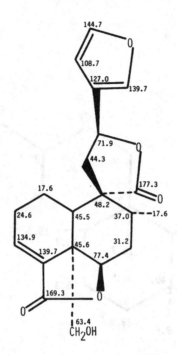

J.L. Marco, B. Rodriguez, C. Pascual, G. Savona and F. Piozzi, <u>Phytochemistry</u>, 22 (3), 727 (1983).

SONDERIANIN

Source: *Croton sonderianus* Muell. Arg.

Mol. formula: $C_{21}H_{26}O_5$

Mol. wt.: 358

Solvent: $CDCl_3$

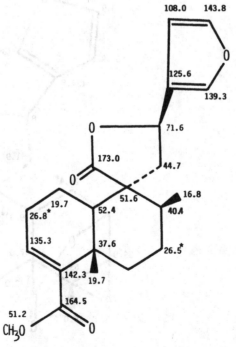

A.A. Craveiro, E.R. Silveira, R.B. Filho and I.P. Mascarenhas, <u>Phytochemistry</u>, 20 (4), 852 (1981).

1,2-GNAPHALIN

Source: *Teucrium gnaphalodes* L'Her.

Mol. formula: $C_{20}H_{24}O_6$

Mol. wt.: 360

Solvent: $CDCl_3$

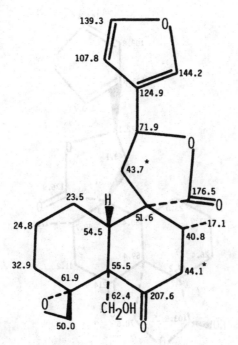

J. Fayos, M. Martinez-Ripoll, M. Paternostro, F. Piozzi, B. Rodriguez and G. Savona, <u>J. Org. Chem.</u>, **44**(26), 4992 (1979).

||||||||||||||||||||||||||||||||||||||||||||||||||||||||||||||||||||||||||||||||||||||||||||||||

TEUCRIN H$_2$

Source: *Teucrium hyrcanicum*

Mol. formula: $C_{20}H_{24}O_6$

Mol. wt.: 360

Solvent: $CDCl_3$-DMSO-d$_6$ (4:1)

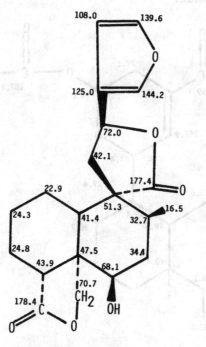

E. Gas-Baitz, L. Radics, G.B. Oganessian and V.A. Mnatsakanian, <u>Phytochemistry</u>, **17** (9), 1967 (1978).

Source: *Teucrium scorodonia* L.

Mol. formula: $C_{20}H_{24}O_6$

Mol. wt.: 360

Solvent: C_5D_5N

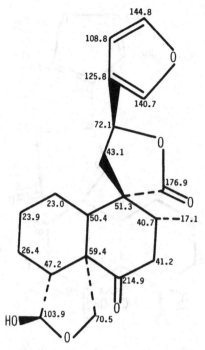

J.L. Marco, B. Rodriguez, C. Pascual, G. Savona and F. Piozzi, <u>Phytochemistry</u>, 22 (3), 727 (1983).

15-ACETOXY-11,15-EPOXY-<u>ENT</u>-CLERODA-4(18), 12-DIEN-1<u>6</u>-<u>AL</u>

Source: *Linaria saxatilis* (L.) Chaz.

Mol. formula: $C_{22}H_{32}O_4$

Mol. wt.: 360

Solvent: $CDCl_3$

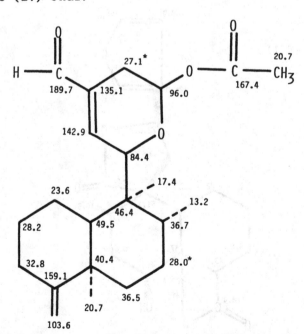

A.S. Feliciano, A.F. Barrero, J.M.M. Del Corral, M. Gordaliza and M. Medarde, <u>Tetrahedron</u>, 41 (4), 671 (1985).

19-DEACETYLTEUSCORODOL

Source: *Teucrium botrys* L.

Mol. formula: $C_{20}H_{26}O_6$

Mol. wt.: 362

Solvent: C_5D_5N

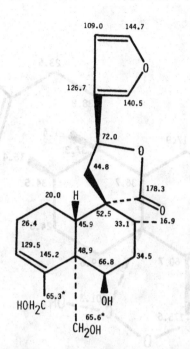

M.C.De La Torre, F. Fernandez-Gadea A. Michavila, B. Rodriguez, F. Piozzi and G. Savona, Phytochemistry, 25(10), 2385(1986).

||||||||||||||||||||||||||||||||||||||||||||||||||||||||||||||||||||||||||||||||||||||||||||||||||||||||||||||||||||||||||

2β,7α-DIHYDROXY-ENT-CLERODAN-3,13-DIEN-18,19:16,15-DIOLIDE

Source: *Salvia melissodora* Lag.

Mol. formula: $C_{20}H_{26}O_6$

Mol. wt.: 362

Solvent: DMSO-d_6

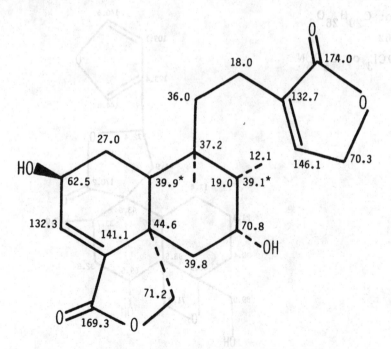

B. Esquivel, L.M. Hernandez, J. Cardenas, T.P. Ramamoorthy and L. Rodriguez-Hahn, Phytochemistry, 28 (2), 561 (1989).

3α,4-EPOXY-19α-HYDROXY-CIS-CLERODA-13(14)-ENE-15,16:18,19-DIOLIDE

Source: *Gutierrezia taxana* (DC) T. and G.

Mol. formula: $C_{20}H_{26}O_6$

Mol. wt.: 362

Solvent: $CDCl_3$

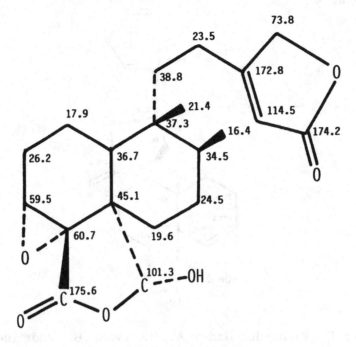

F. Gao and T.J. Mabry, Phytochemistry, 26(1), 209 (1987).

||||||||||||||||||||||||||||||||||||||||||||||||||||||||||||||||||||||||||||||||||||||||||||||||||||||||||

MONTANIN D

Source: *Teucrium montanum* L.

Mol. formula: $C_{20}H_{26}O_6$

Mol. wt.: 362

Solvent: $CDCl_3$ or C_5D_5N

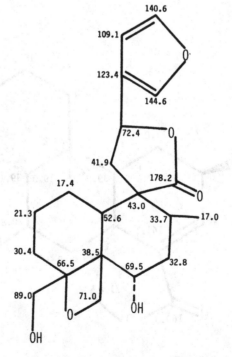

P.Y. Malakov, G.Y. Papanov, N.M. Mollov and S.L. Spassov, Z. Naturforsch., 33b (10), 1142 (1978).

SEMIATRIN

Source: *Salvia semiatratha* Zucc.
Mol. formula: $C_{20}H_{26}O_6$
Mol. wt.: 362
Solvent: DMSO-d$_6$

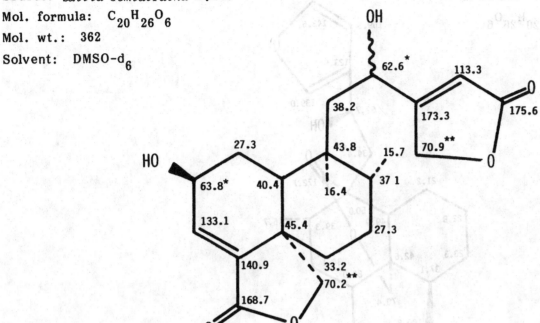

B. Esquivel, M. Hernandez, T.P. Ramamoorthy, J. Cardenas and L. Rodriguez-Hahn,
Phytochemistry, 25(6), 1484 (1986).

||||||||||||||||||||||||||||||||||||||||||||||||||||||||||||||||||||||||||||||||||||||||||||||||||||||||

TEUBOTRIN

Source: *Teucrium botrys* L.
Mol. formula: $C_{20}H_{26}O_6$
Mol. wt.: 362
Solvent: C_5D_5N

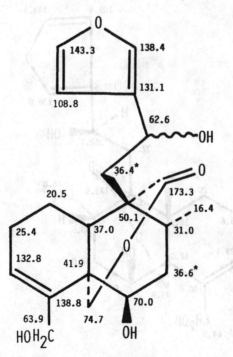

M.C. De La Torre, F. Fernandez-Gadea, A. Michavila, B. Rodriguez, F. Piozzi and G.
Savona, Phytochemistry, 25(10), 2385 (1986)

TEUCHAMAEDRIN C

Source: *Teucrium chamaedrys* L.

Mol. formula: $C_{20}H_{26}O_6$

Mol. wt.: 362

Solvent: C_5D_5N

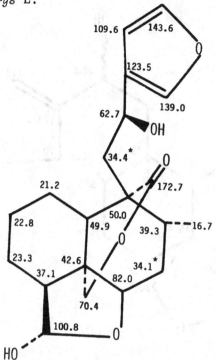

P.Y. Malakov and G.Y. Papanov, Phytochemistry, 24(2), 301 (1985).

TEULAMIFIN B

Source: *Teucrium lamiifolium* and *T. polium*

Mol. formula: $C_{20}H_{26}O_6$

Mol. wt.: 362

Solvent: C_5D_5N

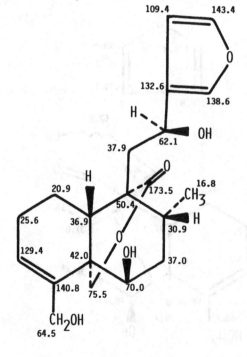

P.Y. Malakov, I.M. Boneva, G.Y. Papanov and S.L. Spassov, Phytochemistry, 27(4), 1141 (1988).

Source: *Teucrium polium* L.

Mol. formula: $C_{20}H_{26}O_6$

Mol. wt.: 362

Solvent: C_5D_5N

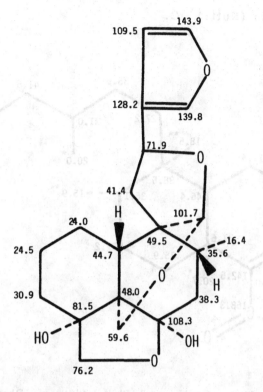

P.Y. Malakov and G.Y. Papanov, <u>Phytochemistry</u>, 22(12), 2791 (1983).

||||||||||||||||||||||||||||||||||||||||||||||||||||||||||||||||||||||||||||||||||||||||||||||||||||||||||||||

3α,4:18,19-DIEPOXY-18β,19α-DIHYDROXY-CIS-CLERODA-13 (14)-ENE-15,16-OLIDE

Source: *Gutierrezia texana* (DC) T. and G.

Mol. formula: $C_{20}H_{28}O_6$

Mol. wt.: 364

Solvent: $CDCl_3$

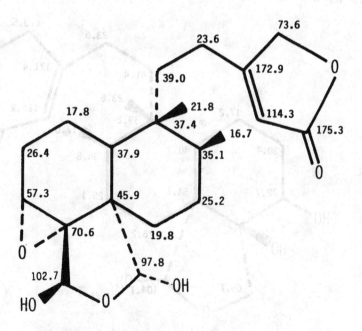

F. Gao and T.J. Mabry, <u>Phytochemistry</u>, 26(1), 209 (1987).

HAPLOCILIATIC ACID (DIMETHYL ESTER)

Source: *Haplopappus ciliatus* (Nutt.) DC.

Mol. formula: $C_{22}H_{36}O_4$

Mol. wt.: 364

Solvent: $CDCl_3$

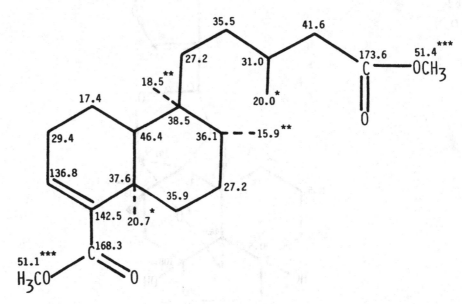

M.L. Bittner, V. Zabel, W.B. Smith and W.H. Watson, Phytochemistry, 17 (10), 1797 (1978).

3α,4β,19α-TRIHYDROXY-18,19-EPOXY-CIS-CLERODA-13 (14)-ENE-15,16-OLIDE

Source: *Gutierrezia texana* (DC) T. and G.

Mol. formula: $C_{20}H_{30}O_6$

Mol. wt.: 366

Solvent: $CDCl_3$

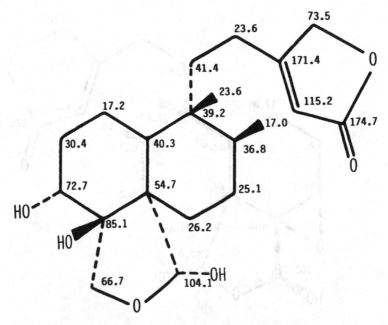

F. Gao and T.J. Mabry, Phytochemistry, 26(1), 209 (1987).

CROVERIN

Source: *Croton verreauxii* Baill.

Mol. formula: $C_{21}H_{22}O_6$

Mol. wt.: 370

Solvent: $CDCl_3$

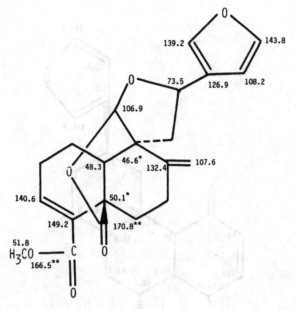

CH_2: 22.7, 26.3, 31.9, 35.3 and 39.1

E. Fujita, M. Node, K. Nishide, M. Sai, K. Fuji, A.T. Mcphail and J.A. Lamberton, <u>J. Chem. Soc. Chem. Commun.</u>, (19), 920 (1980).

FIBRAURIN

Source: *Fibraurea tinctoria* Lour.

Mol. formula: $C_{20}H_{20}O_7$

Mol. wt.: 372

Solvent: DMSO-d_6

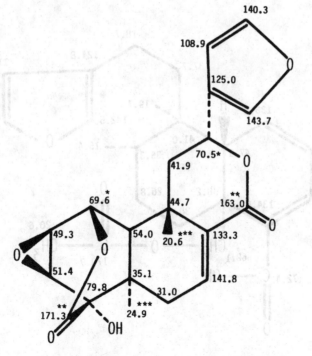

H. Itokawa, K. Mizuno, R. Tajima and K. Takeya, <u>Phytochemistry</u>, 25 (4), 905 (1986).

Source: *Rhynchospermum verticillatum* Reinw.

Mol. formula: $C_{20}H_{20}O_7$

Mol. wt.: 372

Solvent: C_5D_5N

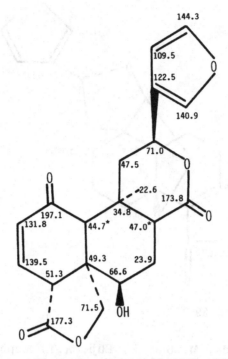

M. Seto, T. Miyase and A. Ueno, <u>Phytochemistry</u>, **26** (12), 3289 (1987).

DITERPENE BF-11a

Source: *Baccharis flabellata*

Mol. formula: $C_{22}H_{28}O_5$

Mol. wt.: 372

Solvent: $CDCl_3$

J.R. Saad, J.G. Davicino and O.S. Giordano, <u>Phytochemistry</u>, 27(6), 1884 (1988).

10α-HYDROXYCOLUMBIN

Source: *Tinospora malabarica* (Miers)
Mol. formula: $C_{20}H_{22}O_7$
Mol. wt.: 374
Solvent: $(CD_3)_2CO$

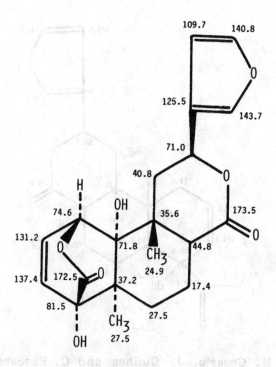

Atta-ur-Rahman and S. Ahmad, <u>Phytochemistry</u>, 27(6), 1882 (1988).

2-KETO-19-HYDROXYTEUSCORDIN

Source: *Teucrium scordium* L.
Mol. formula: $C_{20}H_{22}O_7$
Mol. wt.: 374
Solvent: C_5D_5N

G.Y. Papanov and P.Y. Malakov, <u>Phytochemistry</u>, 24(2), 297 (1985).

CORDATIN

Source: *Aparisthmium cordatum*

Mol. formula: $C_{21}H_{26}O_6$

Mol. wt.: 374

Solvent: $CDCl_3$

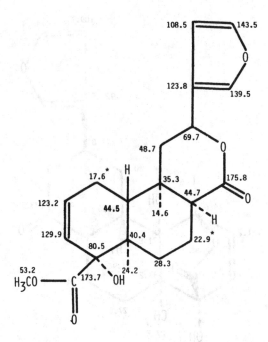

H. Dadoun, A.H. Muller, M. Cesario, J. Guilhem and C. Pascard, <u>Phytochemistry</u>, 26(7), 2108 (1987).

LASIANTHIN

Source: *Salvia lasiantha*

Mol. formula: $C_{22}H_{30}O_5$

Mol. wt.: 374

Solvent: $CDCl_3$

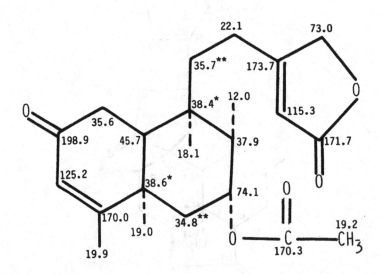

A.-D. Sanchez, B. Esquivel, A. Pera, J. Cardenas, M. Soriano-Garcia, A. Toscano and L. Rodriguez-Hahn, <u>Phytochemistry</u>, 26(2), 479 (1987).

BORAPETOL A

Source: *Tinospora tuberculata* Beumee (syn. *T. crispa* Diers)

Mol. formula: $C_{20}H_{24}O_7$

Mol. wt.: 376

Solvent: $CDCl_3$

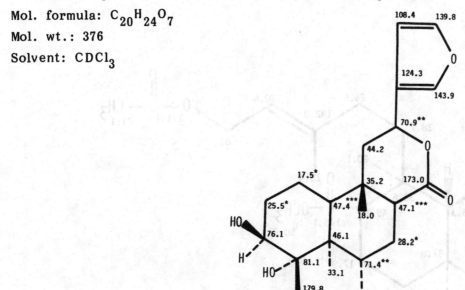

For chemical shifts in DMSO-d_6 see:

N. Fukuda, M. Yonemitsu, T. Kimura, S. Hachiyama, K. Miyahara and T. Kawasaki, Chem. Pharm. Bull. (Tokyo), **33** (10), 4438 (1985).

||||||||||||||||||||||||||||||||||||||||||||||||||||||||||||||||||||||||||||||||||||||||||||||||||||||||||||

DIHYDROTEUGIN

Source: *Teucrium chamaedrys* L.

Mol. formula: $C_{20}H_{24}O_7$

Mol. wt.: 376

Solvent: C_5D_5N

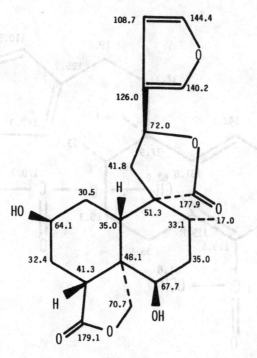

G. Savona, M.C. Garcia-Alvarez and B. Rodriguez, <u>Phytochemistry</u>, 21(3), 721 (1982).

BACCHASALICYLIC ACID (METHYL ESTER ACETATE)

Source: *Baccharis salicifolia* Pers.

Mol. formula: $C_{23}H_{36}O_4$

Mol. wt.: 376

Solvent: $CDCl_3$

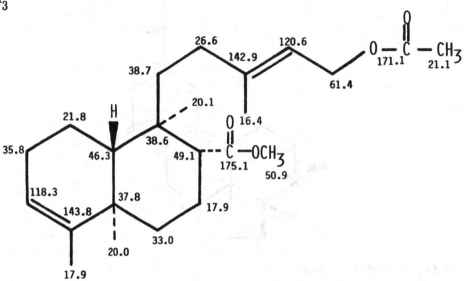

C. Zdero, F. Bohlmann, R.M. King and H. Robinson, <u>Phytochemistry</u>, 25 (12), 2841 (1986).

||||||||||||||||||||||||||||||||||||||||||||||||||||||||||||||||||||||||||||||||||||||||||||||||||||||||||||||

DITERPENE BF-Ia (ACETATE)

Source: *Baccharis flabellata*

Mol. formula: $C_{23}H_{30}O_5$

Mol. wt.: 386

Solvent: $CDCl_3$

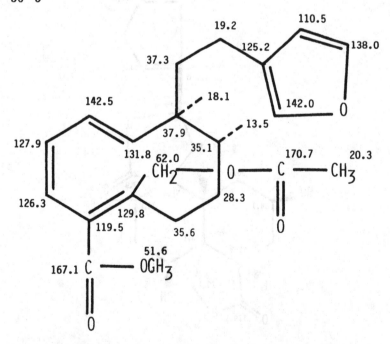

J.R. Saad, J.G. Davicino and O.S. Giordano, <u>Phytochemistry</u>, 27(6), 1884 (1988).

6-HYDROXYFIBRAURIN

Source: *Arcangelisia flava* Merr.

Mol. formula: $C_{20}H_{20}O_8$

Mol. wt.: 388

Solvent: DMSO-d$_6$

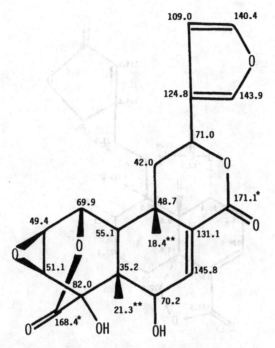

T. Kunii, K. Kagei, Y. Kawakami, Y. Nagai, Y. Nezu and T. Sato, Chem. Pharm. Bull. (Tokyo), 33(2), 479 (1985).

8-EPIDIOSBULBIN E ACETATE

Source: *Dioscorea bulbifera* L.

Mol. formula: $C_{21}H_{24}O_7$

Mol. wt.: 388

Solvent: CDCl$_3$

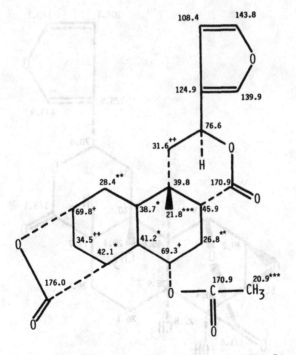

R.D.H. Murray, Z.D. Jorge, N.H. Khan, M. Shahjahan and M. Quaisuddin, Phytochemistry, 23 (3), 623 (1984).

Source: *Baccharis rhomboidalis* Remy

Mol. formula: $C_{22}H_{28}O_6$

Mol. wt.: 388

Solvent: $CDCl_3$

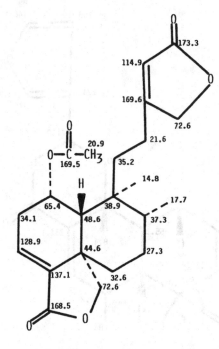

A. San-Martin, J. Rovirosa, C. Labbe, A. Givovich, M. Mahu and M. Castillo, Phyto-chemistry, 25(6), 1393 (1986).

DITERPENE TC-I

Source: *Tinospora cordifolia* Miers

Mol. formula: $C_{20}H_{22}O_8$

Mol. wt.: 390

Solvent: DMSO-d_6

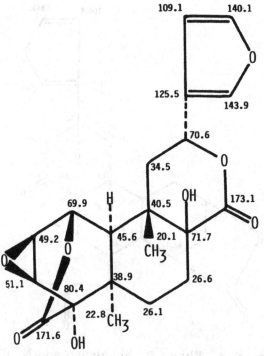

R.K. Bhatt, J.B. Hanuman and B.K. Sabata, Phytochemistry, 27(4), 1212 (1988).

DITERPENE TC-II

Source: *Tinospora cordifolia* Miers
Mol. formula: $C_{20}H_{22}O_8$
Mol. wt.: 390
Solvent: $(CD_3)_2CO$

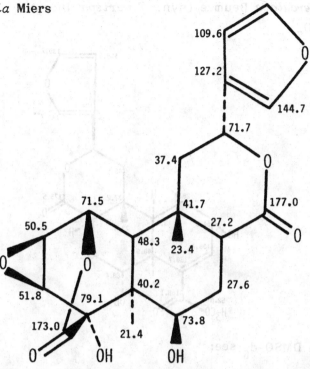

J.B. Hanuman, R.K. Bhatt and B. Sabata, J. Nat. Prod.,(Lloydia), 51 (2), 197 (1988).

||||||||||||||||||||||||||||||||||||||||||||||||||||||||||||||||||||||||||||||||||||||||||||||||

6-HYDROXYARCANGELISIN

Source: *Arcangelisia flava* Merr.
Mol. formula: $C_{20}H_{22}O_8$
Mol. wt.: 390
Solvent: DMSO-d_6

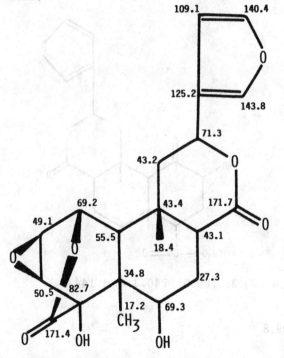

T. Kunii, K. Kagei, Y. Kawakami, Y. Nagai, Y. Nezu and T. Sato, Chem. Pharm. Bull. (Tokyo), 33 (2), 479 (1985).

BORAPETOL B

Source: *Tinospora tuberculata* Beumee (syn. *T. crispa* Diers)

Mol. formula: $C_{21}H_{26}O_7$

Mol. wt.: 390

Solvent: $CDCl_3$

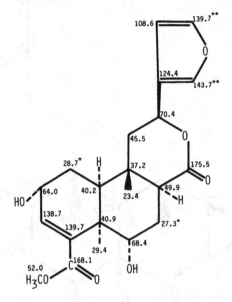

For chemical shifts in DMSO-d_6 see:

N. Fukuda, M. Yonemitsu and T. Kimura, <u>Chem. Pharm. Bull.</u> (Tokyo), **34** (7), 2868 (1986).

||||||||||||||||||||||||||||||||||||||||||||||||||||||||||||||||||||||||||||||||||||||||||||||||||||||||||||||||||||||||||||||||||||

DIVINORIN B

Source: *Salvia divinorum* (Epling and Jativa-M.)

Mol. formula: $C_{21}H_{26}O_7$

Mol. wt.: 390

Solvent: C_5D_5N

CH_3: 15.3, 16.5 and 51.5

CH_2: 18.9, 35.8 and 43.5

CH: 51.2, 53.6, 63.2, 72.0, 75.3, 109.3, 140.3 and 144.1

C: 38.3, 42.4 and 126.6

C=O: 171.4, 172.6 and 209.8

L.J. Valdes, W.M. Butler, G.M. Hatfield, A.G. Paul and M. Koreeda, <u>J. Org. Chem.</u>, **49** (24), 4716 (1984).

FRUTICOLONE

Source: *Teucrium fruticans*

Mol. formula: $C_{22}H_{30}O_6$

Mol. wt.: 390

Solvent: $CDCl_3$

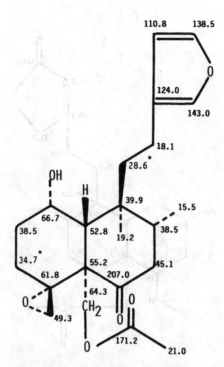

G. Savona, S. Passannanti, M.P. Paternostro, F. Piozzi, J.R. Hanson, P.B. Hitchcock and M. Siverns, J. Chem. Soc. Perkin Trans. I, (4), 356 (1978).

ISOFRUTICOLONE

Source: *Teucrium fruticans*

Mol. formula: $C_{22}H_{30}O_6$

Mol. wt.: 390

Solvent: $CDCl_3$

G. Savona, S. Passannanti, M.P. Paternostro, F. Piozzi, J.R. Hanson, P.B. Hitchcock and M. Siverns, J. Chem. Soc. Perkin Trans. I, (4), 356 (1978).

BLININ

Source: *Conyza blinii* Levl.

Mol. formula: $C_{22}H_{32}O_6$

Mol. wt.: 392

Solvent: $CDCl_3$

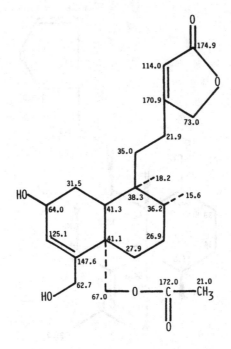

C.-R. Yang, Z.-T. He, X.-C. Li, Q.-T. Zheng, C.-H. He, J. Yang and T. Morita, Phytochemistry, **28** (11), 3131 (1989).

||||||||||||||||||||||||||||||||||||||||||||||||||||||||||||||||||||||||||||||||||||||||||||||||||||||||

DITERPENE BR-V

Source: *Baccharis rhomboidalis* Remy

Mol. formula: $C_{24}H_{40}O_4$

Mol. wt.: 392

Solvent: $CDCl_3$

A. San-Martin, J. Rovirosa, C. Labbe, A. Givovich, M. Mahu and M. Castillo, Phytochemistry, 25(6), 1391 (1986).

SALVIGENOLIDE

Source: *Salvia fulgens* Cav.
Mol. formula: $C_{22}H_{22}O_7$
Mol. wt.: 398
Solvent: $CDCl_3$

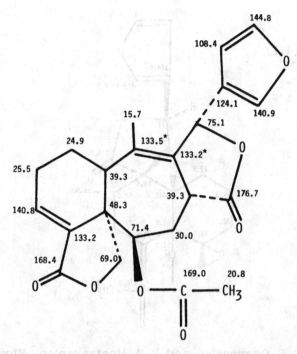

B. Esquivel, J. Cardenas, A. Toscano, M. Soriano-Garcia and L. Rodriguez-Hahn, <u>Tetrahedron</u>, **41(16)**, 3213 (1985).

||||||||||||||||||||||||||||||||||||||||||||||||||||||||||||||||||||||||||||||||||||||||||||||||||||||||||||||||||||||||||||||||||||||||||||||||||||||||

19-ACETYLGNAPHALIN

Source: *Teucrium gnaphalodes* L'Her.
Mol. formula: $C_{22}H_{26}O_7$
Mol. wt.: 402
Solvent: $CDCl_3$

J. Fayos, M. Martinez-Ripoll, M. Paternostro, F. Piozzi, B. Rodriguez and G. Savona, <u>J. Org. Chem.</u>, **44(26)**, 4992 (1979).

211

TEUCRIN H3

Source: *Teucrium hyrcanicum*

Mol. formula: $C_{22}H_{26}O_7$

Mol. wt.: 402

Solvent: $CDCl_3$-DMSO-d_6 (4:1)

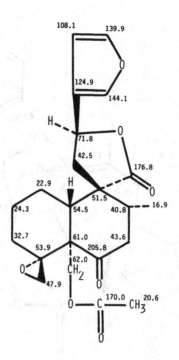

E.Gacs-Baitz, L.Radics, G.B.Oganessian and V.A.Mnatsakanian, <u>Phytochemistry</u>, 17(11), 1967 (1978).

TEUPYRENONE

Source: *Teucrium pyrenaicum* L.

Mol. formula: $C_{22}H_{26}O_7$

Mol. wt.: 402

Solvent: $CDCl_3$

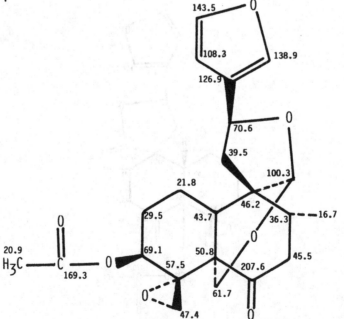

M.C. Garcia-Alvarez, J.L. Marco, B. Rodriguez, G. Savona and F. Piozzi, <u>Phytochemistry</u>, 21(10), 2559 (1982).

212

BARTICULIDIOL (DIACETATE)

Source: *Baccharis incarum* Wedd

Mol. formula: $C_{24}H_{34}O_5$

Mol. wt.: 402

Solvent: $CDCl_3$

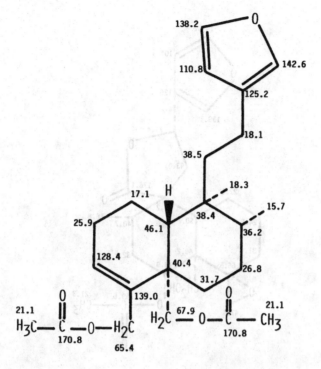

A. San-Martin, A. Givovich annd M. Castillo, <u>Phytochemistry</u>, 25 (1), 264 (1986).

||||||||||||||||||||||||||||||||||||||||||||||||||||||||||||||||||||||||||||||||||||||||||||||||||||||||||||||||

7α-ACETOXY-2β-HYDROXY-<u>ENT</u>-CLERODAN-3,13-DIEN-18,19:16,15-DIOLIDE

Source: *Salvia melissodora* Lag.

Mol. formula: $C_{22}H_{28}O_7$

Mol. wt.: 404

Solvent: $CDCl_3$

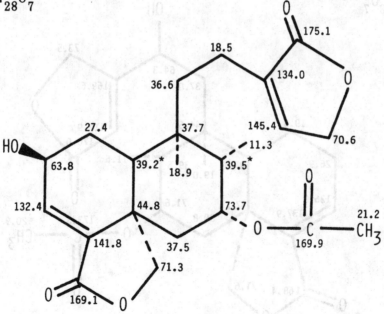

B. Esquivel, L.M. Hernandez, J. Cardenas, T.P. Ramamoorthy and L. Rodriguez-Hahn, <u>Phytochemistry</u>, 28 (2), 561 (1989).

Source: *Teucrium lamiifolium* D'Urv.

Mol. formula: $C_{22}H_{28}O_7$

Mol. wt.: 404

Solvent: $CDCl_3$

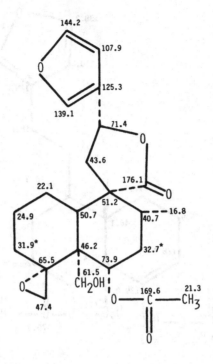

I.M. Boneva, P.Y. Malakov and G.Y. Papanov, <u>Phytochemistry</u>, **27** (1), 295 (1988).

KERLINOLIDE

Source: *Salvia keerlii* Bentham

Mol. formula: $C_{22}H_{28}O_7$

Mol. wt.: 404

Solvent: $DMSO-d_6$

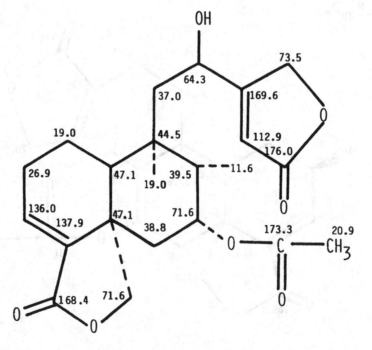

B. Esquivel, A. Mendez, A. Ortega, M. Soriano-Garcia, A. Toscano and L. Rodriguez-Hahn, <u>Phytochemistry</u>, **24**(8), 1769 (1985).

TEUPOLIN-II

Source: *Teucrium lamiifolium* D'urv.

Mol. formula: $C_{22}H_{28}O_7$

Mol. wt.: 404

Solvent: $CDCl_3$

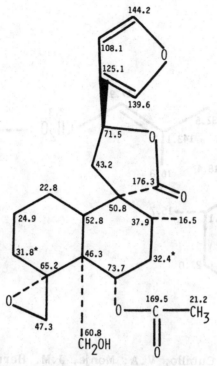

I.M. Boneva, P.Y. Malakov and G.Y. Papanov, <u>Phytochemistry</u>, **27** (1), 295 (1988).

||||||||||||||||||||||||||||||||||||||||||||||||||||||||||||||||||||||||||||||||||||||||||||||||||||||

12-<u>EPI</u> TEUPOLIN-II

Source: *Teucrium lamiifolium* D'urv.

Mol. formula: $C_{22}H_{28}O_7$

Mol. wt.: 404

Solvent: $CDCl_3$

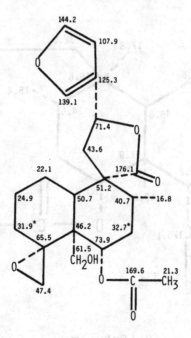

I.M. Boneva, P.Y. Malakov and G.Y. Papanov, <u>Phytochemistry</u>, **27** (1), 295 (1988).

ENT-2-OXO-3,13E-CLERODADIEN-15-YL-METHYLMALONIC ACID DIESTER

Source: *Parentucellia latifolia* (L) Caruel.

Mol. formula: $C_{24}H_{36}O_5$

Mol. wt.: 404

Solvent: $CDCl_3$

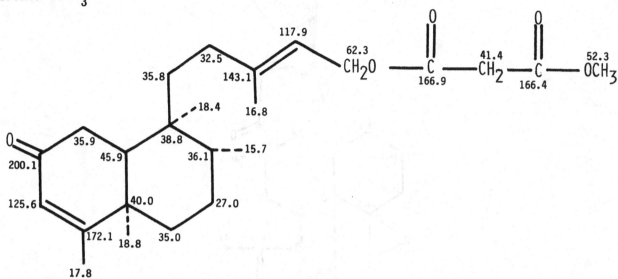

J.G. Urones, I.S. Marcos, L. Cubillo, V.A. Monje, J.M. Hernandez and P. Basabe, Phyto-chemistry, 28(2), 651 (1989).

AJUGARIN-IV

Source: *Ajuga remota*

Mol. formula: $C_{23}H_{34}O_6$

Mol. wt.: 406

Solvent: $CDCl_3$

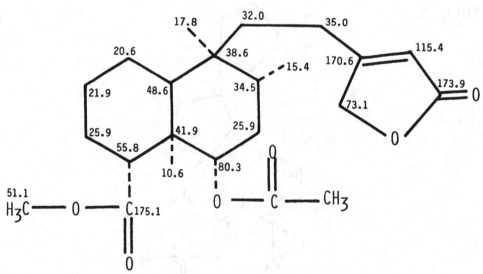

I. Kubo, J.A. Klocke, I. Miura and Y. Fukuyama, J. Chem. Soc. Chem. Commun., (II), 618 (1982).

ENT-3,13E-CLERODADIEN-15-YL-METHYL MALONIC ACID DIESTER

Source: *Parentucellia latifolia* (L) Caruel

Mol. formula: $CH_{24}H_{38}O_4$

Mol. wt.: 390

Solvent: $CDCl_3$

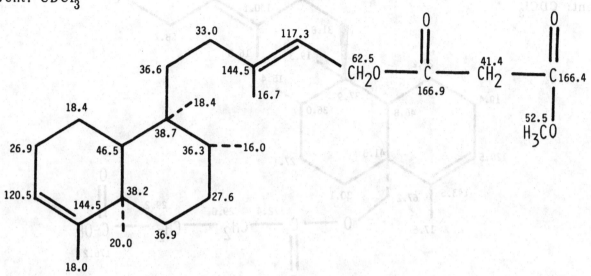

J.G. Urones, I.S. Marcos, L. Cubillo, V.A. Monje, J.M. Hernandez and P. Basabe, <u>Phyto</u>chemistry, 28 (2),651 (1989).

||||||||||||||||||||||||||||||||||||||||||||||||||||||||||||||||||||||||||||||||||||||||||||||||||||||||||||||||||||||||||||||||||

VANCLEVIC ACID A

Source: *Vanclevea stylosa* (Eastw.) Greene

Mol. formula: $C_{24}H_{38}O_5$

Mol. wt.: 406

Solvent: $CDCl_3$

S.D. Jolad, B.N. Timmermann, J.J. Hoffmann, R.B. Bates, F.A. Camou and T.J. Siahaan, <u>Phytochemistry</u>, 27(2), 505 (1988).

VANCLEVIC ACID B

Source: *Vanclevea stylosa* (Eastw.) Greene

Mol. formula: $C_{24}H_{38}O_5$

Mol. wt.: 406

Solvent: $CDCl_3$

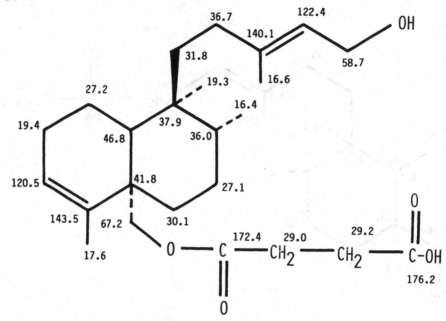

S.D. Jolad, B.N. Timmermann, J.J. Hoffmann, R.B. Bates, F.A. Camou and T.J. Siahaan, Phytochemistry, 27(2), 505 (1988).

TEUPYRIN B

Source: *Teucrium pyrenaicum* L.

Mol. formula: $C_{22}H_{32}O_7$

Mol. wt.: 408

Solvent: $CDCl_3$

P. Fernandez, B. Rodriguez, J.-A. Villegas, A. Perales, G. Savona, F. Piozzi and M. Bruno, Phytochemistry, 25(6), 1405 (1986).

PICROPOLINONE

Source: *Teucrium capitatum* L.

Mol. formula: $C_{22}H_{24}O_8$

Mol. wt.: 416

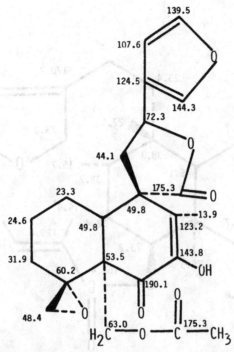

C. Marquez, R.M. Rabanal, S. Valverde, L. Eguren, A. Perales and J. Fayos, <u>Tetrahedron Lett.</u>, **21**(52), 5039 (1980).

||||||||||||||||||||||||||||||||||||||||||||||||||||||||||||||||||||||||||||||||||||||||||||||||||||||||||||||||||||||||||

SOLIDAGOLACTONE-VII

Source: *Solidago altissima* L.

Mol. formula: $C_{25}H_{36}O_5$

Mol. wt.: 416

Solvent: $CDCl_3$

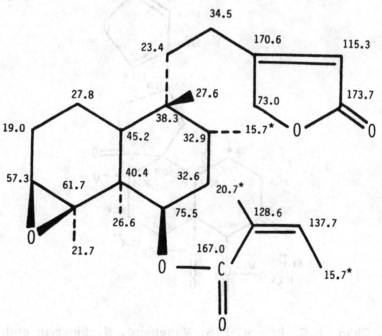

S. Manabe and C. Nishino, <u>Tetrahedron</u>, **42**(13), 3461 (1986).

SOLIDAGOLACTONE–VIII

Source: *Solidago altissima* L.

Mol. formula: $C_{25}H_{36}O_5$

Mol. wt.: 416

Solvent: $CDCl_3$

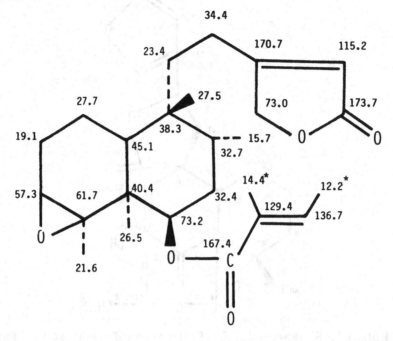

S. Manabe and C. Nishino, <u>Tetrahedron</u>, **42**(13), 3461 (1986).

CORYLIFURAN

Source: *Croton corylifolius* Lam

Mol. formula: $C_{22}H_{26}O_8$

Mol. wt.: 418

Solvent: $CDCl_3$

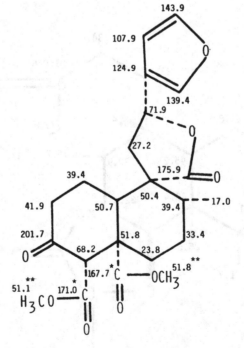

B.A. Burke, W.R. Chan, E.C. Prince, P.S. Manchand, N. Eickman and J. Clardy, <u>Tetrahedron</u>, **32**(15), 1881 (1976).

7-DEACETYLCAPITATIN

Source: *Teucrium polium* Linn.

Mol. formula: $C_{22}H_{26}O_8$

Mol. wt.: 418

Solvent: C_5D_5N

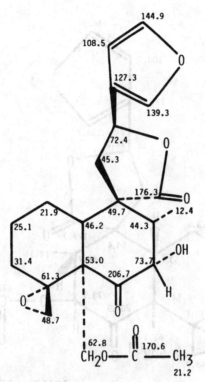

P.Fernandez, B.Rodriguez, G.Savona and F.Piozzi, <u>Phytochemistry</u>, 25(1), 181 (1986).

||||||||||||||||||||||||||||||||||||||||||||||||||||||||||||||||||||||||||||||||||||||||||||||||||||||||||||||||||||||||||

PICROPOLIN

Source: *Teucrium capitatum* L.

Mol. formula: $C_{22}H_{26}O_8$

Mol. wt.: 418

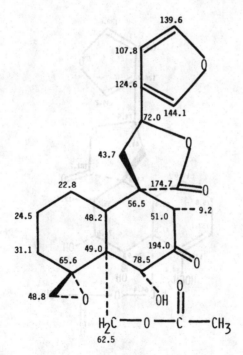

C. Marquez, R.M. Rabanal, S. Valverde, L. Eguren, A. Perales and J. Fayos, <u>Tetrahedron Lett.</u>, 21(52), 5039 (1980).

TEUPOLIN-IV

Source: *Teucrium polium* L.

Mol. formula: $C_{22}H_{26}O_8$

Mol. wt.: 418

Solvent: $CDCl_3$

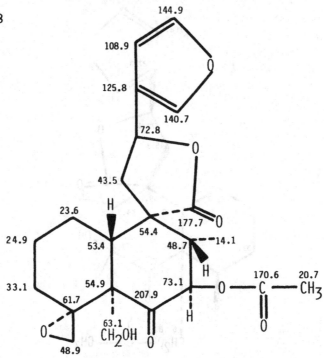

P.Y. Malakov and G.Y. Papanov, <u>Phytochemistry</u>, 22(12), 2791 (1983).

||||||||||||||||||||||||||||||||||||||||||||||||||||||||||||||||||||||||||||||||||||||||||||||||||||||||||

LOLIN

Source: *Teucrium capitatum* L.

Mol. formula: $C_{22}H_{28}O_8$

Mol. wt.: 420

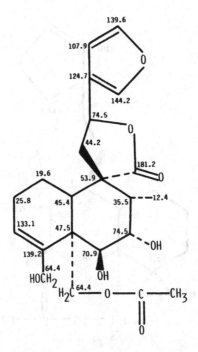

C. Marquez, R.M. Rabanal, S. Valverde, L. Eguren, A. Perales and J. Fayos, <u>Tetrahedron Lett.</u>, 22(29), 2823 (1981).

TEUFLAVIN

Source: *Teucrium flavum* L.

Mol. formula: $C_{22}H_{28}O_8$

Mol. wt.: 420

Solvent: C_5D_5N

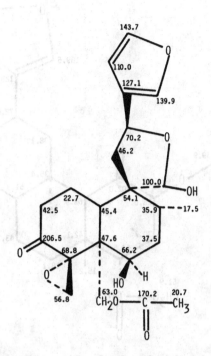

G. Savona, F. Piozzi, O. Servettaz, B. Rodriguez, F.F. Gadea and M. Martin-Lomas, Phytochemistry, 23 (4), 843 (1984).

‖‖

TEUMARIN

Source: *Teucrium marum* L.

Mol. formula: $C_{22}H_{28}O_8$

Mol. wt.: 420

Solvent: $CDCl_3$

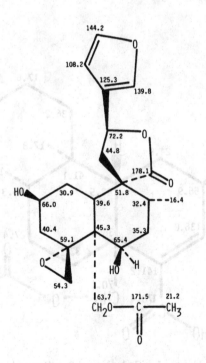

G. Savona, F. Piozzi, O. Servettaz, F. Fernandez-Gadea and B. Rodriguez, Phytochemistry, 23 (3), 611 (1984).

SALVINORIN

Source: *Salvia divinorum*

Mol. formula: $C_{23}H_{28}O_8$

Mol. wt.: 432

Solvent: $CDCl_3$

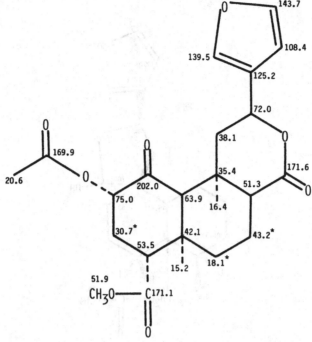

A. Ortega, J.F. Blount and P.S. Manchand, J. Chem. Soc. Perkin Trans. 1, (10), 2505 (1982).

|||||||||||||||||||||||||||||||||||||||||||||||||||||||||||||||||||||||||||||||||||||||||||||||||||||||||||||||||

2β-HYDROXYHAUTRIWAIC ACID (DIACETATE)

Source: *Baccharis sarothroides* A.Gray

Mol. formula: $C_{24}H_{32}O_7$

Mol. wt.: 432

Solvent: $CDCl_3$

F.J. Arriaga-Giner, E. Wollenweber, I. Schober, P. Dostal and S. Braun, Phytochemistry, 25(3), 719 (1986).

AJUGARIN I

Source: *Ajuga remota*

Mol. formula: $C_{24}H_{34}O_7$

Mol. wt.: 434

Solvent: $CDCl_3$

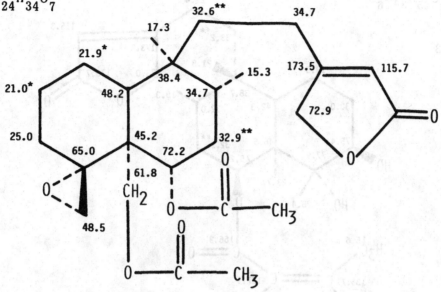

C=O: 169.7, 169.9 and 170.7

I. Kubo, Y.-W. Lee, V. Balogh-Nair, K. Nakanishi and A. Chapya, <u>J. Chem. Soc. Chem. Commun.</u>, (22), 949 (1976).

6,19-DIACETYLTEUMASSILIN

Source: *Teucrium massiliense* L.

Mol. formula: $C_{24}H_{34}O_7$

Mol. wt.: 434

Solvent: $CDCl_3$

G. Savona, M. Bruno, F. Piozzi, O. Servettaz and B. Rodriguez, <u>Phytochemistry</u>, 23 (4), 849 (1984).

DITERPENE SV-VIa

Source: *Solidago virgaurea* L.

Mol. formula: $C_{25}H_{38}O_6$

Mol. wt.: 434

Solvent: $CDCl_3$

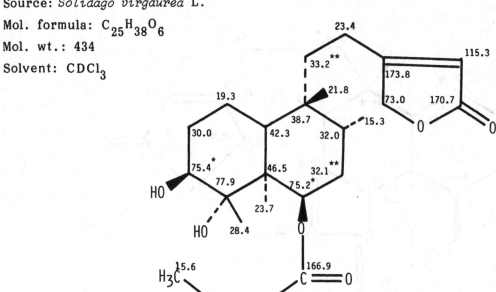

A. Goswami, R.N. Barua, R.P. Sharma, J.N. Baruah, P. Kulanthaivel and W. Herz, Phytochemistry, 23(4), 837 (1984).

||||||||||||||||||||||||||||||||||||||||||||||||||||||||||||||||||||||||||||||||||||||||||||||||||||||||||||||||||||||||||||

DITERPENE SV-VIc

Source: *Solidago virgaurea* L.

Mol. formula: $C_{25}H_{38}O_6$

Mol. wt.: 434

Solvent: $CDCl_3$

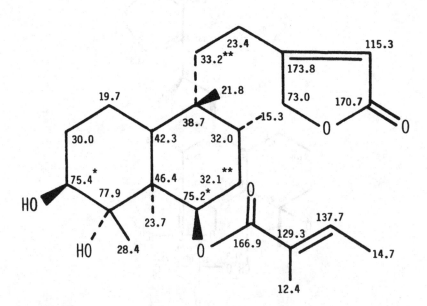

A. Goswami, R.N. Barua, R.P. Sharma, J.N. Baruah, P. Kulanthaivel and W. Herz, Phytochemistry, 23(4), 837 (1984).

GNAPHALIDIN

Source: *Teucrium gnaphalodes* L'Her.

Mol. formula: $C_{24}H_{30}O_8$

Mol. wt.: 446

Solvent: $CDCl_3$

J. Fayos, M. Martinez-Ripoll, M. Paternostro, F. Piozzi, B. Rodriguez and G. Savona, <u>J. Org. Chem.</u>, **44**(26), 4992 (1979).

|||||||||||||||||||||||||||||||||||||||||||||||||||||||||||||||||||||||||||||||||||||||||||||||||||||||||||||||||||||||||||||||||||||||||

MONTANIN C

Source: *Teucrium montanum* L.

Mol. formula: $C_{24}H_{30}O_8$

Mol. wt.: 446

Solvent: $CDCl_3$

P.Y. Malakov, G.Y. Papanov, N.M. Mollov and S.L. Spassov, <u>Z. Naturforsch.</u>, **33b**, 789 (1978).

ISOLINARITRIOL TRIACETATE

Source: *Linaria saxatilis* (L.) Chaz.

Mol. formula: $C_{26}H_{40}O_6$

Mol. wt.: 448

Solvent: $CDCl_3$

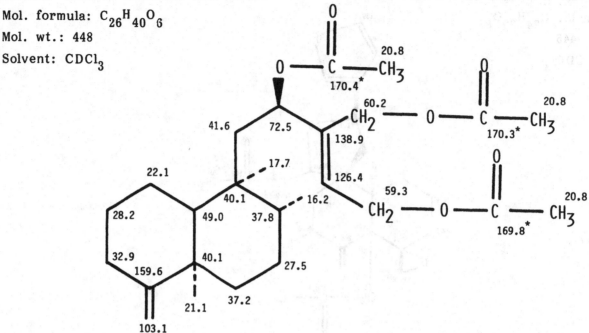

A.S. Feliciano, A.F. Barrero, J.M.M. Del Corral, M. Gordaliza and M. Medarde, <u>Tetrahedron</u>, **41**(4), 671 (1985).

TEUCRETOL

Source: *Teucrium creticum* L. (syn. *T. rosmarinifolium* Lam.)

Mol. formula: $C_{24}H_{34}O_8$

Mol. wt.: 450

Solvent: $CDCl_3$

G. Savona, F. Piozzi, M. Bruno, G. Dominguez, B. Rodriguez and O. Servettaz, <u>Phytochemistry</u>, **26** (12), 3285 (1987).

DITERPENE BR-VI

Source: *Baccharis rhomboidalis* Remy

Mol. formula: $C_{26}H_{42}O_6$

Mol. wt.: 450

Solvent: $CDCl_3$

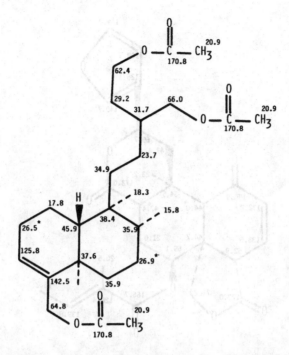

A. San-Martin, J. Rovirosa, C. Labbe, A. Givovich, M. Mahu and M. Castillo, <u>Phyto-chemistry</u>, 25(6), 1393 (1986).

TAFRICANIN A

Source: *Teucrium africanum*

Mol. formula: $C_{22}H_{25}ClO_8$

Mol. wt.: 452

Solvent: $CDCl_3$

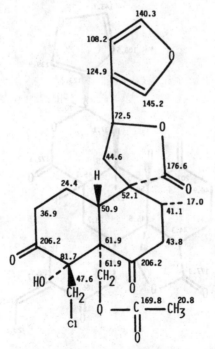

J.R. Hanson, D.E.A. Rivett, S.V. Ley and D.J. Williams, <u>J. Chem. Soc. Perkin Trans. I</u>, (4), 1005 (1982).

Source: *Rhynchospermum verticillatum* Reinw.

Mol. formula: $C_{25}H_{26}O_8$

Mol. wt.: 454

Solvent: C_5D_5N

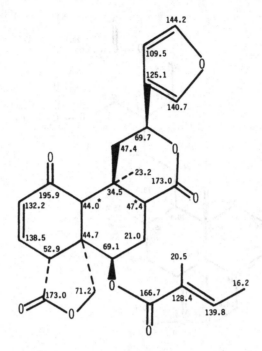

M. Seto, T. Miyase and A. Ueno, <u>Phytochemistry</u>, **26** (12), 3289 (1987).

||||||||||||||||||||||||||||||||||||||||||||||||||||||||||||||||||||||||||||||||||||||||||||||||||||||||||||||||||||||||||||

6α-ANGELOYLOXY-1-OXO-2,3-DIHYDROSALVIARIN

Source: *Aster alpinus* L.

Mol. formula: $C_{25}H_{28}O_8$

Mol. wt.: 456

Solvent: $CDCl_3$

F. Bohlmann, J. Jakupovic, M. Hashemi-Nejad and S. Huneck, <u>Phytochemistry</u>, **24**(3), 608 (1985).

SPLENDIDIN

Source: *Salvia splendens*
Mol. formula: $C_{24}H_{26}O_9$
Mol. wt.: 458

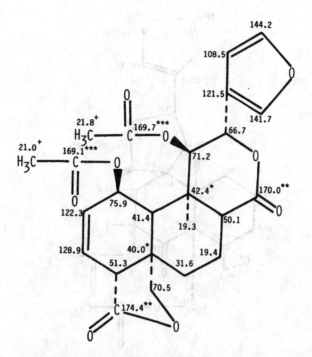

G. Savona, M.P. Paternostro, F. Piozzi and J.R. Hanson, J. Chem. Soc. Perkin Trans. I, (2), 533 (1979).

||||||||||||||||||||||||||||||||||||||||||||||||||||||||||||||||||||||||||||||||||||||||||||||||||||||||||||||||||

19-ACETYLTEUPOLIN IV

Source: *Teucrium polium* L.
Mol. formula: $C_{24}H_{28}O_9$
Mol. wt.: 460
Solvent: C_5D_5N

M.C. De La Torre, F. Piozzi, A.-F. Rizk, B. Rodriguez and G. Savona, Phytochemistry, 25(9), 2239 (1986).

CAPITATIN

Source: *Teucrium capitatum* L.

Mol. formula: $C_{24}H_{28}O_9$

Mol. wt.: 460

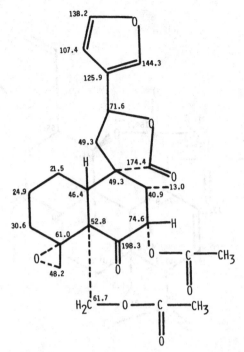

C. Marquez, R.M. Rabanal, S. Valverde, L. Eguren, A. Perales and J. Fayos, <u>Tetrahedron Lett.</u>, **21**(52), 5039 (1980).

ERIOCEPHALIN

Source: *Teucrium eriocephalum* WK.

Mol. formula: $C_{24}H_{30}O_9$

Mol. wt.: 462

Solvent: $CDCl_3$

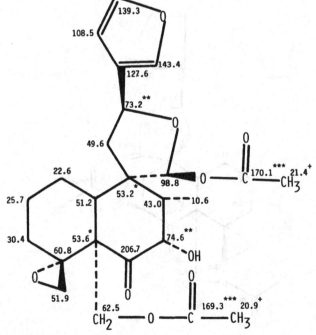

J. Fayos, M. Martinez-Ripoll, M. Paternostro, F. Piozzi, B. Rodriguez and G. Savona, <u>J. Org. Chem.</u>, **44**(26), 4992 (1979).

20-EPI-ISOERIOCEPHALIN

Source: *Teucrium polium* Linn

Mol. formula: $C_{24}H_{30}O_9$

Mol. wt.: 462

Solvent: $CDCl_3$

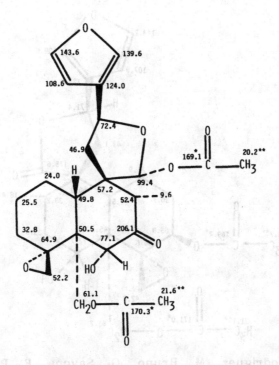

P.Fernandez, B.Rodriguez, G.Savona and F.Piozzi, <u>Phytochemistry</u>, 25(1), 181 (1986).

TEUCAPITATIN

Source: *Teucrium capitatum* L.

Mol. formula: $C_{24}H_{30}O_9$

Mol. wt.: 462

C. Marquez, R.M. Rabanal, S. Valverde, L. Eguren, A. Perales and J. Fayos, <u>Tetrahedron Lett.</u>, 21(52), 5039 (1980).

TEUMICROPODIN

Source: *Teucrium micropodioides* Rouy

Mol. formula: $C_{24}H_{30}O_9$

Mol. wt.: 462

Solvent: $CDCl_3$

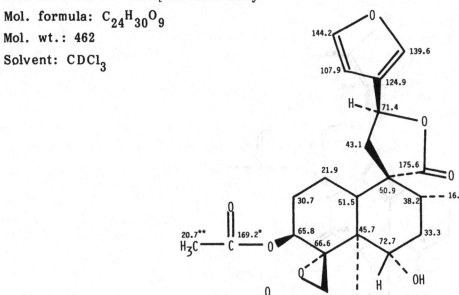

M.C. De La Torre, B. Rodriguez, M. Bruno, G. Savona, F. Piozzi and O. Servettaz, Phytochemistry, 27(1), 213 (1988).

TEUPYRIN A

Source: *Teucrium pyrenaicum* L.

Mol. formula: $C_{24}H_{30}O_9$

Mol. wt.: 462

Solvent: $CDCl_3$

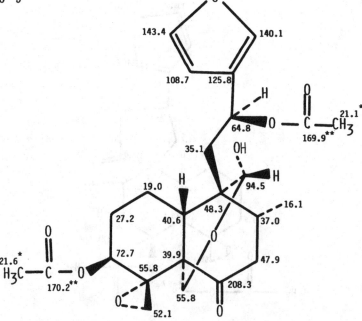

P. Fernandez, B. Rodriguez, J.-A. Villegas, A. Perales, G. Savona, F. Piozzi and M. Bruno, Phytochemistry, 25(6), 1405 (1986).

2β,18ξ,19ξ-TRIACETOXY-4ξ,17α,19α,20β-(-)-CLERODA-13(16), 14-DIENE-18,19-OXIDE

Source: *Monodora brevipes* Benth.

Mol. formula: $C_{26}H_{38}O_7$

Mol. wt.: 462

Solvent: $CDCl_3$

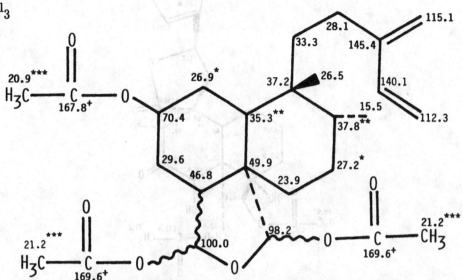

J.T. Etse, A.I. Gray, D.W. Thomas and P.G. Waterman, <u>Phytochemistry</u>, 28 (9), 2489 (1989).

19-O-α-L-ARABINOPYRANOSYL-CIS-CLERODA-3,13(14)-DIENE-15, 16-OLIDE-19-OIC ESTER

Source: *Gutierrezia texana* (DC) T. and G.

Mol. formula: $C_{25}H_{36}O_8$

Mol. wt.: 464

Solvent: $CDCl_3$

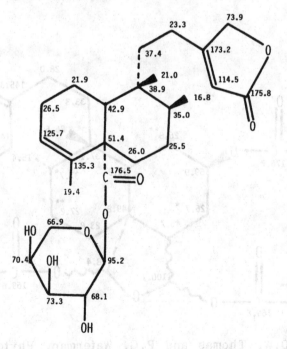

F. Gao and T.J. Mabry, <u>Phytochemistry</u>, 26(1), 209 (1987).

Source: *Teucrium polium* Linn

Mol. formula: $C_{24}H_{30}O_{10}$

Mol. wt.: 478

Solvent: $CDCl_3$

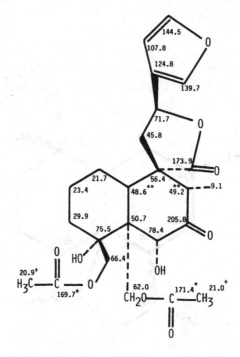

P.Fernandez, B.Rodriguez, G.Savona and F.Piozzi, <u>Phytochemistry</u>, **25**(1), 181 (1986).

2β-ISOBUTOXY-18ξ,19ξ-DIACETOXY-4ξ,17α,19α,20β-(-)-CLERODA-13(16), 14-DIENE-18,19-OXIDE

Source: *Monodora brevipes* Benth.

Mol. formula: $C_{28}H_{42}O_7$

Mol. wt.: 490

Solvent: $CDCl_3$

J.T. Etse, A.I. Gray, D.W. Thomas and P.G. Waterman, <u>Phytochemistry</u>, **28** (9), 2489 (1989).

TAFRICANIN B

Source: *Teucrium africanum*

Mol. formula: $C_{24}H_{29}ClO_9$

Mol. wt.: 496

Solvent: $CDCl_3$

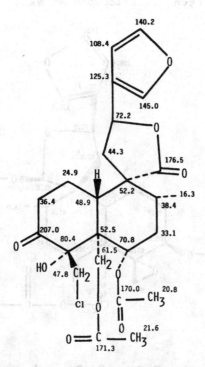

J.R. Hanson, D.E.A. Rivett, S.V.Ley and D.J. Williams, J. Chem. Soc. Perkin Trans.I, (4), 1005 (1982).

||||||||||||||||||||||||||||||||||||||||||||||||||||||||||||||||||||||||||||||||||||||||||||||||||||||||

TEULANIGIN

Source: *Teucrium lanigerum* Lag. (syn. *T. eriocephalum* Wk.)

Mol. formula: $C_{26}H_{32}O_{10}$

Mol. wt.: 504

Solvent: $CDCl_3$

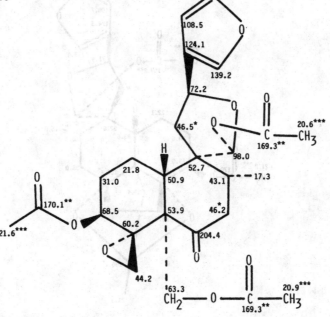

J.A. Hueso-Rodriguez, F. Fernandez-Gadea, C. Pascual, B. Rodriguez, G. Savona and F. Piozzi, Phytochemistry, **25** (1), 175 (1986).

Source: *Teucrium lanigerum* Lag. (syn. *T. eriocephalum* Wk.)

Mol. formula: $C_{26}H_{32}O_{10}$

Mol. wt.: 504

Solvent: $CDCl_3$

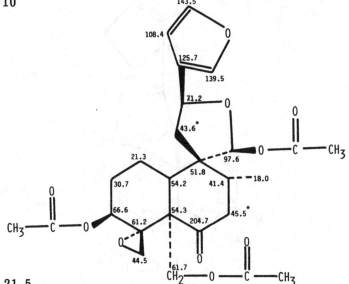

CH$_3$: 20.5, 20.9 and 21.5

C=O: 169.2, 169.3 and 170.4

J.A. Hueso-Rodriguez, F. Fernandez-Gadea, C. Pascual, B. Rodriguez, G. Savona and F. Piozzi, <u>Phytochemistry</u>, 25(1), 175 (1986).

|||||||||||||||||||||||||||||||||||||||||||||||||||||||||||||||||||||||||||||||||||||||||||||||||||||||||||||||||||||||||||||||||||||||||

TEULANIGERIDIN

Source: *Teucrium lanigerum* Lag.

Mol. formula: $C_{26}H_{32}O_{10}$

Mol. wt.: 504

Solvent: $CDCl_3$

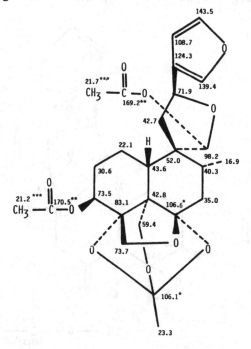

J.A. Hueso-Rodriguez, F. Fernandez-Gadea, C. Pascual, B. Rodriguez, G. Savona and F. Piozzi, <u>Phytochemistry</u>, 25 (1), 175 (1986).

TEUPYREININ

Source: *Teucrium pyrenaicum* L.

Mol. formula: $C_{26}H_{32}O_{10}$

Mol. wt.: 504

Solvent: $CDCl_3$

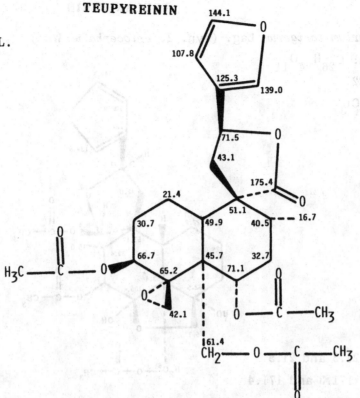

CH_3: 20.8, 21.0 and 21.0

C=O: 169.0, 169.6 and 170.4

M.C. Garcia-Alvarez, J.L. Marco, B. Rodriguez, G. Savona and F. Piozzi, <u>Phytochemistry</u>, **21** (10), 2559 (1982).

||||||||||||||||||||||||||||||||||||||||||||||||||||||||||||||||||||||||||||||||||||||||||||||||||||||||||||||||||||||||||||

DITERPENE TC-I

Source: *Tinospora cordifolia* Miers

Mol. formula: $C_{26}H_{34}O_{11}$

Mol. wt.: 522

Solvent: DMSO-d_6

R.K. Bhatt and B.K. Sabata, <u>Phytochemistry</u>, **28** (9), 2419 (1989).

TEULANIGERIN

Source: *Teucrium lanigerum* Lag. (syn. *T. eriocephalum* Wk.)

Mol. formula: $C_{26}H_{34}O_{11}$

Mol. wt.: 522

Solvent: $CDCl_3$

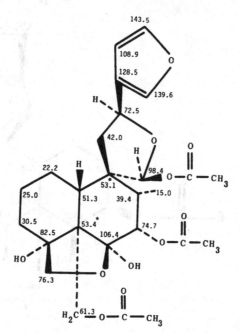

CH_3: 20.7, 21.3 and 21.5

C=O: 169.0, 171.3 and 171.4

J.A. Hueso-Rodriguez, F. Fernandez-Gadea, C. Pascual, B. Rodriguez, G. Savona and F. Piozzi, <u>Phytochemistry</u>, **25**(1), 175 (1986).

||||||||||||||||||||||||||||||||||||||||||||||||||||||||||||||||||||||||||||||||||||||||||||||||||||||||||||||||

FIBRAURINOSIDE

Source: *Fibraurea tinctoria* Lour.

Mol. formula: $C_{26}H_{30}O_{12}$

Mol. wt.: 534

Solvent: CD_3OD

H. Itokawa, K. Mizuno, R. Tajima and K. Takeya, <u>Phytochemistry</u>, **25** (4), 905 (1986).

DITERPENE CS-I

Source: *Casearia sylvestris* SW.

Mol. formula: $C_{29}H_{42}O_9$

Mol. wt.: 534

Solvent: $CDCl_3$

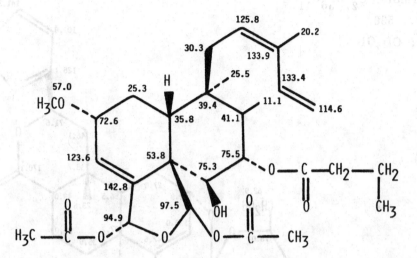

CH$_3$: 13.7, 21.2 and 21.3

CH$_2$: 18.6 and 36.3

C=O: 169.0, 170.2 and 174.3

H. Itokawa, N. Totsuka, K. Takeya, K. Watanabe and E. Obata, <u>Chem. Pharm. Bull.</u> (Tokyo), **36**(4), 1585 (1988).

PALMATOSIDE A

Source: *Jateorhiza palmata* Miers

Mol. formula: $C_{26}H_{32}O_{12}$

Mol. wt.: 536

Solvent: DMSO-d$_6$

M. Yonemitsu, N. Fukuda, T. Kimura and T. Komori, <u>Liebigs Ann. Chem.</u>, 1327 (1986).

TINOPHYLLOLOSIDE

Source: *Fibraurea tinctoria* Lour.

Mol. formula: $C_{27}H_{36}O_{11}$

Mol. wt.: 536

Solvent: CD_3OD

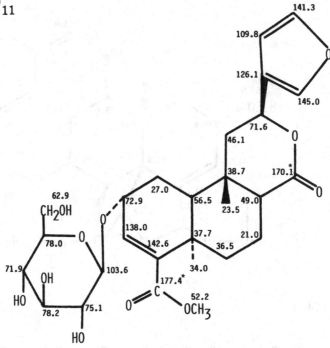

H. Itokawa, K.Mizuno, R. Tajima and K.Takeya, <u>Phytochemistry</u>, 25(4), 905 (1986).

||||||||||||||||||||||||||||||||||||||||||||||||||||||||||||||||||||||||||||||||||||||||||

BORAPETOSIDE A

Source: *Tinospora tuberculata* Beumee

Mol. formula: $C_{26}H_{34}O_{12}$

Mol. wt.: 538

Solvent: DMSO-d_6

N. Fukuda, M. Yonemitsu, T. Kimura, S. Hachiyama, K. Miyahara and T. Kawasaki, <u>Chem. Pharm. Bull.</u> (Tokyo), 33 (10), 4438 (1985).

IVAIN-I

Source: *Ajuga iva*

Mol. formula: $C_{28}H_{42}O_{10}$

Mol. wt.: 538

F. Camps, J. Coll and A. Cortel, <u>Chem. Lett.</u>, 1053 (1982).

||||||||||||||||||||||||||||||||||||||||||||||||||||||||||||||||||||||||||||||||||||||||||||||||||||||||||||||||

TEUSALVIN E (TETRAACETATE)

Source: *Teucrium salviastrum* Schreber

Mol. formula: $C_{28}H_{34}O_{11}$

Mol. wt.: 546

Solvent: $CDCl_3$

CH_3: 21.0, 21.1, 21.2 and 21.3
C=O: 169.7, 169.8, 169.9 and 170.1

M.C. De La Torre, C. Pascual, B. Rodriguez, F. Piozzi, G. Savona and A. Perales, <u>Phytochemistry</u>, **25** (6), 1397 (1986).

TEUPYREINIDIN

Source: *Teucrium pyrenaicum* L.

Mol. formula: $C_{28}H_{36}O_{11}$

Mol. wt.: 548

Solvent: $CDCl_3$

CH_3: 20.8, 20.9, 21.0 and 21.8

C=O: 169.2, 169.5, 169.7 and 170.9

M.C. Garcia-Alvarez, J.L. Marco, B. Rodriguez, G. Savona and F. Piozzi, <u>Phytochemistry</u>, 21(10), 2559 (1982).

AJUGAMARIN

Source: *Ajuga nipponensis* Makino

Mol. formula: $C_{29}H_{40}O_{10}$

Mol. wt.: 548

Solvent: $CDCl_3$

H. Shimomura, Y. Sashida, K. Ogawa and Y. Iitaka, <u>Tetrahedron Lett.</u>, 22(14), 1367 (1981).

AJUGAREPTANSONE A

Source: *Ajuga reptans*
Mol. formula: $C_{29}H_{40}O_{10}$
Mol. wt.: 548
Solvent: $CDCl_3$

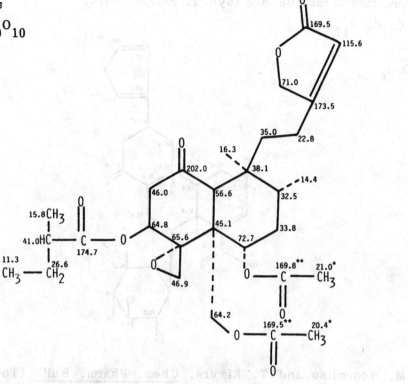

F. Camps, J. Coll and A. Cortel, Chem. Lett., 1093 (1981).

||||||||||||||||||||||||||||||||||||||||||||||||||||||||||||||||||||||||||||||||||||||||||||||||||||||||||||

DIHYDROAJUGAMARIN

Source: *Ajuga nipponensis* Makino
Mol. formula: $C_{29}H_{42}O_{10}$
Mol. wt.: 550
Solvent: $CDCl_3$

H. Shimomura, Y. Sashida, K. Ogawa and Y. Iitaka, Chem. Pharm. Bull. (Tokyo), 31 (7), 2192 (1983).

BORAPETOSIDE B

Source: *Tinospora tuberculata* Beumee (syn. *T. crispa* Diers)
Mol. formula: $C_{27}H_{36}O_{12}$
Mol. wt.: 552
Solvent: DMSO-d_6

N. Fukuda, M. Yonemitsu and T. Kimura, <u>Chem. Pharm. Bull.</u> (Tokyo), **34** (7), 2868 (1986).

||||||||||||||||||||||||||||||||||||||||||||||||||||||||||||||||||||||||||||||||||||||||||||||||||||||||||||||||||||||||||||||||

AJUGAREPTANSIN

Source: *Ajuga reptans*
Mol. formula: $C_{29}H_{44}O_{10}$
Mol. wt.: 552
Solvent: $CDCl_3$

F. Camps, J. Coll and A. Cortel, <u>Chem. Lett.</u>, 1093 (1981).

IVAIN-IV

Source: *Ajuga iva*
Mol. formula: $C_{29}H_{44}O_{10}$
Mol. wt.: 552

F. Camps, J. Coll and A. Cortel, <u>Chem. Lett.</u>, 1053 (1982).

||||||||||||||||||||||||||||||||||||||||||||||||||||||||||||||||||||||||||||||||||||||||||||||||||||||||||

DITERPENE CS-II

Source: *Casearia sylvestris* SW.
Mol. formula: $C_{31}H_{44}O_{10}$
Mol. wt.: 576
Solvent: $CDCl_3$

CH_3: 13.8, 21.0, 21.2 and 21.4
CH_2: 18.4 and 36.1
C=O: 169.1, 170.1, 170.1 and 172.5

H. Itokawa, N. Totsuka, K. Takeya, K. Watanabe and E. Obata, <u>Chem. Pharm. Bull.</u> (Tokyo), **36**(4), 1585 (1988).

Source: *Ajuga pseudoiva* (L.) Schreber
Mol. formula: $C_{30}H_{44}O_{11}$
Mol. wt.: 580
Solvent: $CDCl_3$

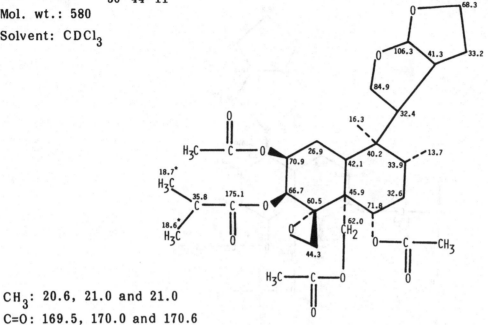

CH_3: 20.6, 21.0 and 21.0
C=O: 169.5, 170.0 and 170.6

F. Camps, J. Coll and O. Dargallo, <u>Phytochemistry</u>, 23 (2), 387 (1984).

PIMARANE

Basic skeleton:

8α-ISOPIMARA-9(11)-15-DIENE

Source: *Dacrydium biforme* Hook.

Mol. formula: $C_{20}H_{32}$

Mol. wt.: 272

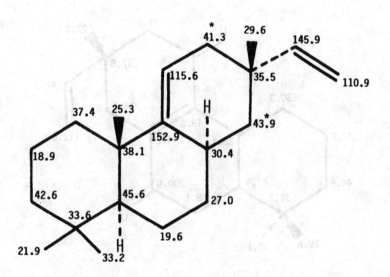

A.R. Hayaman, N.B. Perry and R.T. Weavers, <u>Phytochemistry</u>, 25(3), 649 (1986).

||||||||||||||||||||||||||||||||||||||||||||||||||||||||||||||||||||||||||||||||||||||||||||||||||||||||||||||||||||||||||||

PIMARADIENE

Source: *Erythroxylon monogynum*

Mol. formula: $C_{20}H_{32}$

Mol. wt.: 272

Solvent: $CDCl_3$

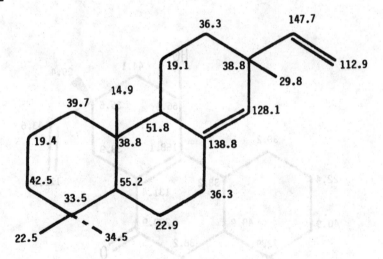

1. E. Wenkert and B.L. Buckwalter, <u>J. Am. Chem.Soc.</u>, 94(12), 4367 (1972).

2. A.H. Kapadi, R.R. Sobti and S. Dev, <u>Tetrahedron Lett.</u>, (31), 2729

   (1965).

20-NOR-8(9),15-ISOPIMARADIEN-7,11-DIONE

Source: *Vellozia pusilla* Pohl
Mol. formula: $C_{19}H_{26}O_2$
Mol. wt.: 286
Solvent: $CDCl_3$

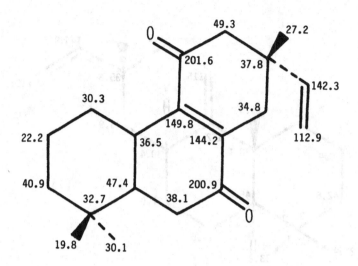

A.C. Pinto, L.M.M. Valente and R.S. Da Silva, <u>Phytochemistry</u>, **27** (12), 3913 (1988).

11β-HYDROXY-20-NOR-8(9),15-ISOPIMARDIEN-7-ONE

Source: *Vellozia pusilla* Pohl
Mol. formula: $C_{19}H_{28}O_2$
Mol. wt.: 288
Solvent: $CDCl_3$

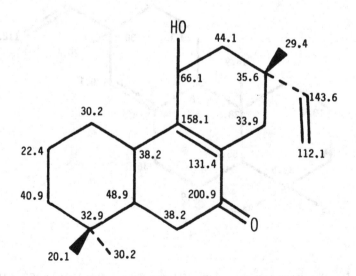

A.C. Pinto, L.M.M. Valente and R.S. Da Silva, <u>Phytochemistry</u>, **27** (12), 3913 (1988).

17-HYDROXY-9-Epi, ENT-7,15-ISOPIMARADIENE

Source: *Calceolaria foliosa* Phil

Mol. formula: $C_{20}H_{32}O$

Mol. wt.: 288

Solvent: $CDCl_3$

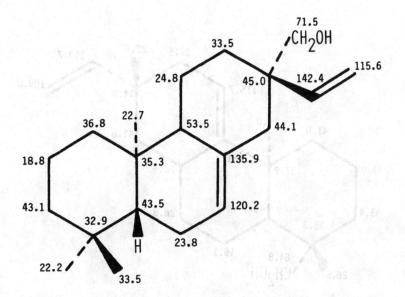

M.C. Chamy, M. Piovano, J.A. Garbarino, V. Gambaro and C. Miranda, <u>Phytochemistry</u>, 28 (2), 571 (1989).

||||||||||||||||||||||||||||||||||||||||||||||||||||||||||||||||||||||||||||||||||||||||||||||||||||||||||||||||||||||||||

ISOPIMARAN-19-OL

Source: *Fritillaria thunbergii* Miq.

Mol. formula: $C_{20}H_{32}O$

Mol. wt.: 288

Solvent: $CDCl_3$

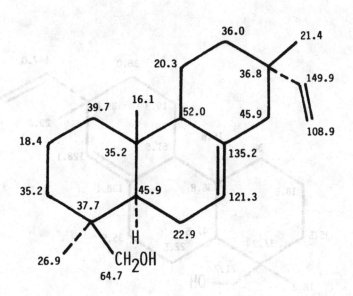

J. Kitajima, T. Komori and T. Kawasaki, Chem. Pharm. Bull., (Tokyo), 30(11), 3912 (1982).

(-)-PIMARA-9(11),15-DIEN-19-OL

Source: *Acanthopanax koreanum* Nakai

Mol. formula: $C_{20}H_{32}O$

Mol. wt.: 288

Solvent: $CDCl_3$

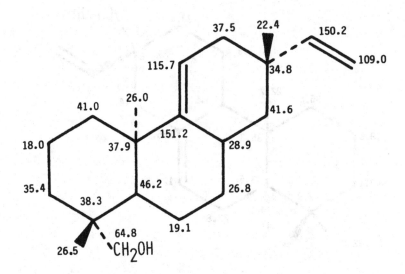

Y.H. Kim, Bo S. Chung and U. Sankawa, <u>J. Nat. Prod.</u>, (Lloydia), **51** (6), 1080 (1988).

|||||||||||||||||||||||||||||||||||||||||||||||||||||||||||||||||||||||||||||||||||||||||||||||||||||||

PIMAROL

Source: *Aralia cordata* Thunb

Mol. formula: $C_{20}H_{32}O$

Mol. wt.: 288

Solvent: CCl_4

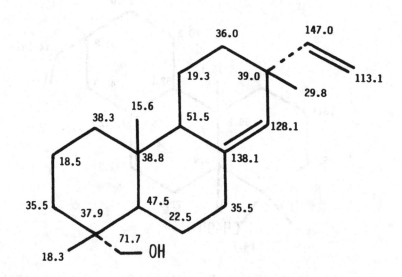

E. Wenkert and B.L. Buckwalter, <u>J. Am. Chem. Soc.</u>, 94(12), 4367 (1972).

BIS-NOR-DITERPENE VB-I

Source: *Vellozia bicolor* L.B. Smith
Mol. formula: $C_{18}H_{28}O_3$
Mol. wt.: 292
Solvent: C_5D_5N

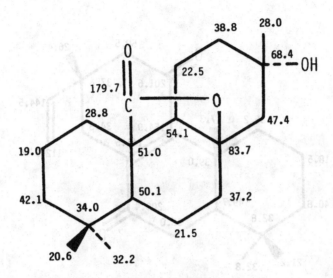

A.C. Pinto, W.S. Garcez, W.E. Hull, A. Neszmelyi and G. Lukacs, J. Chem. Soc. Chem. Commun., (8), 464 (1983).

||||||||||||||||||||||||||||||||||||||||||||||||||||||||||||||||||||||||||||||||||||||||||||||||

7,11-DIOXOPIMAR-8(9),15 DIEN

Source: *Vellozia compacta* Martius ex Schultes
Mol. formula: $C_{20}H_{28}O_2$
Mol. wt.: 300
Solvent: $CDCl_3$

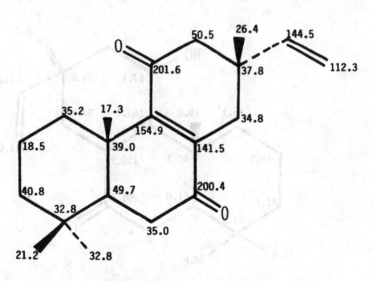

A.C. Pinto and C. Borges, Phytochemistry, 22(9), 2011 (1983).

Source: *Vellozia pusilla* Pohl

Mol. formula: $C_{20}H_{28}O_2$

Mol. wt.: 300

Solvent: $CDCl_3$

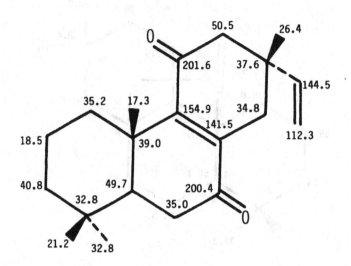

A.C. Pinto, L.M.M. Valente and R.S. Da Silva, <u>Phytochemistry</u>, **27** (12), 3913 (1988).

||||||||||||||||||||||||||||||||||||||||||||||||||||||||||||||||||||||||||||||||||||||||||||||||||||||||||||||

11β-HYDROXY-8(9),15-ISOPIMAREN-7-ONE

Source: *Vellozia nanuzae* L.B. Smith and Ayensu

Mol. formula: $C_{20}H_{30}O_2$

Mol. wt.: 302

Solvent: $CDCl_3$

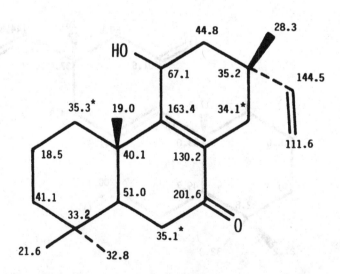

A.C. Pinto, R.S. Da Silva and L.M.M. Valente, <u>Phytochemistry</u>, **27**(12), 3909 (1988).

11β-HYDROXY-7-OXOPIMAR-8(9),15-DIEN

Source: *Vellozia compacta* Martius ex Schultes
Mol. formula: $C_{20}H_{30}O_2$
Mol. wt.: 302
Solvent: $CDCl_3$

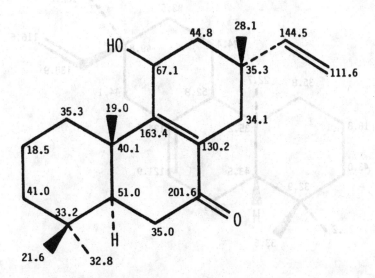

A.C. Pinto and C. Borges, Phytochemistry, 22 (9), 2011 (1983).

12β-HYDROXY-7-OXOPIMAR-8(9),15-DIEN

Source: *Vellozia compacta* Martus ex Schultes
Mol. formula: $C_{20}H_{30}O_2$
Mol. wt.: 302
Solvent: $CDCl_3$

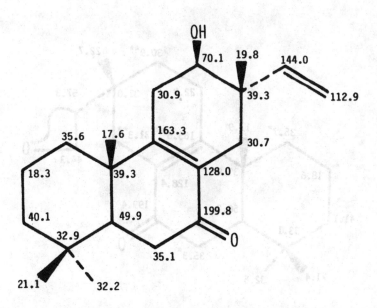

A.C. Pinto and C. Borges, Phytochemistry, 22(9), 2011 (1983).

9-EPI-ENT-7,15-ISOPIMARADIENE-17-OIC ACID

Source: *Calceolaria foliosa* Phil

Mol. formula: $C_{20}H_{30}O_2$

Mol. wt.: 302

Solvent: $CDCl_3$

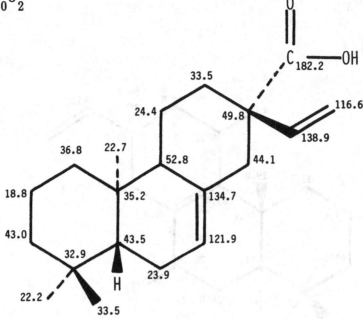

M.C. Chamy, M. piovano, J.A. Garbarino, V. Gambaro and C. Miranda, <u>Phytochemistry</u>, **28** (2), 571 (1989).

||||||||||||||||||||||||||||||||||||||||||||||||||||||||||||||||||||||||||||||||||||||||||||||||||||||||||||

NANUZONE

Source: *Vellozia nanuzae* L.B.Smith and Ayensu

Mol. formula: $C_{20}H_{30}O_2$

Mol. wt.: 302

Solvent: $CDCl_3$

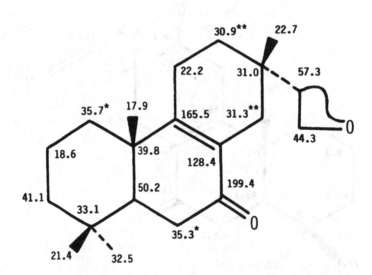

A.C. Pinto, R.S. Da Silva and L.M.M. Valente, <u>Phytochemistry</u>, 27(12), 3909 (1988).

Source: *Vellozia pusilla* Pohl
Mol. formula: $C_{20}H_{30}O_2$
Mol. wt.: 302
Solvent: $CDCl_3$

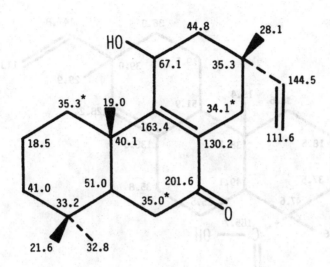

A.C. Pinto. L.M.M. Valente and R.S. Da Silva, Phytochemistry, **27** (12), 3913 (1988).

IIIIIIIIIIIIIIIIIIIIIIIIIIIIIIIIIIIIIIIIIIIIIIIIIIIIIIIIIIIIIIIIIIIIIIIIIIIIIIIIIIIIIIIIIIIIIIIIIIIIIIIIIIIIIIIIIIIIIIIIIIIIIIIIIIIIIIIIIII

(-)-PIMARA-9(11),15-DIEN-19-OIC ACID

Source: *Acanthopanax koreanum* Nakai
Mol. formula: $C_{20}H_{30}O_2$
Mol. wt.: 302
Solvent: $CDCl_3$

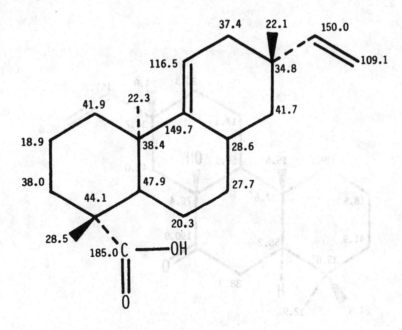

Y.H. Kim, Bo S. Chung and U. Sankawa, J. Nat. Prod., (Lloydia), **51** (6), 1080 (1988).

PIMARIC ACID

Source: *Aralia cordata* Thunb.

Mol. formula: $C_{20}H_{30}O_2$

Mol. wt.: 302

Solvent: $CDCl_3$

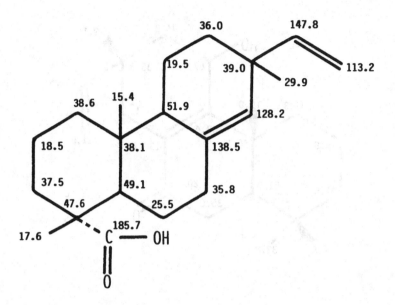

E. Wenkert and B.L. Buckwalter, J. Am. Chem. Soc., 94 (12), 4367 (1972).

||||||||||||||||||||||||||||||||||||||||||||||||||||||||||||||||||||||||||||||||||||||||||||||||||||||||||||||

COMPACTONE

Source: *Vellozia compacta* Martius ex Schultes

Mol. formula: $C_{20}H_{32}O_2$

Mol. wt.: 304

Solvent: $CDCl_3$

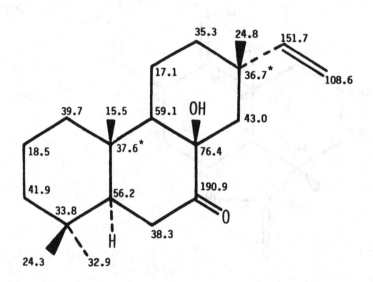

A.C. Pinto, A.J.R. Silva, L.M.U. Mayer and R. Braz F, Phytochemistry, 18(12), 2036 (1979).

DITERPENE OV-Ib

Source: *Oospora virescens* (Link) Wallr.

Mol. formula: $C_{20}H_{32}O_2$

Mol. wt.: 304

Solvent: $CDCl_3$

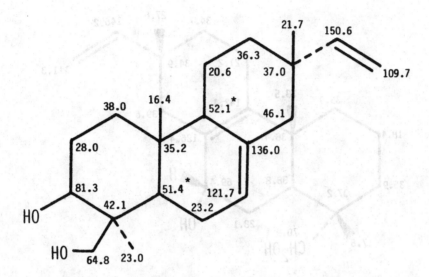

J. Polonsky, Z. Baskevitch, N. Cagnoli-Bellavita, P. Ceccherelli, B.L. Buckwalter and E. Wenkert, J. Am. Chem. Soc., **94**(12), 4369 (1972).

DITERPENE OV-IIIb

Source: *Oospora virescens* (Link) Wallr.

Mol. formula: $C_{20}H_{32}O_2$

Mol. wt.: 304

Solvent: $CDCl_3$

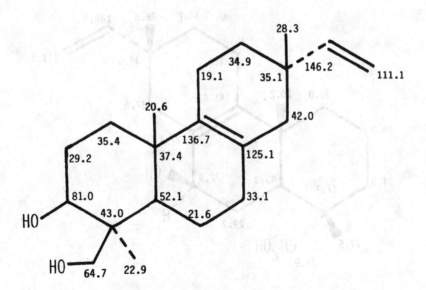

J. Polonsky, Z. Baskevitch, N. Cagnoli-Bellavita, P. Ceccherelli, B.L. Buckwalter and E. Wenkert, J. Am. Chem. Soc., **94**(12), 4369 (1972).

ISOPIMARA-8,15-DIEN-7α,18-DIOL

Source: *Nepeta tuberosa*
Mol. formula: $C_{20}H_{32}O_2$
Mol. wt.: 304

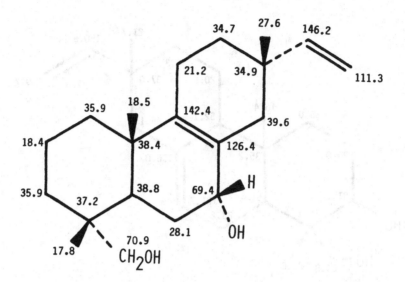

J.G. Urones, I.S. Marcos, J.F. Ferreras and P.B. Barcala, <u>Phytochemistry</u>, 27(2), 523 (1988).

|||||||||||||||||||||||||||||||||||||||||||||||||||||||||||||||||||||||||||||||||||||||||||||||||||||||||||||||||||||||||

ISOPIMARA-8,15-DIEN-7β,18-DIOL

Source: *Nepeta tuberosa*
Mol. formula: $C_{20}H_{32}O_2$
Mol. wt.: 304

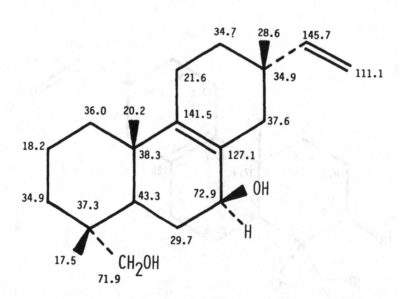

J.G. Urones, I.S. Marcos, J.F. Ferreras and P.B. Barcala, <u>Phytochemistry</u>, 27(2), 523 (1988).

ENT-8(14),15-PIMARADIENE-3β,19-DIOL

Source: *Gochnatia glutinosa* Don
Mol. formula: $C_{20}H_{32}O_2$
Mol. wt.: 304
Solvent: $CDCl_3$

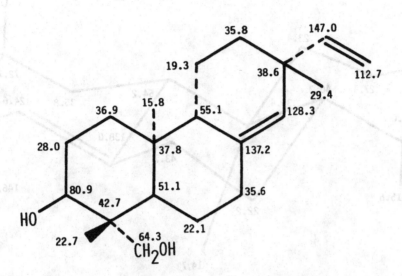

E.E. Garcia, E. Guerreiro and P. Joseph-Nathan, <u>Phytochemistry</u>, **24**(12), 3059 (1985).

8(14),15-SANDARACOPIMARADIENE-2α,18-DIOL

Source: *Tetradenia riparia* (Hochst) Codd.
Mol. formula: $C_{20}H_{32}O_2$
Mol. wt.: 304
Solvent: $CDCl_3$

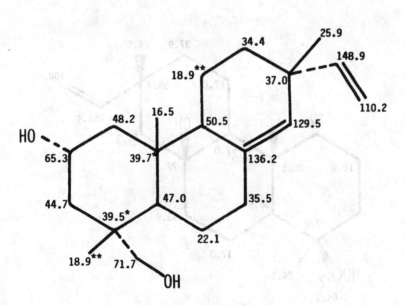

L.V. Puyvelde, N.De. Kimpe, F. Borremans, W. Zhang and N. Schamp, <u>Phytochemistry</u>,
26(2), 493 (1987).

YUCALEXIN P-21

Source: *Manihot esculenta* Crantz
Mol. formula: $C_{20}H_{32}O_2$
Mol. wt.: 304
Solvent: $CDCl_3$

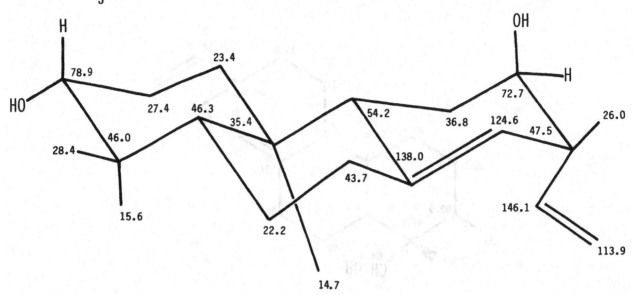

T. Sakai and Y. Nakagawa, <u>Phytochemistry</u>, **27** (12), 3769 (1988).

||||||||||||||||||||||||||||||||||||||||||||||||||||||||||||||||||||||||||||||||||||||||||||||||||||||||||||||||||||

AKHDARDIOL

Source: *Polemonium viscosum*
Mol. formula: $C_{20}H_{34}O_2$
Mol. wt.: 306
Solvent: $CDCl_3$

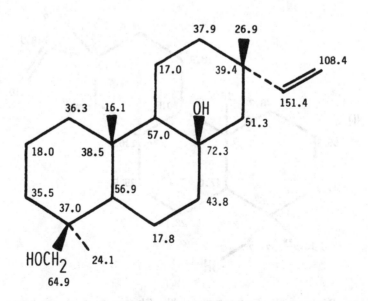

D.B. Stierle, A.A. Stierle and R.D. Larsen, <u>Phytochemistry</u>, 27(2), 517 (1988).

COMPACTOL

Source: *Vellozia compacta* Martius ex Schultes

Mol. formula: $C_{20}H_{34}O_2$

Mol. wt.: 306

Solvent: $CDCl_3$

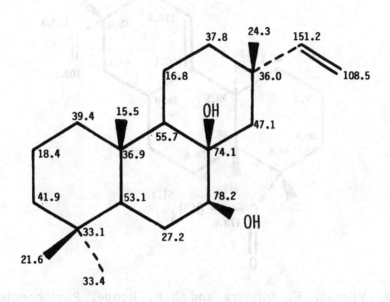

A.C. Pinto and C. Borges, <u>Phytochemistry</u>, **22**(9), 2011 (1983).

||

MOMILACTONE A

Source: *Oryza sativa* L. cv Koshihikari

Mol. formula: $C_{20}H_{26}O_3$

Mol. wt.: 314

Solvent: $CDCl_3$

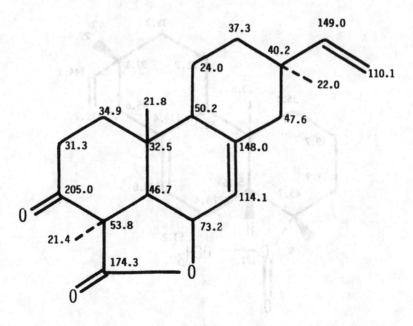

T. Kato, M. Tsunakawa, N. Sasaki, H. Aizawa, K. Fujita, Y. Kitahara and N. Takahashi, <u>Phytochemistry</u>, **16**(1), 45 (1977).

DITERPENE MT-Ia

Source: *Mikania triangularis* Baker

Mol. formula: $C_{21}H_{32}O_2$

Mol. wt.: 316

Solvent: $CDCl_3$

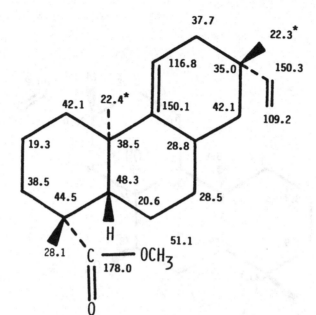

F.S. Knudsen, W. Vilegas, F. Oliveira and N.F. Roque, Phytochemistry, 25(5), 1240 (1986).

‖‖

DITERPENE MT-II

Source: *Mikania triangularis* Baker

Mol. formula: $C_{21}H_{32}O_2$

Mol. wt.: 316

Solvent: $CDCl_3$

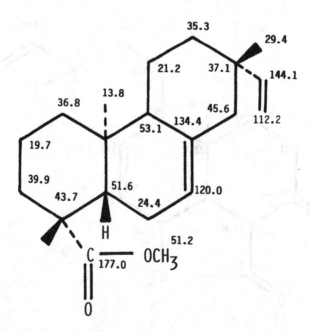

F.S. Knudsen, W. Vilegas, F. Oliveira and N.F. Roque, Phytochemistry, 25(5), 1240 (1986).

12β-HYDROXY-7, 11-DIOXOPIMAR-8,(9),15-DIEN

Source: *Vellozia compacta* Martius ex Schultes
Mol. formula: $C_{20}H_{28}O_3$
Mol. wt.: 316
Solvent: $CDCl_3$

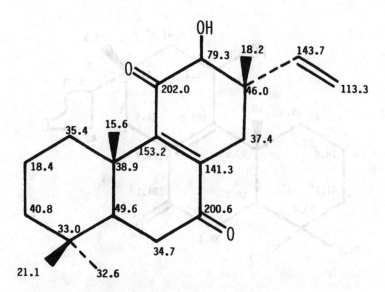

A.C. Pinto and C. Borges, <u>Phytochemistry</u>, 22(9), 2011 (1983).

||||||||||||||||||||||||||||||||||||||||||||||||||||||||||||||||||||||||||||||||||||||||||||||||||||||||||

ISOPIMARAN-19-OIC ACID (METHYL ESTER)

Source: *Fritillaria thunbergii* Miq.
Mol. formula: $C_{21}H_{32}O_2$
Mol. wt.: 316
Solvent: $CDCl_3$

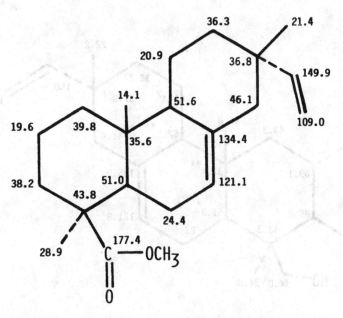

J. Kitajima, T. Komori and T. Kawasaki, <u>Chem. Pharm. Bull.</u>, (Tokyo) 30(11), 3912 (1982).

Source: *Vellozia nanuzae* L.B. Smith and Ayensu
Mol. formula: $C_{20}H_{30}O_3$
Mol. wt.: 318
Solvent: $CDCl_3$

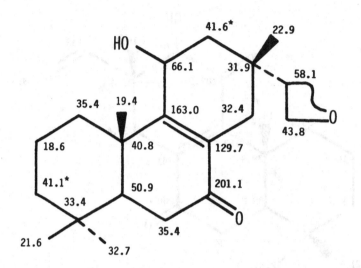

A.C. Pinto, R.S. Da Silva and L.M.M. Valente, <u>Phytochemistry</u>, **27**(12), 3909 (1988).

|||||||||||||||||||||||||||||||||||||||||||||||||||||||||||||||||||||||||||||||||||||||||||||||||||||||||||||||||||||||||||||||||||||||||||||

DITERPENE OV-Ia

Source: *Oospora virescens* (Link) Wallr.
Mol. formula: $C_{20}H_{32}O_3$
Mol. wt.: 320
Solvent: $CDCl_3$

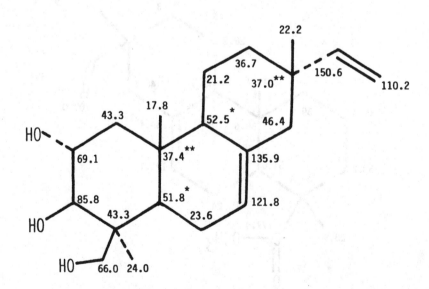

J. Polonsky, Z. Baskevitch, N. Cagnoli-Bellavita, P. Ceccherelli, B.L. Buckwalter and E. Wenkert, <u>J. Am. Chem. Soc.</u>, **94**(12), 4369 (1972).

DITERPENE OV-IIIa

Source: *Oospora virescens* (Link) Wallr.

Mol. formula: $C_{20}H_{32}O_3$

Mol. wt.: 320

Solvent: $CDCl_3$

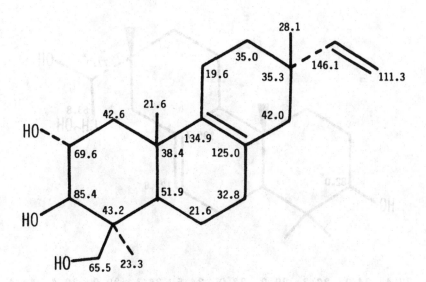

J. Polonsky, Z. Baskevitch, N. Cagnoli-Bellavita, P. Ceccherelli, B.L. Buckwalter and E. Wenkert, J. Am. Chem. Soc., **94**(12), 4369 (1972).

AGATHIS DITERPENTRIOL S-VI

Source: *Agathis* sp.

Mol. formula: $C_{20}H_{34}O_3$

Mol. wt.: 322

Solvent: CD_3OD

15.3, 16.2, 17.7, 20.9, 24.2, 28.2, 29.0, 30.6, 34.4, 36.3, 38.3, 39.2, 39.6, 44.2 and 53.5

N. Fujita, T. Yoshimoto and M. Samejima, Mokuzai Gakkaishi, **30** (3), 264 (1984).

Source: *Agathis* sp.
Mol. formula: $C_{20}H_{34}O_3$
Mol. wt.: 322
Solvent: CD_3OD

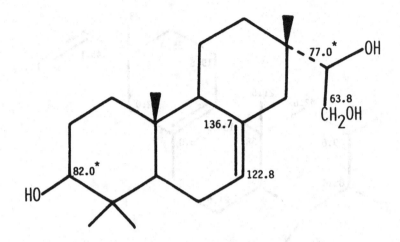

15.3, 17.7, 20.8, 23.4, 24.1, 26.3, 29.0, 33.0, 34.5, 36.1, 38.0, 38.4, 44.4, 45.4 and 53.2

N. Fujita, T. Yoshimoto and M. Samejima, <u>Mokuzai Gakkaishi</u>, **30** (3), 264 (1984).

||||||||||||||||||||||||||||||||||||||||||||||||||||||||||||||||||||||||||||||||||||||||||||||||||||||||

COMPACTOTRIOL

Source: *Vellozia compacta* Martius ex Schultes
Mol. formula: $C_{20}H_{34}O_3$
Mol. wt.: 322
Solvent: C_5D_5N

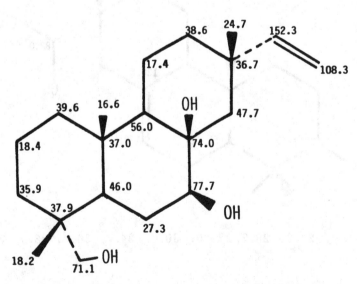

A.C. Pinto and C. Borges, <u>Phytochemistry</u>, 22(9), 2011 (1983).

ISOAKHDARTRIOL

Source: *Amaracus akhdarensis* (Ietswaart et Boulos) Brullo et Furnari (syn. *Origanum akhdarensis*)

Mol. formula: $C_{20}H_{34}O_3$

Mol. wt.: 322

Solvent: $CDCl_3$

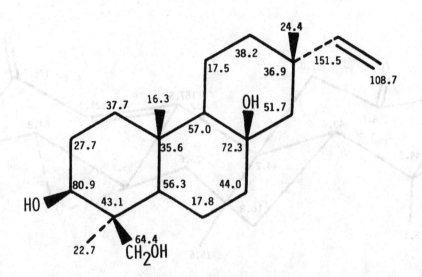

F. Piozzi, M. Paternostro and S. Passannanti, <u>Phytochemistry</u>, 24(5), 1113 (1985).

|||||||||||||||||||||||||||||||||||||||||||||||||||||||||||||||||||||||||||||||||||||||||||||||||||||||||||||||

MOMILACTONE B

Source: *Oryza sativa* L.cv Koshihikari

Mol. formula: $C_{20}H_{26}O_4$

Mol. wt.: 330

Solvent: $CDCl_3$

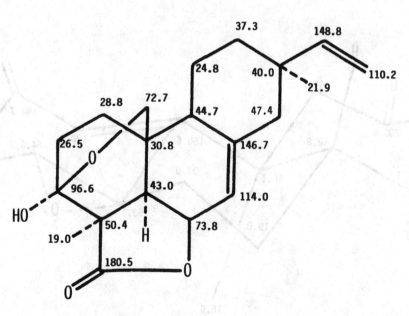

T. Kato, M. Tsunakawa, N. Sasaki, H. Aizawa, K. Fujita, Y. Kitahara and N. Takahashi, <u>Phytochemistry</u>, 16(1), 45 (1977).

YUCALEXIN P-8

Source: *Manihot esculenta* Crantz
Mol. formula: $C_{20}H_{26}O_4$
Mol. wt.: 330
Solvent: $CDCl_3$

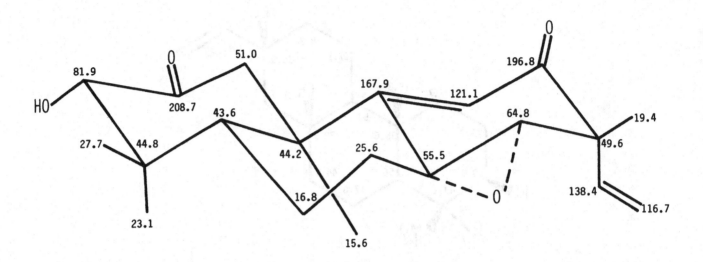

T. Sakai and Y. Nakagawa, <u>Phytochemistry</u>, 27 (12), 3769 (1988).

YUCALEXIN P-10

Source: *Manihot esculenta* Crantz
Mol. formula: $C_{20}H_{26}O_4$
Mol. wt.: 330
Solvent: $CDCl_3$

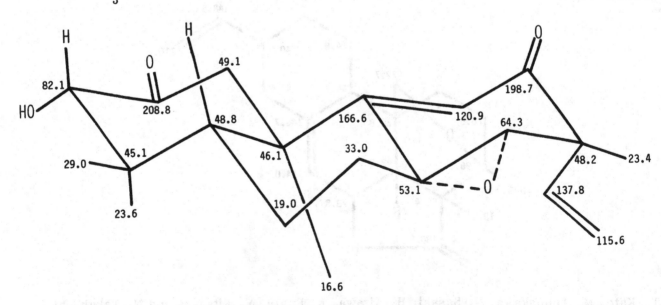

T. Sakai and Y. Nakagawa, <u>Phytochemistry</u>, 27 (12), 3769 (1988).

Source: *Manihot esculenta* Crantz

Mol. formula: $C_{20}H_{26}O_4$

Mol. wt.: 330

Solvent: $CDCl_3$

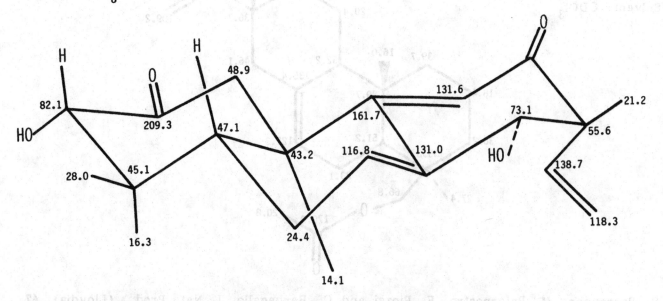

T. Sakai and Y. Nakagawa, <u>Phytochemistry</u>, 27 (12), 3769 (1988).

|||||||||||||||||||||||||||||||||||||||||||||||||||||||||||||||||||||||||||||||||||||||||||||||||||||||||||||||||||

17-ACETOXY-9-<u>EPI</u>-<u>ENT</u>-7,15-ISOPIMARADIENE

Source: *Calceolaria foliosa* Phil

Mol. formula: $C_{22}H_{34}O_2$

Mol. wt.: 330

Solvent: $CDCl_3$

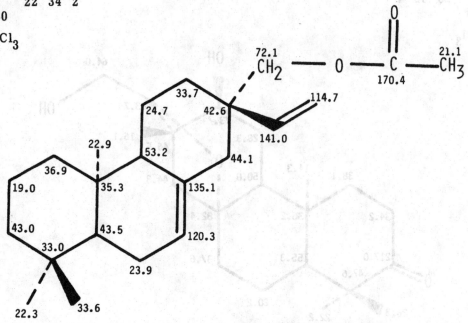

M.C. Chamy, M. Piovano, J.A. Garbarino, V. Gambaro and C. Miranda, <u>Phytochemistry</u>, 28(2), 571 (1989).

AKHDARENOL(ACETATE)

Source: *Amaracus akhdarensis* (Ietswaart et Boulos) Brullo et Furnari (syn. *Origanum akhdarensis*)

Mol. formula: $C_{22}H_{34}O_2$

Mol. wt.: 330

Solvent: $CDCl_3$

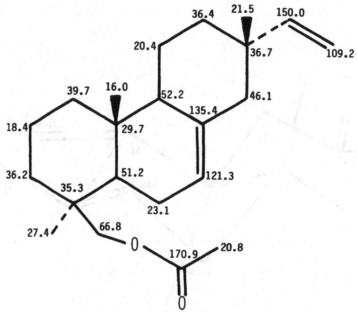

S. Passannanti, M. Paternostro, F. Piozzi and C. Barbagallo, <u>J. Nat. Prod.</u>, (Lloydia), 47 (5), 885 (1984).

IIIIIIIIIIIIIIIIIIIIIIIIIIIIIIIIIIIIIIIIIIIIIIIIIIIIIIIIIIIIIIIIIIIIIIIIIIIIIIIIIIIIIIIIIIIIIIIIIIIIIIIIIII

<u>ENT</u>-8,15<u>R</u>-EPOXY-3-OXOPIMARA-12α,16-DIOL

Source: *Liatris laevigata* Nutt.

Mol. formula: $C_{20}H_{32}O_4$

Mol. wt.: 336

Solvent: $CDCl_3$

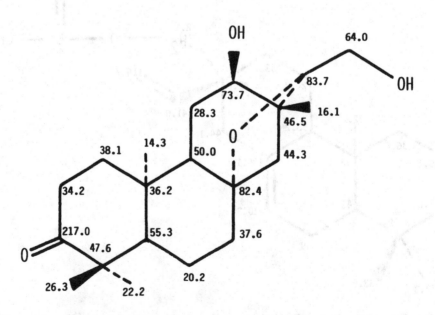

W. Herz and P. Kulanthaivel, <u>Phytochemistry</u>, 22 (3), 715 (1983).

HALLOL

Source: *Podocarpus hallii* Kirk

Mol. formula: $C_{20}H_{34}O_4$

Mol. wt.: 338

Solvent: C_5D_5N

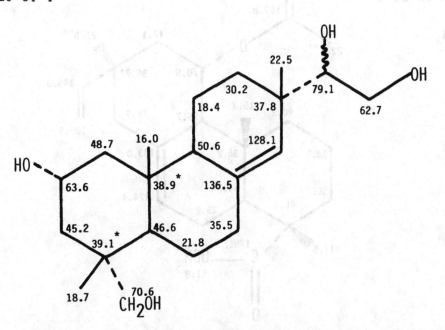

R.C. Cambie, I.R. Burfitt, T.E. Goodwin and E. Wenkert, J. Org. Chem., 40(25), 3789 (1975).

||||||||||||||||||||||||||||||||||||||||||||||||||||||||||||||||||||||||||||||||||||||||||||||||||||||||||||||||||||||||||||

7α-ACETOXYSANDARACOPIMARIC ACID

Source: *Salvia microphylla* H.B.K. (syn. *S. grehamii* Benth)

Mol. formula: $C_{22}H_{32}O_4$

Mol. wt.: 360

Solvent: $CDCl_3$

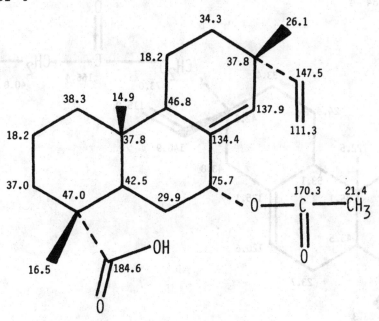

B. Esquivel, N. del Socorro Martinez, J. Cardenas, T.P. Ramamoorthy and L. Rodriguez-Hahn, Planta Med.,55 (1), 62 (1989).

CLEONIOIC ACID (METHYL ESTER)

Source: *Cleonia lusitanica*

Mol. formula: $C_{23}H_{34}O_4$

Mol. wt.: 374

Solvent: $CDCl_3$

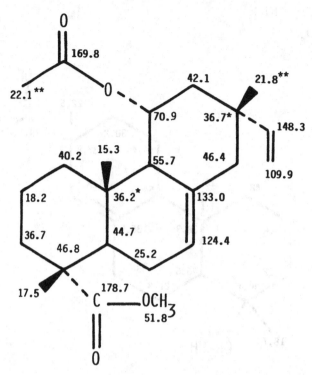

M.C. Garcia-Alvarez, M. Paternostro, F. Piozzi, B. Rodriguez and G. Savona, <u>Phytochemistry</u>, **18**(11), 1835 (1979).

||||||||||||||||||||||||||||||||||||||||||||||||||||||||||||||||||||||||||||||||||||||||||||||||||||||||||

17-MALONYLOXY-9-EPI-<u>ENT</u>-7,15-ISOPIMARADIENE

Source: *Calceolaria foliosa* Phil

Mol. formula: $C_{23}H_{34}O_4$

Mol. wt.: 374

Solvent: $CDCl_3$

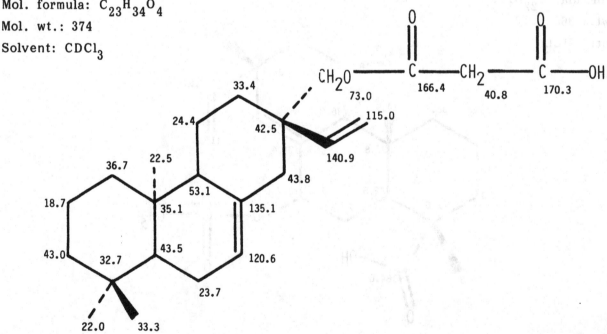

M.C. Chamy, M. Piovano, J.A. Garbarino, V. Gambaro and C. Miranda, <u>Phytochemistry</u>, **28**(2), 571 (1989).

18-MALONYLOXY-9-EPI-ENT-ISOPIMAROL

Source: *Calceolaria glandulosa* Poepp. ex Benth.

Mol. formula: $C_{23}H_{34}O_4$

Mol. wt.: 374

Solvent: $CDCl_3$

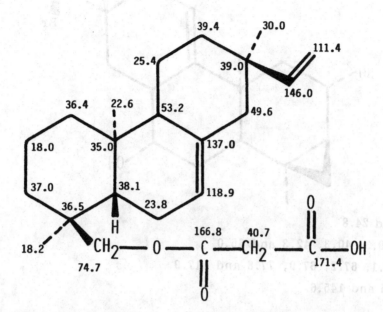

M. Piovano, V. Gambaro, M.C. Chamy, J.A. Garbarino, M. Nicoletti, J. Guilhem and C. Pascard, <u>Phytochemistry</u>, 27(4), 1145 (1988).

||||||||||||||||||||||||||||||||||||||||||||||||||||||||||||||||||||||||||||||||||||||||||||||||||||||||

ISOPIMARA-8,15-DIEN-7α,18-DIOL (DIACETATE)

Source: *Nepeta tuberosa*

Mol. formula: $C_{24}H_{36}O_4$

Mol. wt.: 388

J.G. Urones, I.S. Marcos, J.F. Ferreras and P.B. Barcala, <u>Phytochemistry</u>, 27(2), 523 (1988).

15-BROMO-2,7,16-TRIHYDROXY-9(11)-PARGUERENE

Source: *Laurencia obtusa* (Hudson) Lamouroux

Mol. formula: $C_{20}H_{31}BrO_3$

Mol. wt.: 399

Solvent: CD_3OD

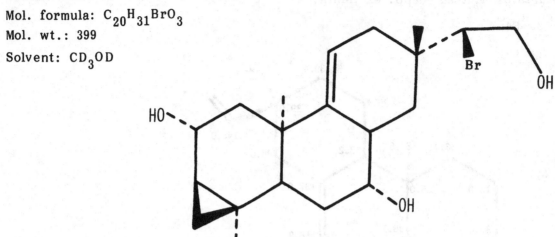

CH_3: 20.8, 23.9 and 24.8

CH_2: 22.6, 35.5, 39.3, 40.3, 42.3 and 65.0

CH: 27.5, 39.4, 48.1, 67.2, 67.9, 77.8 and 117.9

C: 18.4, 36.3, 38.5 and 145.6

T. Suzuki, S. Takeda, N. Hayama, I. Tanaka and K. Komiyama, <u>Chem. Lett.</u>, (6), 969 (1989).

||||||||||||||||||||||||||||||||||||||||||||||||||||||||||||||||||||||||||||||||||||||||||||||||

AKHDARTRIOL (DIACETATE)

Source: *Amaracus akhdarensis* (Ietswaart et Boulos) Brullo et Furnari (syn. *Origanum akhdarensis*)

Mol. formula: $C_{24}H_{38}O_5$

Mol. wt.: 406

Solvent: $CDCl_3$

S. Passannanti, M. Paternostro, F. Piozzi and C. Barbagallo, <u>J. Nat. Prod.</u>, (Lloydia), **47** (5), 885 (1984).

7α-ACETYL-ISOPIMARA-8(14),15-DIEN-18-YL (METHYL MALONATE)

Source: *Nepeta tuberosa*
Mol. formula: $C_{26}H_{38}O_6$
Mol. wt.: 446

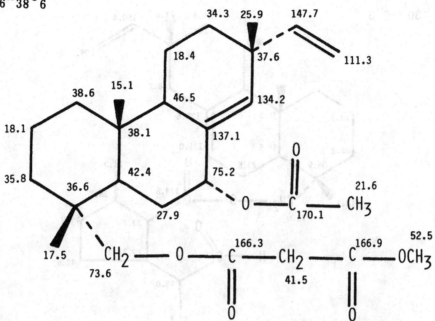

J.G. Urones, I.S. Marcos, J.F. Ferreras and P.B. Barcala, <u>Phytochemistry</u>, 27(2), 523 (1988).

|||||||||||||||||||||||||||||||||||||||||||||||||||||||||||||||||||||||||||||||||||||||||||||||||||||||||||

ISOPIMARYL-2β-ACETYL-5α-METHYL-CYCLOPENTA-β-CARBOXYLATE

Source: *Nepeta tuberosa*
Mol. formula: $C_{29}H_{44}O_3$
Mol. wt.: 440

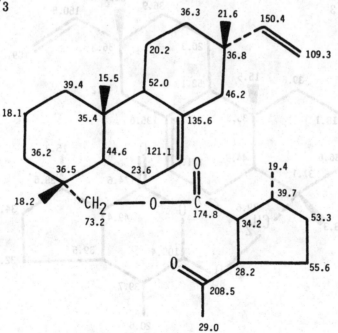

J.G. Urones, P.B. Barcala, I.S. Marcos, J.F. Ferreras and A.F. Rodriguez, <u>Phytochemistry</u>, 27(6), 1783 (1988).

ISOPIMARYL-2β[(1'-METHYL-2'-AL)ETHYL]-5α-METHYL-CYCLOPENTA-β-CARBOXYLATE

Source: *Nepeta tuberosa*

Mol. formula: $C_{30}H_{46}O_3$

Mol. wt.: 454

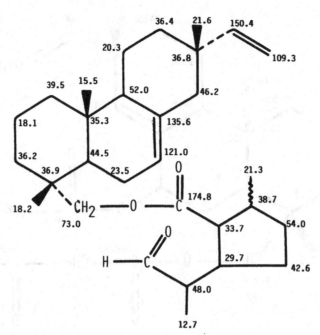

J.G. Urones, P.B. Barcala, I.S. Marcos, J.F. Ferreras and A.F. Rodriguez, <u>Phytochemistry</u>, 27(6), 1783 (1988).

3α-ISOPIMARYLOXY-4α, 4aα, 7α, 7aα-DIHYDRONEPETALACTONE

Source: *Nepeta tuberosa*

Mol. formula: $C_{30}H_{46}O_3$

Mol. wt.: 454

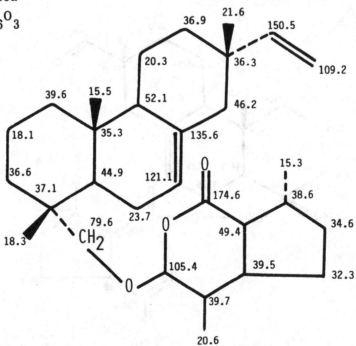

J.G. Urones, P.B. Barcala, I.S. Marcos, J.F. Ferreras and A.F. Rodriguez, <u>Phyto-chemistry</u>, 27(6), 1783 (1988).

3,15-DIBROMO-7,16-DIHYDROXYISOPIMAR-9(11)-ENE

Source: *Laurencia perforata*

Mol. formula: $C_{20}H_{32}Br_2O_2$

Mol. wt.: 464

Solvent: $CDCl_3$

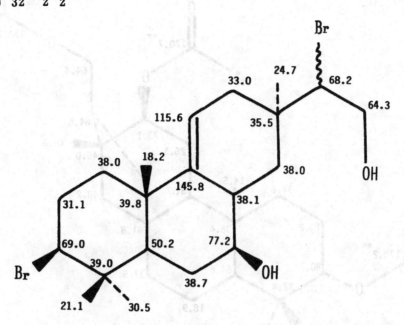

A.G. Gonzalez, J.F. Ciccio, A.P. Rivera and J.D. Martin, <u>J. Org. Chem.</u>, **50**(8), 1261 (1985).

||||||||||||||||||||||||||||||||||||||||||||||||||||||||||||||||||||||||||||||||||||||||||||||||||||||||||||||||||||||||||||||||

<u>ENT-8, 15R-EPOXYPIMARA-3β, 12α,16-TRIOL (TRIACETATE)</u>

Source: *Liatris laevigata* Nutt.

Mol. formula: $C_{26}H_{40}O_7$

Mol. wt.: 464

Solvent: $CDCl_3$

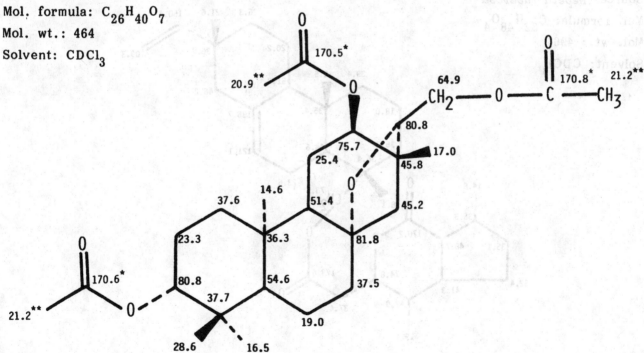

W. Herz and P. Kulanthaivel, <u>Phytochemistry</u>, **22**(3), 715 (1983).

279

ENT-8,15S-EPOXYPIMARA-3β,12α,16-TRIOL (TRIACETATE)

Source: *Liatris laevigata* Nutt.

Mol. formula: $C_{26}H_{40}O_7$

Mol. wt.: 464

Solvent: $CDCl_3$

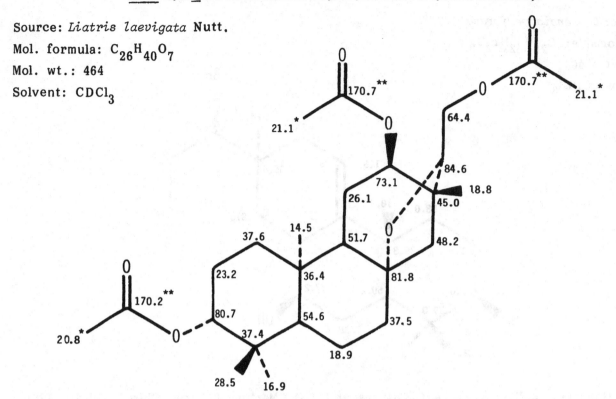

W. Herz and P. Kulanthaivel, <u>Phytochemistry</u>, **22**(3), 715 (1983).

||||||||||||||||||||||||||||||||||||||||||||||||||||||||||||||||||||||||||||||||||||||||||||||||||||||||||||||

ISOPIMARYL-3α-METHYLENECARBOXYLATE-4α,4aα, 7α, 7aα-DIHYDRONEPETALACTONE

Source: *Nepeta tuberosa*

Mol. formula: $C_{32}H_{48}O_4$

Mol. wt.: 496

Solvent: $CDCl_3$

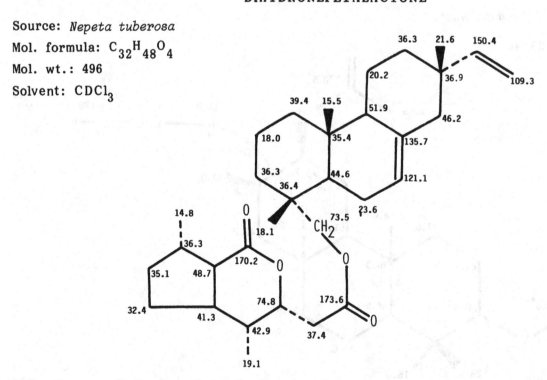

J.G. Urones, P.B. Barcala, I.S. Marcos, J.F. Ferreras and A.F. Rodriguez, <u>Phyto-chemistry</u>, **27**(6), 1783 (1988).

ISOPIMARYL-4β[(3'α-METHYL-2'β-METHOXYCARBONYL) CYCLOPENTHYL]-2-PENTHENOATE

Source: *Nepeta tuberosa*

Mol. formula: $C_{33}H_{50}O_4$

Mol. wt.: 510

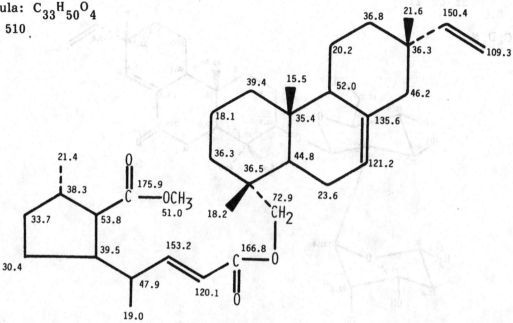

J.G. Urones, P.B. Barcala, I.S. Marcos, J.F. Ferreras and A.F. Rodriguez, <u>Phytochemistry</u>, 27(6), 1783 (1988).

ISOPIMARYL-3-METHYLENECARBOXYLATE-4-HYDROXY-4aα, 7α, 7aα-DIHYDRONEPETALACTONE

Source: *Nepeta tuberosa*

Mol. formula: $C_{32}H_{48}O_5$

Mol. wt.: 512

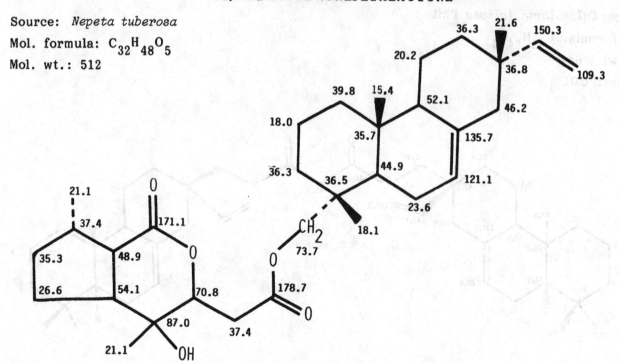

J.G. Urones, P.B. Barcala, I.S. Marcos, J.F. Ferreras and A.F. Rodriguez, <u>Phytochemistry</u>, 27(6), 1783 (1988).

LINIFOLIOSIDE

Source: *Leucas linifolia* Spreng (syn. *L. lavandulaefolia* Rees)

Mol. formula: $C_{32}H_{50}O_{11}$

Mol. wt.: 610

Solvent: C_5D_5N

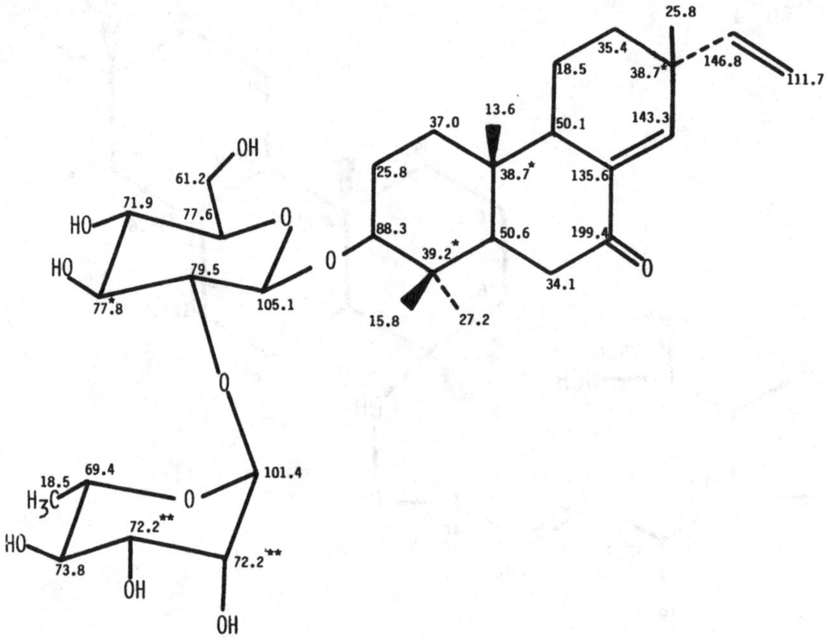

S.B. Mahato and B.C. Pal, <u>Phytochemistry</u>, 25(4), 909 (1986).

FOLIOSATE

Source: *Calceolaria foliosa* Phil

Mol. formula: $C_{43}H_{64}O_4$

Mol. wt.: 644

Solvent: $CDCl_3$

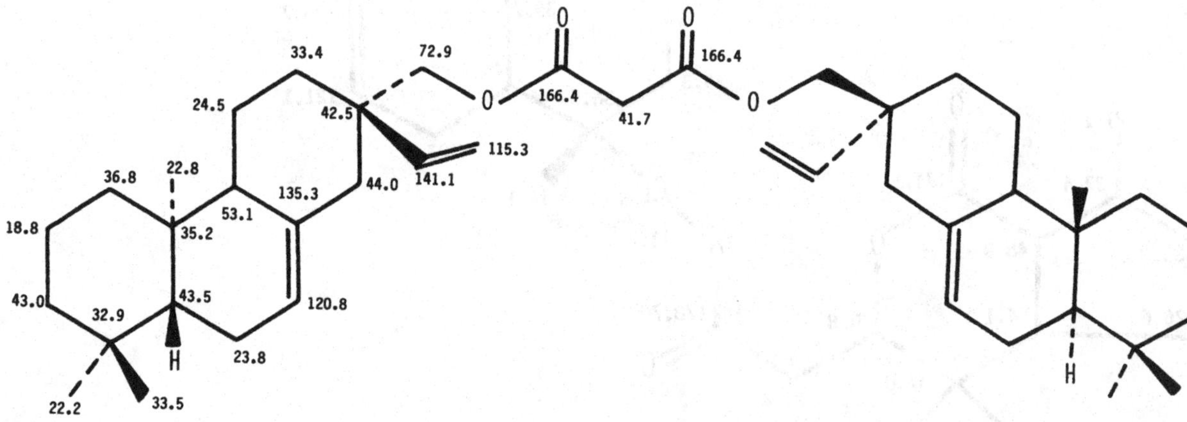

M.C. Chamy, M. Piovano, J.A. Garbarino, V. Gambaro and C. Miranda, <u>Phytochemistry</u>, **28** (2), 571 (1989).

HOOKEROSIDE A

Source: *Scypholepia hookeriana* J.SM. (syn. *Microlepia hookeriana* (Wall.) Presl)

Mol. formula: $C_{39}H_{64}O_{14}$

Mol. wt.: 756

Solvent: C_5D_5N

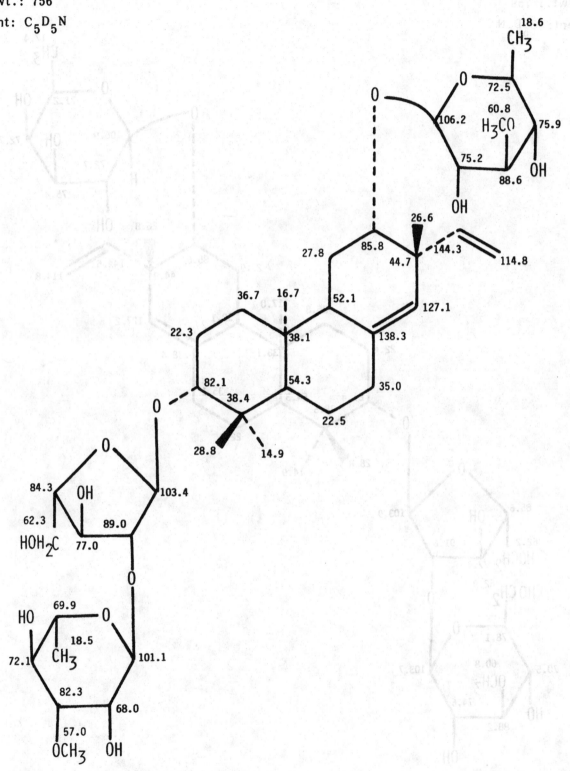

T. Kuraishi, K. Ninomiya, T. Murakami, N. Tanaka, Y. Saiki and C.-M. Chen, Chem. Pharm. Bull. (Tokyo), **32** (12), 4883 (1984).

HOOKEROSIDE C

Source: *Scypholepia hookeriana* J.SM. (syn. *Microlepia hookeriana* (Wall.) Presl)

Mol. formula: $C_{38}H_{62}O_{15}$

Mol. wt.: 758

Solvent: C_5D_5N

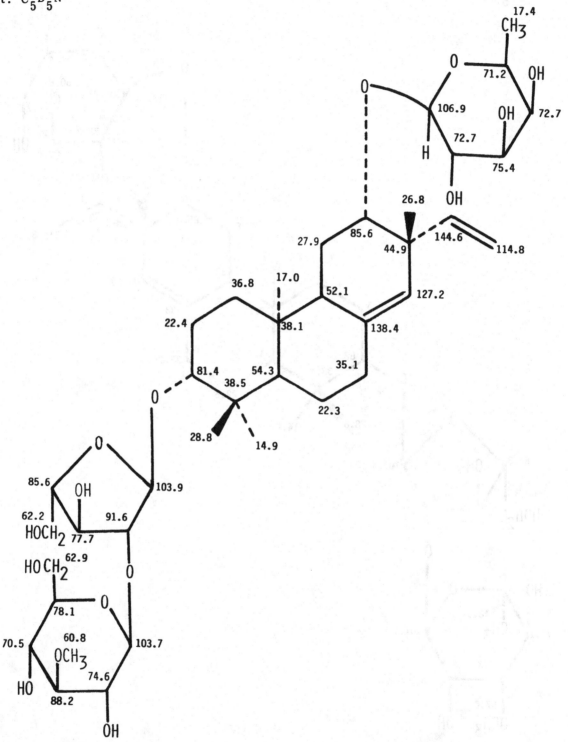

T. Kuraishi, K. Ninomiya, T. Murakami, N. Tanaka, Y. Saiki and C.-M. Chen, Chem. Pharm. Bull. (Tokyo), 32 (12), 4883 (1984).

HOOKEROSIDE D

Source: *Scypholepia hookeriana* J.SM. (syn.*Microlepia hookeriana* (Wall.) Presl)

Mol. formula: $C_{43}H_{70}O_{18}$

Mol. wt.: 874

Solvent: C_5D_5N

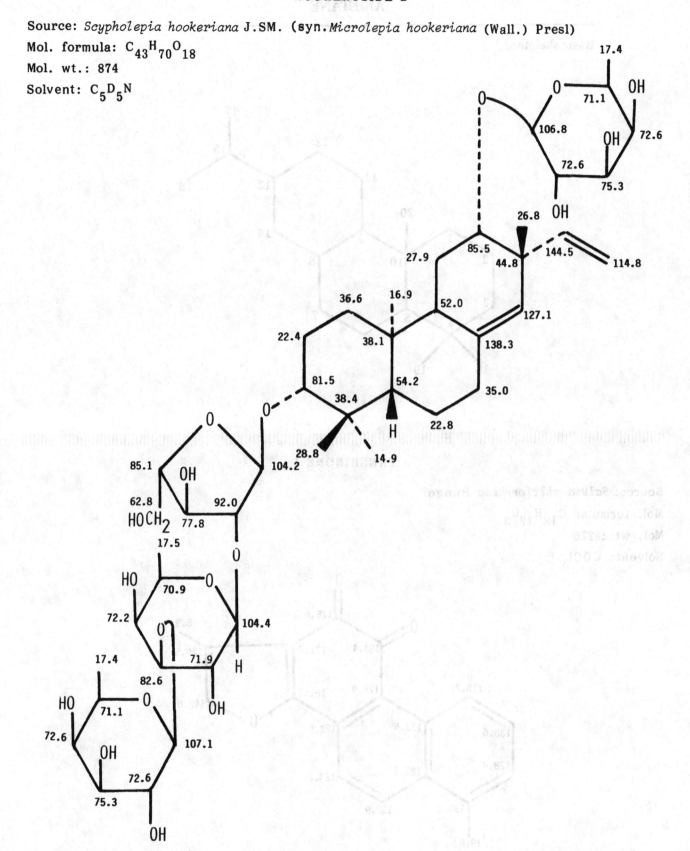

T. Kuraishi, K. Ninomiya, T. Murakami, N. Tanaka, Y. Saiki and C.-M. Chen, <u>Chem. Pharm. Bull.</u> (Tokyo), 32(12), 4883 (1984).

Basic skeleton:

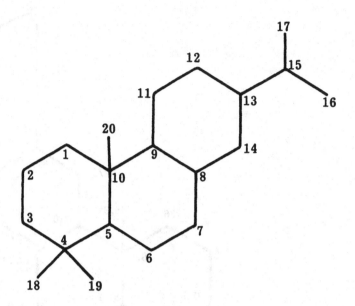

||||||||||||||||||||||||||||||||||||||||||||||||||||||||||||||||||||||||||||||||||||||||||||||||||||||||||||||||||||

TANSHINONE I

Source: *Salvia miltiorrhiza* Bunge

Mol. formula: $C_{18}H_{12}O_3$

Mol. wt.: 276

Solvent: $CDCl_3$

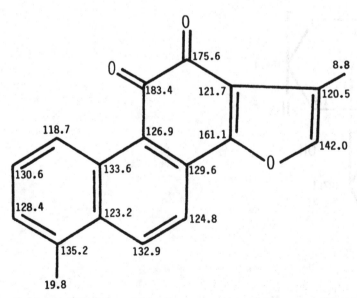

H.-W. Luo, J. Ji, M.-Y. Wu, Z.-G. Yong, M. Niwa and Y. Hirata, Chem. Pharm. Bull. (Tokyo), 34 (8), 3166 (1986).

AR-ABIETATRIEN-12,16-OXIDE

Source: *Thujopsis dolabrata* Sieb. et Zucc.

Mol. formula: $C_{20}H_{28}O$

Mol. wt.: 284

Solvent: $CDCl_3$

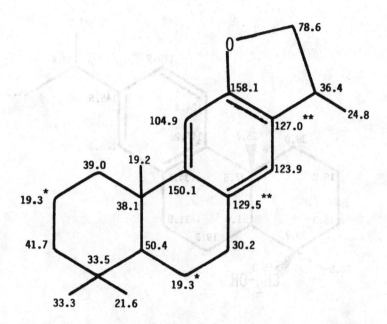

S. Hasegawa and Y. Hirose, <u>Phytochemistry</u>, 21(3), 643 (1982).

||||||||||||||||||||||||||||||||||||||||||||||||||||||||||||||||||||||||||||||||||||||||||||||||||||||||||||||||||||||

ABIETATRIEN-3β-OL

Source: *Nepeta tuberosa*

Mol. formula: $C_{20}H_{30}O$

Mol. wt.: 286

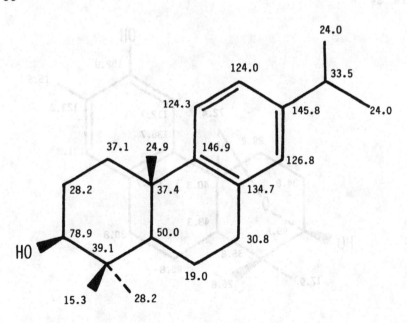

J.G. Urones, I.S. Marcos, J.F. Ferreras and P.B. Barcala, <u>Phytochemistry</u>, 27(2), 523 (1988).

287

DEHYDROABIETINOL

Source: *Calceolaria ascendens* Lind.

Mol. formula: $C_{20}H_{30}O$

Mol. wt.: 286

Solvent: $CDCl_3$

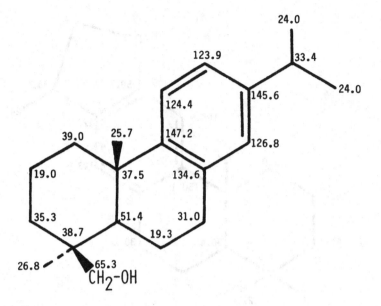

M.C. Chamy, M. Piovano, V. Gambaro, J.A. Garbarino and M. Nicoletti, <u>Phytochemistry</u>, **26** (6), 1763 (1987).

|||||||||||||||||||||||||||||||||||||||||||||||||||||||||||||||||||||||||||||||||||||||||||||||||||||||||||||||||||||||||||

MICRANDROL D

Source: *Micrandropsis scleroxylon*

Mol. formula: $C_{18}H_{24}O_3$

Mol. wt.: 288

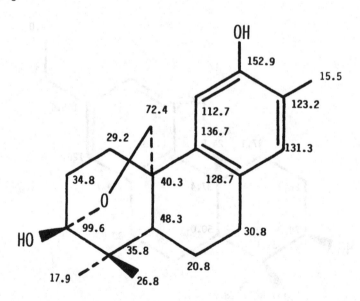

M.A. De Alvarenga, J.J.Da Silva, H.E. Gottlieb and O.R. Gottlieb, <u>Phytochemistry</u>, 20 (5), 1159 (1981).

CRYPTOTANSHINONE

Source: *Salvia miltiorrhiza* Bunge

Mol. formula: $C_{19}H_{20}O_3$

Mol. wt.: 296

Solvent: $CDCl_3$

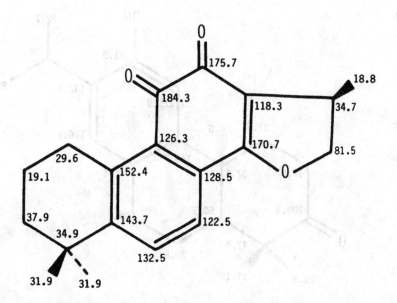

Y. Ikeshiro, I. Mase and Y. Tomita, <u>Phytochemistry</u>, **28** (11), 3139 (1989).

|||||||||||||||||||||||||||||||||||||||||||||||||||||||||||||||||||||||||||||||||||||||||||||||||||||||||||||||||||||||

HYPARGENIN D

Source: *Salvia hypargeia* Fisch. et Mey.

Mol. formula: $C_{20}H_{26}O_2$

Mol. wt.: 298

Solvent: $CDCl_3$

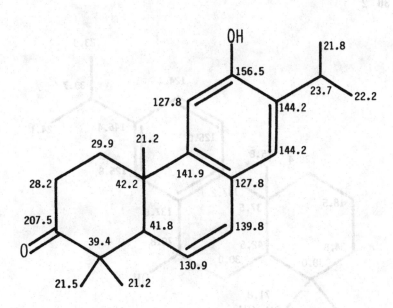

A. Ulubelen, N. Evren, E. Tuzlaci and C. Johansson <u>J. Nat. Prod.</u>, (Lloydia), **51** (6), 1178 (1988)

Source: *Juniperus thurifera*

Mol. formula: $C_{20}H_{28}O_2$

Mol. wt.: 300

Solvent: $CDCl_3$

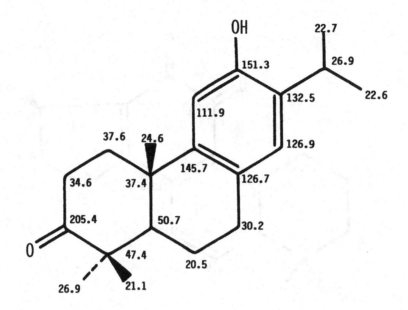

A. San Feliciano, M. Medarde, J.L. Lopez, J.M.M. Del Corral, P. Puebla and A.F. Barrero, Phytochemistry, 27(7), 2241 (1988).

7β-18-DIHYDROXYDEHYDROABIETANOL

Source: *Cedrus deodara* Loud.

Mol. formula: $C_{20}H_{30}O_2$

Mol. wt.: 302

Solvent: $CDCl_3$

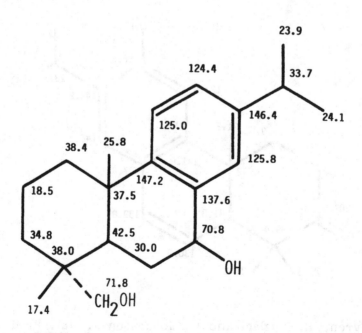

T. Ohmoto, M.Saito and K. Yamaguchi, Chem. Pharm. Bull. (Tokyo), 35(6), 2443 (1987).

Source: *Thujopsis dalabrata* Sieb et Zucc.

Mol. formula: $C_{20}H_{30}O_2$

Mol. wt.: 302

Solvent: $CDCl_3$

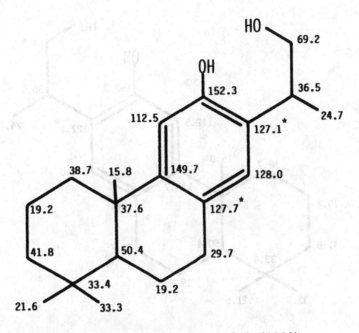

S. Hasegawa and Y. Hirose, Phytochemistry, 21(3), 643 (1982).

||||||||||||||||||||||||||||||||||||||||||||||||||||||||||||||||||||||||||||||||||||||||||||||||||||||||||||||||||||||||||||||||||||||||||||||||

HINOKIOL

Source: *Juniperus thurifera*

Mol. formula: $C_{20}H_{30}O_2$

Mol. wt.: 302

Solvent: DMSO-d_6

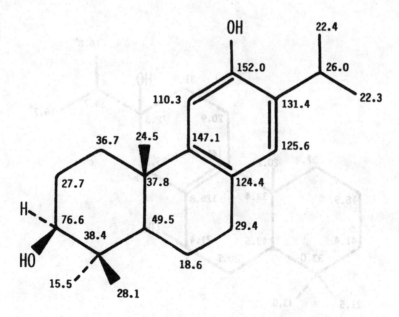

A. San Feliciano, M. Medarde, J.L. Lopez, J.M.M. Del Corral, P. Puebla and A.F. Barrero, Phytochemistry, 27(7), 2241 (1988).

16-HYDROXYFERRUGINOL

Source: *Thujopsis dolabrata* Sieb. et. Zucc.

Mol. formula: $C_{20}H_{30}O_2$

Mol. wt.: 302

Solvent: $CDCl_3$

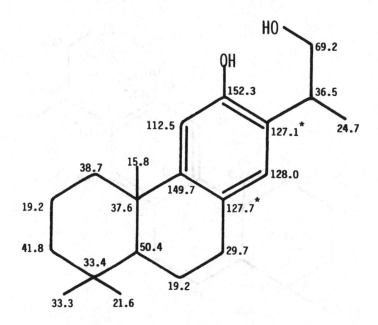

S. Hasegawa and Y. Hirose, <u>Phytochemistry</u>, **21**(3), 643 (1982).

||||||||||||||||||||||||||||||||||||||||||||||||||||||||||||||||||||||||||||||||||||||||||||||||||||||

IBOZOL

Source: *Iboza riparia* N.E. Brown

Mol. formula: $C_{20}H_{34}O_2$

Mol. wt.: 306

Solvent: $CDCl_3$

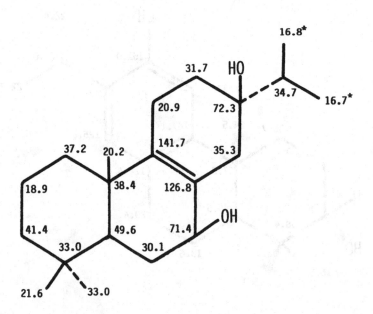

R. Zelnik, E. Rabenhorst, A.K. Matida, H.E. Gottlieb, D. Lavie and S. Panizza, <u>Phyto-chemistry</u>, **17** (9), 1795 (1978).

Source: *Salvia miltiorrhiza* Bunge
Mol. formula: $C_{19}H_{20}O_4$
Mol. wt.: 312
Solvent: $CDCl_3$

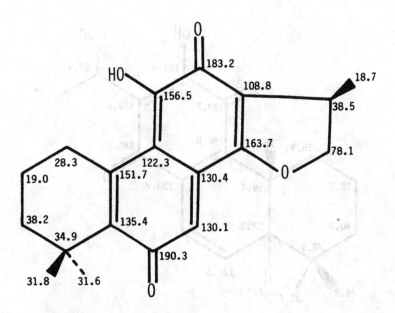

Y. Ikeshiro, I. Mase and Y. Tomita, <u>Phytochemistry</u>, **28** (11), 3139 (1989).

MILTIONONE I

Source: *Salvia miltiorrhiza* Bunge
Mol. formula: $C_{19}H_{20}O_4$
Mol. wt.: 312
Solvent: $CDCl_3$

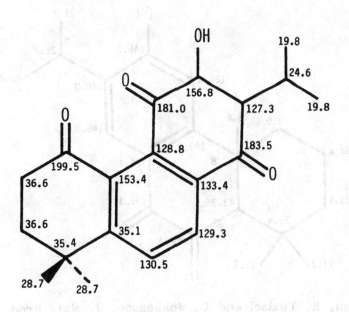

Y. Ikeshiro, I. Mase and Y. Tomita, <u>Phytochemistry</u>, **28** (11), 3139 (1989).

6,7-DIDEHYDROROYLEANONE

Source: *Plectranthus* sp.
Mol. formula: $C_{20}H_{26}O_3$
Mol. wt.: 314
Solvent: $CDCl_3$

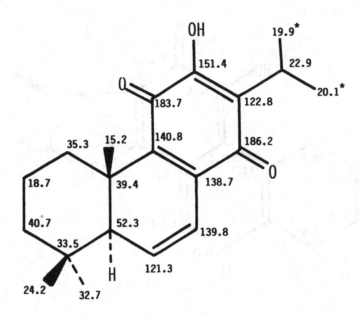

R. Ruedi, <u>Helv. Chim. Acta</u>, 67(4), 1116 (1984).

HYPARGENIN C

Source: *Salvia hypargeia* Fisch. et Mey.
Mol. formula: $C_{20}H_{26}O_3$
Mol. wt.: 314
Solvent: $CDCl_3$

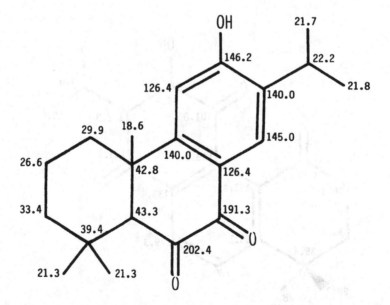

A. Ulubelen, N. Evren, E. Tuzlaci and C. Johansson, <u>J. Nat. Prod.</u>, (Lloydia), 51 (6), 1178 (1988).

NIMOSONE

Source: *Azadirachta indica* A. Juss.

Mol. formula: $C_{20}H_{26}O_3$

Mol. wt.: 314

Solvent: $CDCl_3$

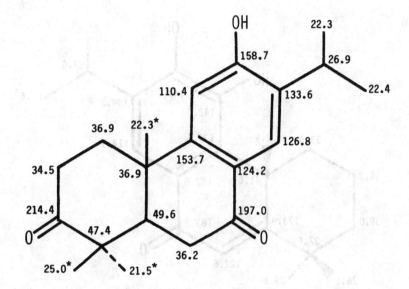

I. Ara, B.S. Siddiqui, S. Faizi and S. Siddiqui, J. Nat. Prod., (Lloydia), 51 (6), 1054 (1988).

PSEUDOJOLKINOLIDE A

Source: *Euphorbia pallasii* Turcz (syn. *E. fisheriana* Stendel)

Mol. formula: $C_{20}H_{26}O_3$

Mol. wt.: 314

Solvent: $CDCl_3$

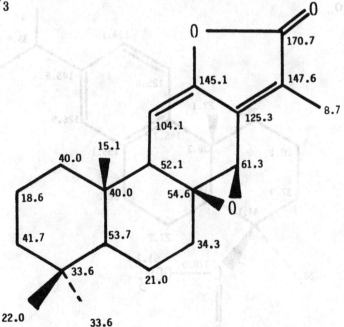

A.I. Sirchina, M.F. Larin and A.A. Semenov, Chem. Nat. Compds., 21 (3), 315 (1985); Khim. Prir. Soedin., 21 (3), 337 (1985).

Source: *Salvia prionitis* Hance

Mol. formula: $C_{20}H_{26}O_3$

Mol. wt.: 314

Solvent: DMSO-d_6

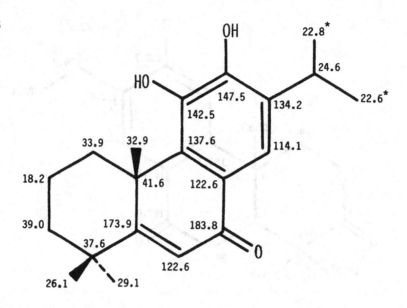

L.-Z. Lin, G. Blasko and G.A. Cordell, <u>Phytochemistry</u>, 28, (1), 177 (1989).

4-<u>EPI</u>-DEHYDROABIETIC ACID (METHYL ESTER)

Source: *Calceolaria ascendens* Lind.

Mol. formula: $C_{21}H_{30}O_2$

Mol. wt.: 314

Solvent: $CDCl_3$

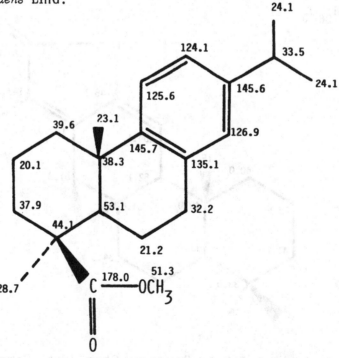

M.C. Chamy, M. Piovano, V. Gambaro, J.A. Garbarino and M. Nicoletti, <u>Phytochemistry</u>, 26 (6), 1763 (1987).

NIMBIDIOL (MONOACETATE)

Source: *Azadirachta indica* A.Juss

Mol. formula: $C_{19}H_{24}O_4$

Mol. wt.: 316

Solvent: $CDCl_3$

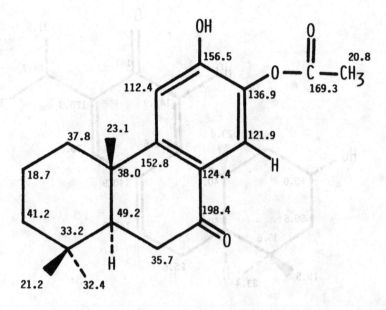

P.L. Majumder, D.C. Maiti, W. Kraus and M. Bokel, <u>Phytochemistry</u>, 26(11), 3021 (1987).

||||||||||||||||||||||||||||||||||||||||||||||||||||||||||||||||||||||||||||||||||||||||||||||||||||||||||||||||||||||||||||||||

DEHYDROABIETIC ACID

Source: *Salvia tomentosa* Mill.

Mol. formula: $C_{20}H_{28}O_3$

Mol. wt.: 316

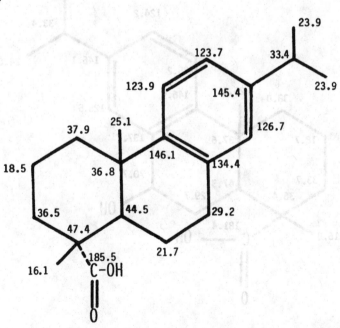

A. Ulubelen, M. Miski and T.J. Mabry, <u>J. Nat. Prod.</u> (Lloydia), 44 (1), 119 (1981).

6-DEOXO-2α-HYDROXYTAXODIONE

Source: *Salvia texana*

Mol. formula: $C_{20}H_{28}O_3$

Mol. wt.: 316

Solvent: $CDCl_3$

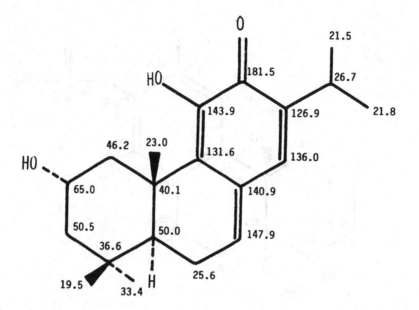

A.G. Gonzalez, Z.E. Aguiar, J.G. Luis, A.G. Ravelo and X. Dominguez, <u>Phytochemistry</u>, 27(6), 1777 (1988).

||||||||||||||||||||||||||||||||||||||||||||||||||||||||||||||||||||||||||||||||||||||||||||||||||||||||||||||||||||

7β-HYDROXYDEHYDROABIETIC ACID

Source: *Cedrus deodara* Loud.

Mol. formula: $C_{20}H_{28}O_3$

Mol. wt.: 316

Solvent: $CDCl_3$

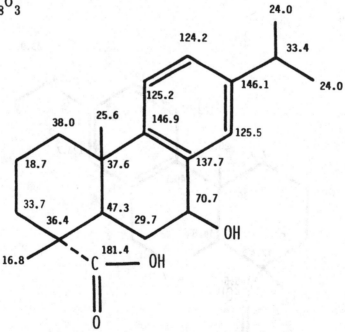

T. Ohmoto, M. Saito and K. Yamaguchi, <u>Chem. Pharm. Bull.</u>, (Tokyo), 35(6), 2443 (1987).

LAMBERTIC ACID

Source: *Dacrycarpus imbricatus* (Bl. de Laub.)

Mol. formula: $C_{20}H_{28}O_3$

Mol. wt.: 316

Solvent: $CDCl_3$

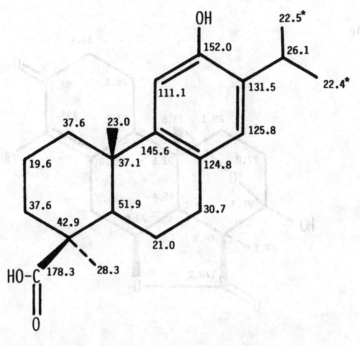

R.C. Cambie, R.E. Cox, K.D. Croft and D. Sidwel, <u>Phytochemistry</u>, 22(5), 1163 (1983).

||||||||||||||||||||||||||||||||||||||||||||||||||||||||||||||||||||||||||||||||||||||||||||||||||||||||||||||||||||||||||||||||||

15-HYDROXYABIETIC ACID

Source: *Cedrus deodara* Loud.

Mol. formula: $C_{20}H_{30}O_3$

Mol. wt.: 318

Solvent: $CDCl_3$

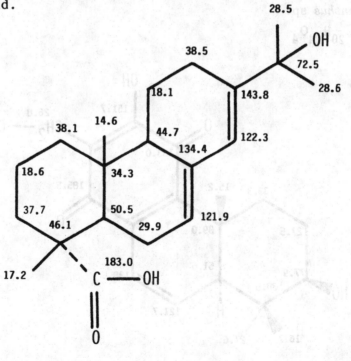

T. Ohmoto, M. Saito and K. Yamaguchi, <u>Chem. Pharm. Bull.</u> (Tokyo), 35(6), 2443 (1987).

HUMIRIANTHENOLIDE D

Source: *Humirianthera rupestris* Ducle
Mol. formula: $C_{17}H_{20}O_6$
Mol. wt.: 320
Solvent: DMSO-d_6

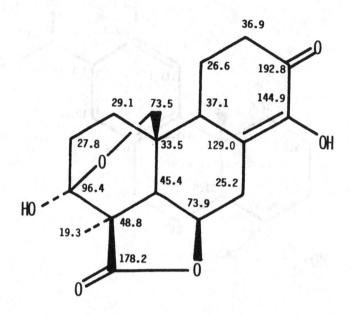

M.D.G.B. Zoghbi, N.F. Roque and H.E. Gottlieb, <u>Phytochemistry</u>, **20**(7), 1669 (1981).

||||||||||||||||||||||||||||||||||||||||||||||||||||||||||||||||||||||||||||||||||||||||||||||||||||||||||||||||||||||||||||||||||||

PLECTRANTHONE F

Source: *Plectranthus* sp.
Mol. formula: $C_{20}H_{24}O_4$
Mol. wt.: 328
Solvent: $CDCl_3$

A.C. Alder, P. Ruedi, R. Prewo, J.H. Bieri and C.H. Eugster, <u>Helv. Chim. Acta</u>, **69**(6), 1395 (1986).

3β-HYDROXY-8,11,13(14),15-ABIETATETRAEN-18-OIC ACID (METHYL ESTER)

Source: *Salvia tomentosa* Mill

Mol. formula: $C_{21}H_{28}O_3$

Mol. wt.: 328

Solvent: $CDCl_3$

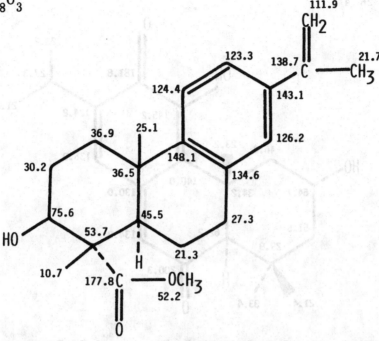

A. Ulubelen, M. Miski and T.J. Mabry, <u>J. Nat. Prod.</u>, (Lloydia), 44(1), 119 (1981).

||||||||||||||||||||||||||||||||||||||||||||||||||||||||||||||||||||||||||||||||||||||||||||||||||||||||||||||||||||

14-DEOXYCOLEON U

Source: *Salvia phlomoides* Asso.

Mol. formula: $C_{20}H_{26}O_4$

Mol. wt.: 330

Solvent: C_5D_5N

J.A. Hueso-Rodriguez, M.L. Jimeno, B. Rodriguez, G. Savona and M. Bruno, <u>Phyto-chemistry</u>, 22(9), 2005 (1983).

2α-HYDROXYTAXODIONE

Source: *Salvia texana*

Mol. formula: $C_{20}H_{26}O_4$

Mol. wt.: 330

Solvent: $CDCl_3$

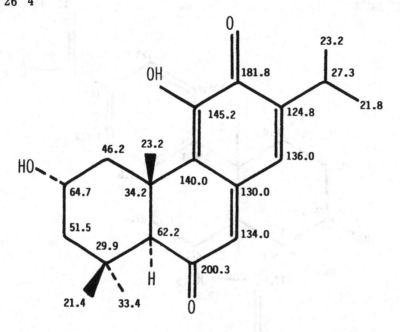

A.G. Gonzalez, Z.E. Aguiar, J.G. Luis, A.G. Ravelo and X. Dominguez, <u>Phytochemistry</u>, 27(6), 1777 (1988).

||||||||||||||||||||||||||||||||||||||||||||||||||||||||||||||||||||||||||||||||||||||||||

HYPARGENIN F

Source: *Salvia hypargeia* Fisch. et Mey.

Mol. formula: $C_{20}H_{26}O_4$

Mol. wt.: 330

Solvent: $CDCl_3$

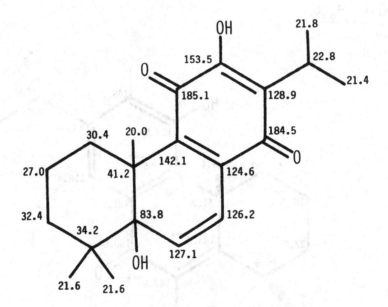

A. Ulubelen, N. Evren, E. Tuzlaci and C. Johansson, <u>J. Nat. Prod.</u>, (Lloydia), 51 (6), 1178 (1988).

ISOCARNOSOL

Source: *Salvia lanigera* Poir
Mol. formula: $C_{20}H_{26}O_4$
Mol. wt.: 330
Solvent: $CDCl_3$

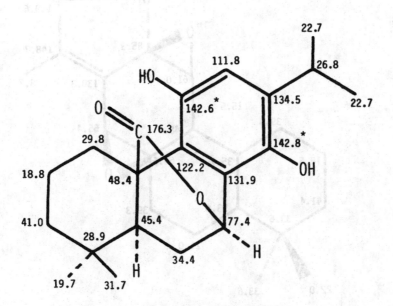

H.M.G. Al-Hazimi, M.S.H. Deep and G.A. Miana, <u>Phytochemistry</u>, 23(4), 919 (1984).

||||||||||||||||||||||||||||||||||||||||||||||||||||||||||||||||||||||||||||||||||||||||||||||||||||||||||||||||||||||||

7-OXOROYLEANON

Source: *Plectranthus* sp.
Mol. formula: $C_{20}H_{26}O_4$
Mol. wt.: 330
Solvent: $CDCl_3$

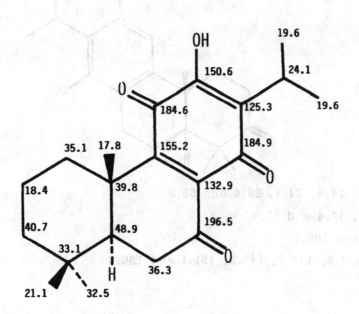

P. Ruedi, <u>Helv. Chim. Acta</u>, 67(4), 1116 (1984).

PSUEDOJOLKINOLIDE B

Source: *Euphorbia pallasii* Turcz (syn. *E. fisheriana* Stendel)

Mol. formula: $C_{20}H_{26}O_4$

Mol. wt.: 330

Solvent: $CDCl_3$

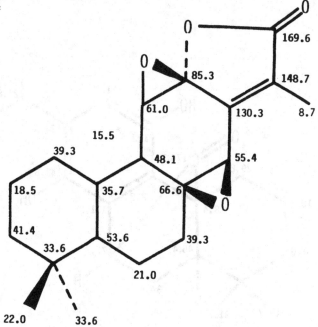

A.I. Sirchina, M.F. Larin and A.A. Semenov, <u>Chem. Nat. Compds.</u>, **21** (3), 315 (1985); <u>Khim. Prir. Soedin.</u>, **21** (3), 337 (1985).

||||||||||||||||||||||||||||||||||||||||||||||||||||||||||||||||||||||||||||||||||||||||||||||||||||||||||||||

NEOTRIPTONOTERPENE

Source: *Tripterygium wilfordii* Hook

Mol. formula: $C_{21}H_{30}O_3$

Mol. wt.: 330

CH_3: 20.4, 20.6, 24.4, 24.4, 26.4 and 55.3

CH_2: 19.2, 24.4, 34.4 and 37.4

CH: 19.2, 49.9 and 100.2

C: 37.0, 46.9, 113.8, 119.2, 145.3, 151.1 and 156.3

C=O: 217.0

B.N. Zhou, D.Y. Zhu, F.X. Deng, C.G. Huang, J.P. Kutney and M. Roberts, <u>Planta Med.</u>, **54**(4), 330 (1988).

HORMINON

Source: *Abyssinian plectranthus*

Mol. formula: $C_{20}H_{28}O_4$

Mol. wt.: 332

Solvent: $CDCl_3$

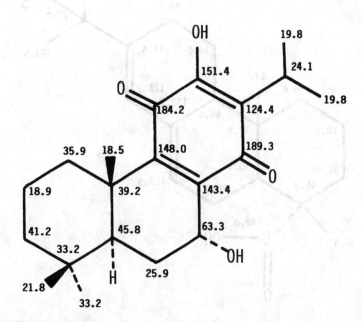

P. Ruedi, <u>Helv. Chim. Acta</u>, 67(4), 1116 (1984).

||||||||||||||||||||||||||||||||||||||||||||||||||||||||||||||||||||||||||||||||||||||||||||||||||||||||||||||||||||||||||||||||||||||||||||||||||||||||

TAXOCHINON

Source: *Abyssinian plectranthus*

Mol. formula: $C_{20}H_{28}O_4$

Mol. wt.: 332

Solvent: $CDCl_3$

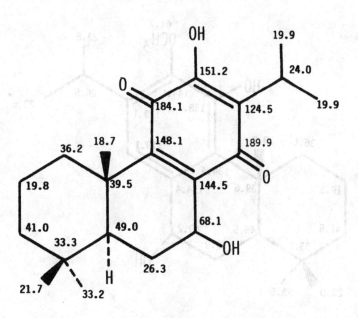

P. Ruedi, <u>Helv. Chim. Acta</u>, 67(4), 1116 (1984).

15-METHOXYABIETIC ACID

Source: *Cedrus deodara* Loud.

Mol. formula: $C_{21}H_{32}O_3$

Mol. wt.: 332

Solvent: $CDCl_3$

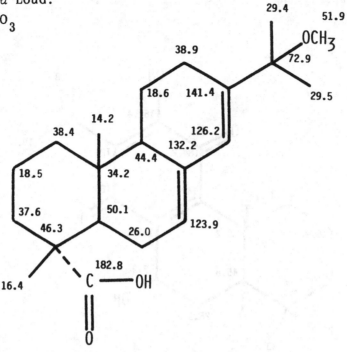

T. Ohmoto, M. Saito and K. Yamaguchi, <u>Chem. Pharm. Bull.</u> (Tokyo), **35** (6), 2443 (1987).

||||||||||||||||||||||||||||||||||||||||||||||||||||||||||||||||||||||||||||||||||||||||||||||||||||||||||||||||||||||||

12-METHOXY-7α,11-DIHYDROXY-DEHYDROABIETANE

Source: *Salvia bicolor*

Mol. formula: $C_{21}H_{32}O_3$

Mol. wt.: 332

Solvent: $CDCl_3$

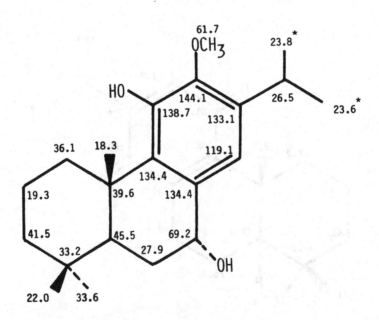

S. Valverde, J. Escudero, J.C. Lopez and R.M. Rabanal, <u>Phytochemistry</u>, **24**(1), 111 (1985).

COLEON-U-QUINONE

Source: *Plectranthus argentatus* S.T. Blake
Mol. formula: $C_{20}H_{24}O_5$
Mol. wt.: 344
Solvent: $CDCl_3$

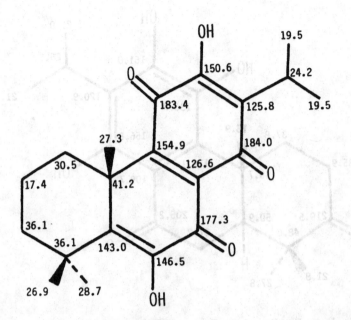

A.C. Alder, P. Ruedi and C.H. Eugster, <u>Helv. Chim. Acta</u>, 67 (6), 1523 (1984).

||||||||||||||||||||||||||||||||||||||||||||||||||||||||||||||||||||||||||||||||||||||||||||||||||||||||||

HUMIRIANTHENOLIDE C

Source: *Humirianthera rupestris* Ducle
Mol. formula: $C_{19}H_{22}O_6$
Mol. wt.: 346
Solvent: DMSO-d_6

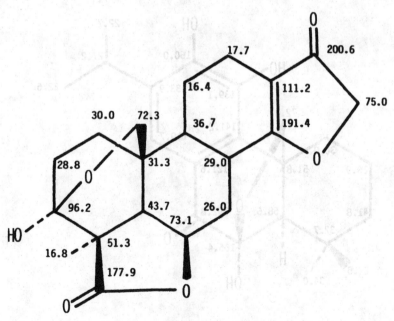

M.D.G.B. Zoghbi, N.F. Roque and H.E. Gottlieb, <u>Phytochemistry</u>, 20(7), 1669 (1981).

CANDELABRONE

Source: *Salvia candelabrum*
Mol. formula: $C_{20}H_{26}O_5$
Mol. wt.: 346
Solvent: CD_3OD

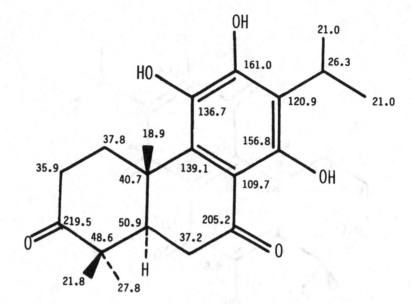

S. Canigueral, J. Iglesias, F. Sanchez-Ferrando and A. Virgilli, <u>Phytochemistry</u>, 27(1), 221 (1988).

CARNOSOLONE

Source: *Coleus carnosus* Hassk.
Mol. formula: $C_{20}H_{26}O_5$
Mol. wt.: 346

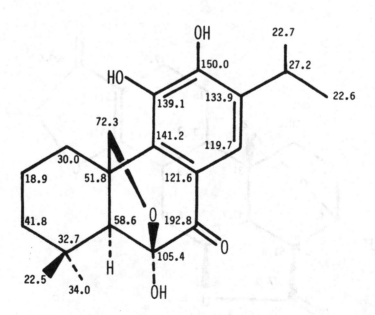

F. Yoshizaki, P. Ruedi and C.H. Eugster, <u>Helv. Chim. Acta.</u>, 62 (8), 2754 (1979).

8α,9α-EPOXY-7-OXOROYLEANONE

Source: *Plectranthus* sp.

Mol. formula: $C_{20}H_{26}O_5$

Mol. wt.: 346

Solvent: $CDCl_3$

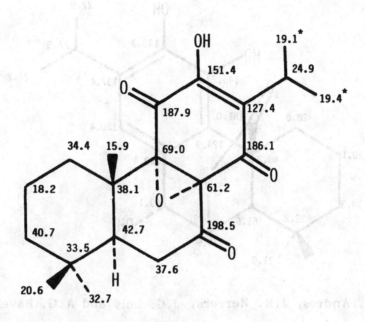

P. Ruedi, <u>Helv. Chim. Acta</u>, 67(4), 1116 (1984).

||||||||||||||||||||||||||||||||||||||||||||||||||||||||||||||||||||||||||||||||||||||||||||||||||||||||||||||||||||||||||||||||||||||||||

16-HYDROXYPSEUDOJOLKINOLIDE B

Source: *Euphorbia pallasii* Turcz (syn. *E. fisheriana* Stendel)

Mol. formula: $C_{20}H_{26}O_5$

Mol. wt.: 346

Solvent: $CDCl_3$

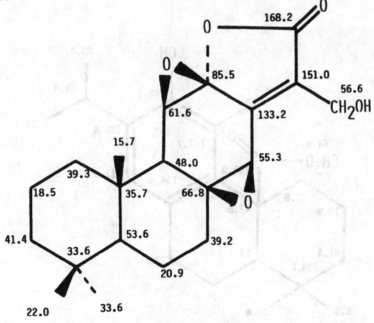

A.I. Sirchina, M.F. Larin and A.A. Semenov, <u>Chem. Nat. Compds.</u>, 21 (3), 315 (1985); <u>Khim. Prir. Soedin.</u>, 21 (3), 337 (1985).

ROSMANOL

Source: *Salvia canariensis* L.

Mol. formula: $C_{20}H_{26}O_5$

Mol. wt.: 346

Solvent: $(CD_3)_2CO$

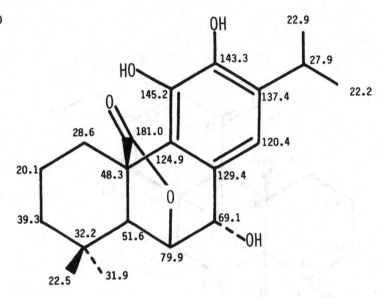

A.G. Gonzalez, L.S. Andres, J.R. Herrera, J.G. Luis and A.G. Ravelo, <u>Can. J. Chem.</u>, 67 (2), 208 (1989).

||||||||||||||||||||||||||||||||||||||||||||||||||||||||||||||||||||||||||||||||||||||||||||||||||||||||||||||||||||||||||||||||||

CARNOSIC ACID (METHYL ESTER)

Source: *Salvia laniger*

Mol. formula: $C_{21}H_{30}O_4$

Mol. wt.: 346

Solvent: $CDCl_3$

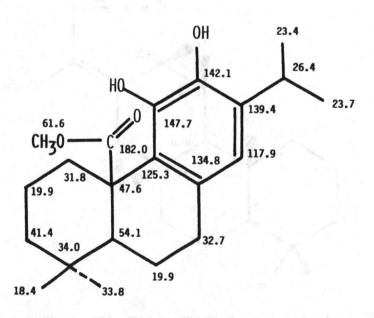

H.M.G. Al-Hazimi, <u>Phytochemistry</u>, 25(5), 1238 (1986).

5,6-DIHYDROCOLEON U

Source: *Plectranthus argentatus* S.T. Blake
Mol. formula: $C_{20}H_{28}O_5$
Mol. wt.: 348
Solvent: C_6D_6

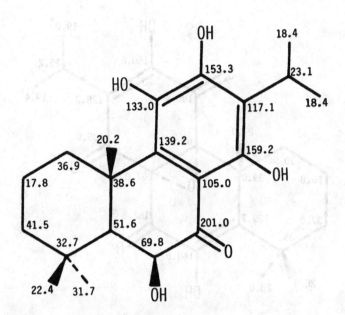

A.C. Alder, P. Ruedi and C.H. Eugster, <u>Helv. Chim. Acta</u>, 67 (6), 1523 (1984).

||||||||||||||||||||||||||||||||||||||||||||||||||||||||||||||||||||||||||||||||||||||||||||||||||||||||||||||||||

2α-7-DIHYDROXYTAXODONE

Source: *Salvia texana*
Mol. formula: $C_{20}H_{28}O_5$
Mol. wt.: 348
Solvent: $CDCl_3$

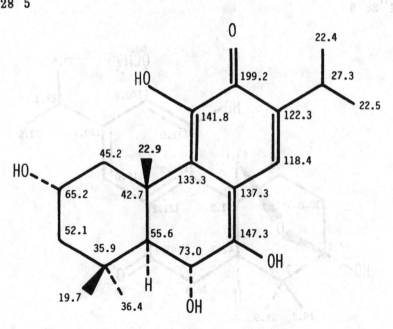

A.G. Gonzalez, Z.E. Aguiar, J.G. Luis, A.G. Ravelo and X. Dominguez, <u>Phytochemistry</u>, 27(6), 1777 (1988).

8α,9α-EPOXYCOLEON-U-CHINON

Source: *Plectranthus argentatus* S.T. Blake
Mol. formula: $C_{20}H_{24}O_6$
Mol. wt.: 360
Solvent: $CDCl_3$

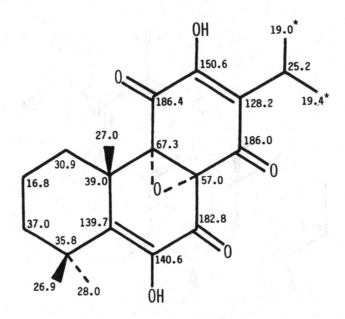

A.C. Alder, P. Ruedi and C.H. Eugster, Helv. Chim. Acta, **67** (6), 1523 (1984).

TAXAMAIRIN C

Source: *Taxus mairei* (Lemee et Levl.) S.Y. Hu
Mol. formula: $C_{21}H_{28}O_5$
Mol. wt.: 360
Solvent: C_5D_5N

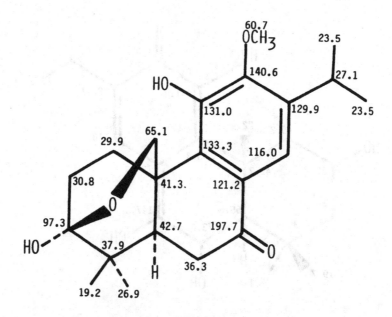

J.-Y. Liang, Z.-D. Min, T. Tanaka, M. Mizuno, M. Iinuma, T. Nakanishi and A. Inada, Acta. Chim. Sinica (Eng. Ed)., (1), 37 (1988).

3β-HYDROXY-DEHYDROABIETIC ACID (METHYL ESTER, ACETATE)

Source: *Salvia oxyodon* Webb et Heldreich

Mol. formula: $C_{23}H_{32}O_4$

Mol. wt.: 372

Solvent: $CDCl_3$

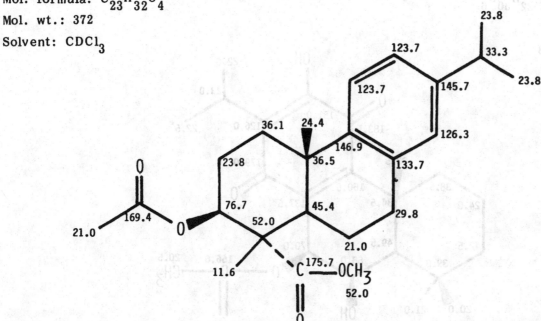

J. Escudero, L. Perez, R.M. Rabanal and S. Valverde, <u>Phytochemistry</u>, 22(2), 585 (1983).

19-MALONYLOXYDEHYDROABIETINOL (METHYL ESTER)

Source: *Calceolaria ascendens* Lindl.

Mol. formula: $C_{24}H_{34}O_4$

Mol. wt.: 386

Solvent: $CDCl_3$

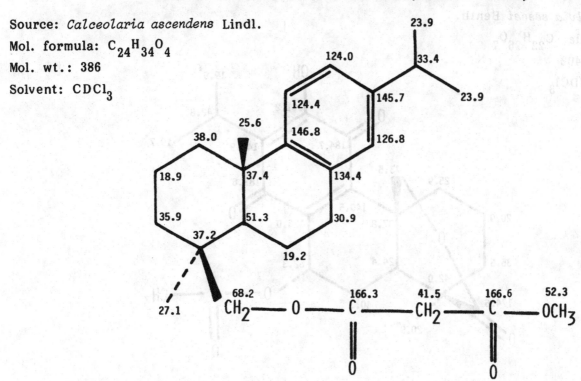

M.C. Chamy, M. Piovano, V. Gambaro, J.A. Garbarino and M. Nicoletti, <u>Phytochemistry</u>, 26(6) 1763 (1987).

7β-ACETOXY-6β-HYDROXYROYLEANONE

Source: *Coleus zeylanicus* (Benth) Cramer (syn. *Plectranthus zeylanicus*)

Mol. formula: $C_{22}H_{30}O_6$

Mol. wt.: 390

Solvent: $CDCl_3$

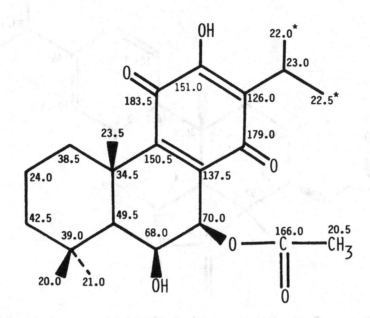

R. Mehrotra, R.A. Vishwakarma and R.S. Thakur, <u>Phytochemistry</u>, 28 (11), 3135 (1989).

SESSEIN

Source: *Salvia sessei* Benth.

Mol. formula: $C_{22}H_{26}O_7$

Mol. wt.: 402

Solvent: $CDCl_3$

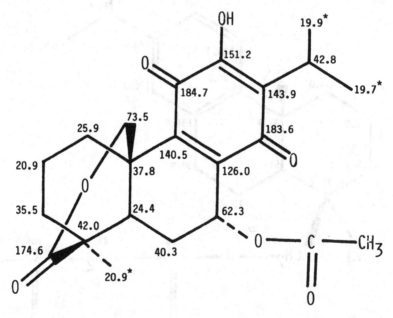

M. Jimenez E., E. Portugal M., A. Lira-Rocha, M. Soriano-Garcia and R.A. Toscano, <u>J. Nat. Prod.</u>, (Lloydia), 51 (2), 243 (1988)

PLECTRANTHONE I

Source: *Plectranthus* sp.

Mol. formula: $C_{22}H_{28}O_7$

Mol. wt.: 404

Solvent: $CDCl_3$

A.C. Alder, P. Ruedi, R. Prewo, J.H. Bieri and C.H. Eugster, Helv. Chim. Acta, 69(6), 1395 (1986).

||||||||||||||||||||||||||||||||||||||||||||||||||||||||||||||||||||||||||||||||||||||||||||||||||||||

LOPHANTHOIDIN E

Source: *Rabdosia lophanthoides* (Buch.-Ham. ex D. Don) Hara

Mol. formula: $C_{22}H_{30}O_7$

Mol. wt.: 406

Solvent: DMSO-d_6

Xu Yunlong, W. Dan, Li Xiaojie and Fu Jian, Phytochemistry, 28 (1), 189 (1989).

3β-ACETOXY-ABIETA-8(14)-EN-18-OIC ACID 9α, 13α-ENDOPEROXIDE (METHYL ESTER, ACETATE)

Source: *Salvia oxyodon* Webb et Heldreich

Mol. formula: $C_{23}H_{34}O_6$

Mol. wt.: 406

Sovlent: $CDCl_3$

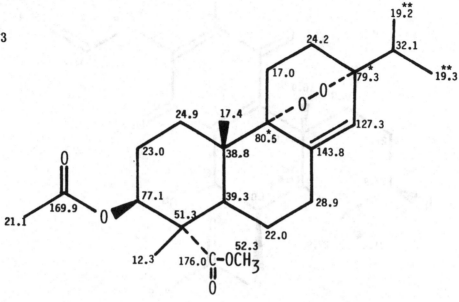

J. Escudero, L. Perez, R.M. Rabanal and S. Valverde, <u>Phytochemistry</u>, 22(2), 585 (1983).

|||||||||||||||||||||||||||||||||||||||||||||||||||||||||||||||||||||||||||||||||||||||||||||||||||||||||||||||||||||||||||||||

LOPHANTHOIDIN A

Source: *Rabdosia lophanthoides* (Buch.-Ham. ex D. Don) Hara

Mol. formula: $C_{23}H_{32}O_7$

Mol. wt.: 420

Solvent: C_5D_5N

Xu Yunlong, W. Dan, Li Xiaojie and Fu Jian, <u>Phytochemistry</u>, 28 (1), 189 (1989).

LOPHANTHOIDIN F

Source: *Rabdosia lophanthoides* (Buch.-Ham. ex D. Don) Hara

Mol. formula: $C_{24}H_{34}O_7$

Mol. wt.: 434

Solvent: DMSO-d_6

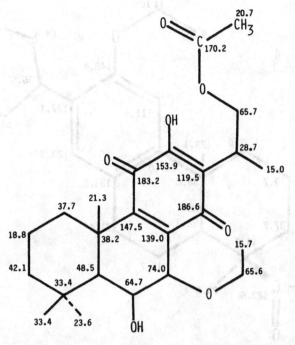

Xu Yunlong, W. Dan, Li Xiaojie and Fu Jian, <u>Phytochemistry</u>, 28 (1), 189 (1989).

LOPHANTHOIDIN B

Source: *Rabdosia lophanthoides* (Buch.-Ham. ex D. Don) Hara

Mol. formula: $C_{24}H_{32}O_8$

Mol. wt.: 448

Solvent: $CDCl_3$

Xu Yunlong, W. Dan, Li Xiaojie and Fu Jian, <u>Phytochemistry</u>, 28 (1), 189 (1989).

PODODACRIC ACID (TRIACETATE)

Source: *Dacrydium nidulum* de Laub.
Mol. formula: $C_{26}H_{34}O_8$
Mol. wt.: 474
Solvent: $CDCl_3$

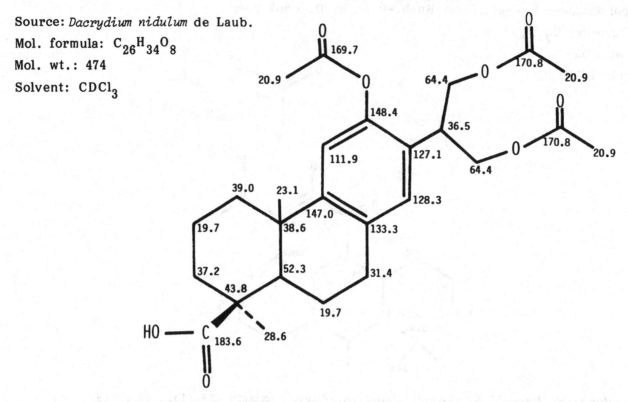

R.C. Cambie, R.E. Cox, K.D. Croft and D. Sidwel, <u>Phytochemistry</u>, 22(5), 1163 (1983).

||||||||||||||||||||||||||||||||||||||||||||||||||||||||||||||||||||||||||||||||||||||||||||||||||||||||||||||||||||||||

Seco-ABIETANE

Basic skeleton:

TANSHINLACTONE

Source: *Salvia miltiorrhiza* Bunge
Mol. formula: $C_{17}H_{12}O_3$
Mol. wt.: 264
Solvent: $CDCl_3$

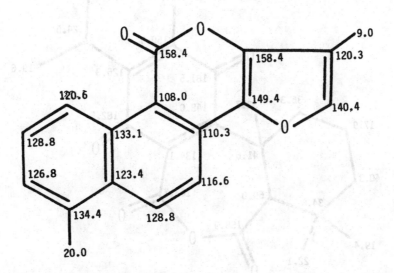

H.-W. Luo, J. Ji, M.-Y. Wu, Z.-G. Yong, M. Niwa and Y. Hirata, Chem. Pharm. Bull. (Tokyo), **34** (8), 3166 (1986).

||||||||||||||||||||||||||||||||||||||||||||||||||||||||||||||||||||||||||||||||||||||||||||||||||||||||||

CARIOCAL

Source: *Coleus barbatus* Bentham
Mol. formula: $C_{20}H_{26}O_5$
Mol. wt.: 346
Solvent: $(CD_3)_2CO$

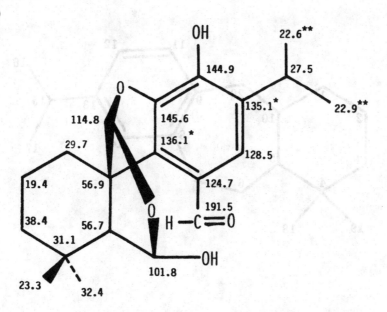

A. Kelecom and T.C. Dos Santos, Tetrahedron Lett., **26**(31), 3659 (1985).

Source: *Plectranthus sanguineus* Britten
Mol. formula: $C_{20}H_{24}O_6$
Mol. wt.: 360
Solvent: $CDCl_3$

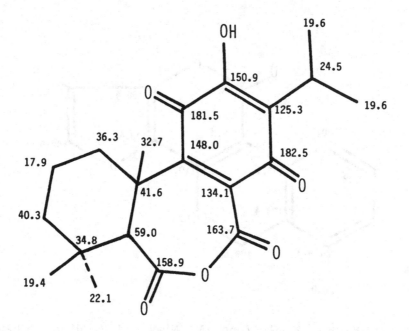

F. Matloubi-Moghadam, P. Ruedi and C.H. Eugster, Helv. Chim. Acta, **70**(4), 975 (1987).

ABEO-ABIETANE

Basic skeleton:

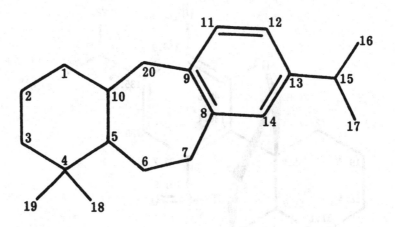

12-DEOXYPISIFERANOL

Source: *Chamaecyparis pisifera* Endl.

Mol. formula: $C_{20}H_{30}O$

Mol. wt.: 286

Solvent: $CDCl_3$

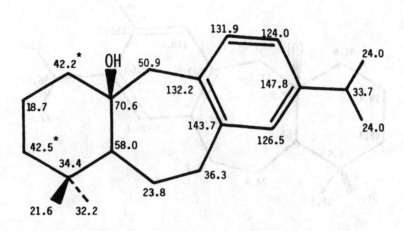

S. Hasegawa, T. Kojima and Y. Hirose, <u>Phytochemistry</u>, 24(7), 1545 (1985).

||||||||||||||||||||||||||||||||||||||||||||||||||||||||||||||||||||||||||||||||||||||||||||||||||||||||||||||||||||||||||||||||||||||||||||||||

BARBATUSOL

Source: *Coleus barbatus* Bentham

Mol. formula: $C_{20}H_{28}O_2$

Mol. wt.: 300

CH_3: 22.5, 22.7, 27.1 and 27.4

CH_2: 23.1, 30.4, 31.2, 34.4 and 35.2

CH: 27.1, 50.6, 117.5 and 120.9

C: 32.0,124.5,131.3,134.3,137.9,139.0 and 140.2

A. Kelecom, <u>Tetrahedron</u>, 39(21), 3603 (1983).

PISIFERANOL

Source: *Chamaecyparis pisifera* Endl.

Mol. formula: $C_{20}H_{30}O_2$

Mol. wt.: 302

Solvent: $CDCl_3$

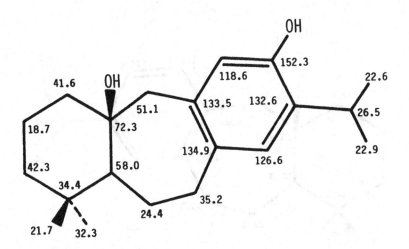

S. Hasegawa, T. Kojima and Y. Hirose, <u>Phytochemistry</u>, 24(7), 1545 (1985).

PISIFERADINOL

Source: *Chamaecyparis pisifera* Endl.

Mol. formula: $C_{20}H_{30}O_3$

Mol. wt.: 318

Solvent: $CDCl_3$

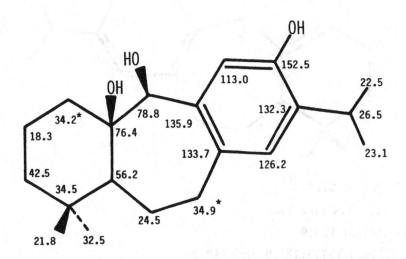

S. Hasegawa, T. Kojima and Y. Hirose, <u>Phytochemistry</u>, 24(7), 1545 (1985).

PISIFERDIOL

Source: *Chamaecyparis pisifera* Endl
Mol. formula: $C_{20}H_{30}O_3$
Mol. wt.: 318
Solvent: $CDCl_3$

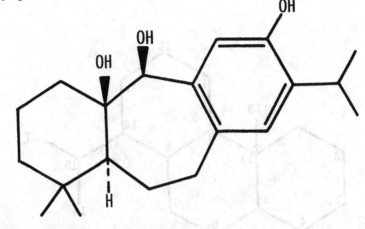

CH$_3$: 21.9, 22.4, 23.1 and 32.5
CH$_2$: 18.4, 24.5, 34.1, 34.9 and 42.5
CH: 26.5, 56.2, 78.9, 112.9 and 126.1
C: 34.5, 76.6, 132.2, 133.6, 135.6 and 152.6

J.W. Ahn, K. Wada and S. Marumo, Tetrahedron, 42(2), 529 (1986).

||||||||||||||||||||||||||||||||||||||||||||||||||||||||||||||||||||||||||||||||||||||||||||||||||||||||||||

TAXAMAIRIN B

Source: *Taxus mairei* (Lemee et levl.) S.Y. Hu
Mol. formula: $C_{22}H_{24}O_4$
Mol. wt.: 352
Solvent: C_5D_5N

J.-Y. Liang, Z.-D. Min, T. Tanaka, M. Mizuno, M. Iinuma, T. Nakanishi and A. Inada, Acta. Chim. Sinica (Eng. Ed)., (1), 37 (1988).

Basic skeleton:

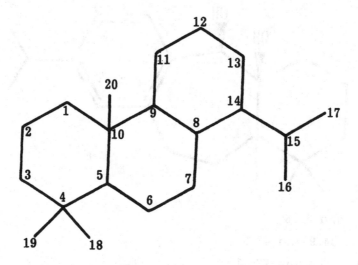

‖‖

19-HYDROXYTOTAROL

Source: *Decussocarpus vitiensis* (Seem.) de Laub. and *Podocarpus gnidioides* Carr.

Mol. formula: $C_{20}H_{30}O_2$

Mol. wt.: 302

Solvent: $CDCl_3$

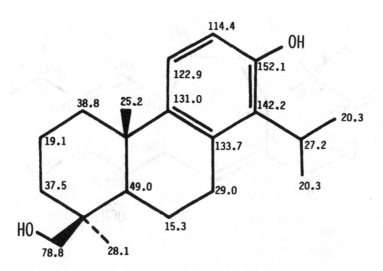

R.C. Cambie, R.E. Cox, K.D. Croft and D. Sidwell, Phytochemistry, **25** (5), 1163 (1983).

Basic skeleton:

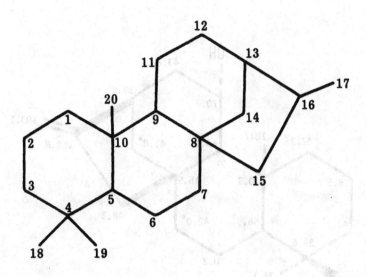

||||||||||||||||||||||||||||||||||||||||||||||||||||||||||||||||||||||||||||||||||||||||||||||||||||||||||||||||

ENT-16β,17-EPOXY-KAURANE

Source: *Fritillaria thunbergii* Miq.

Mol. formula: $C_{20}H_{32}O$

Mol. wt.: 288

Solvent: $CDCl_3$

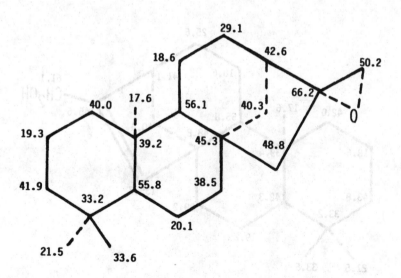

J. Kitajima, T. Komori and T. Kawasaki, Chem. Pharm. Bull. (Tokyo), 30(11), 3912 (1982).

ENT-11α-HYDROXYKAUR-16-ENE

Source: *Vellozia caput-ardeae* L.B.Smith and Ayensu

Mol. formula: $C_{20}H_{32}O$

Mol. wt.: 288

Solvent: $CDCl_3$

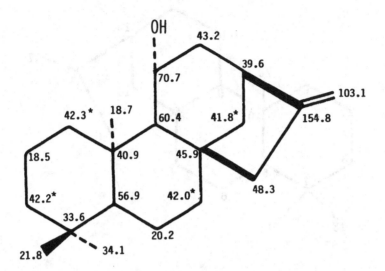

A.C.Pinto, R.Pinchin and S.K.Do Prado, Phytochemistry, 22(9), 2017 (1983).

||||||||||||||||||||||||||||||||||||||||||||||||||||||||||||||||||||||||||||||||||||||||||||||||||||||||||||||||||||||||||||||||||||||||||

ENT-KAUR-15-EN-17-OL

Source: *Fritillaria thunbergii* Miq.

Mol. formula: $C_{20}H_{32}O$

Mol. wt.: 288

Solvent: $CDCl_3$

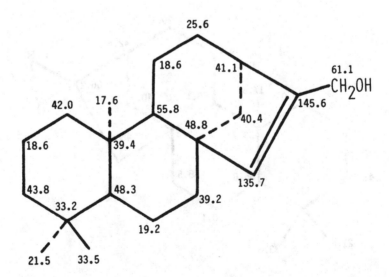

J. Kitajima, T. Komori and T. Kawasaki, Chem. Pharm. Bull. (Tokyo), 30 (11), 3912 (1982).

ENT-12,16-CYCLOKAURANOIC ACID

Source: *Helianthus annuus* L.

Mol. formula: $C_{20}H_{28}O_2$

Mol. wt.: 300

Solvent: $CDCl_3$

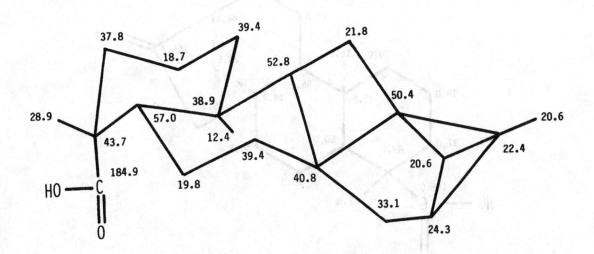

M.H. Beale, J.R. Bearder, J. MacMillan, A. Matsuo and B.O. Phinney, <u>Phytochemistry</u>, 22(4), 875 (1983).

‖‖‖

TETRACHYRIN

Source: *Tetrachyron orizabaensis* Sch. Bip ex Klatt

Mol. formula: $C_{20}H_{28}O_2$

Mol. wt.: 300

Solvent: $CDCl_3$

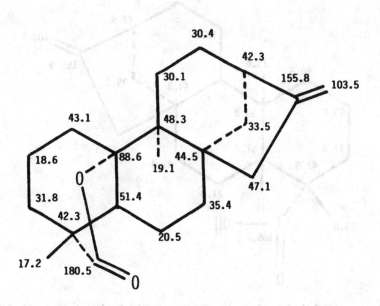

N. Ohno, T.J. Mabry, V. Zabel and W.H. Watson, <u>Phytochemistry</u>, 18 (10), 1687 (1979).

Source: *Croton argyrophylloides*
Mol. formula: $C_{20}H_{30}O_2$
Mol. wt.: 302
Solvent: $CDCl_3$

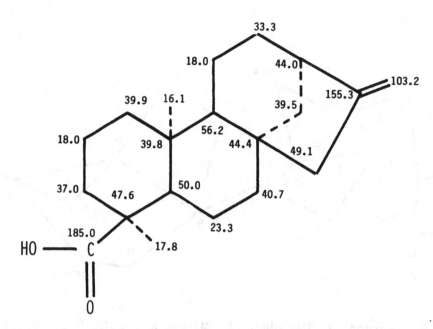

F.J.Q. Monte, E.M.G. Dantas and R. Braz F., Phytochemistry, 27(10), 3209 (1988).

||||||||||||||||||||||||||||||||||||||||||||||||||||||||||||||||||||||||||||||||||||||||||||||||||||||||||||||||||

(-)-KAUR-16-EN-19-OIC ACID

Source: *Tetrachyron orizabaensis* Sch. Bip ex Klatt
Mol. formula: $C_{20}H_{30}O_2$
Mol. wt.: 302
Solvent: $CDCl_3$

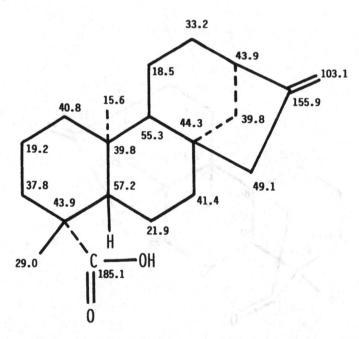

N. Ohno, T.J. Mabry, V. Zabel and W.H. Watson, Phytochemistry, 18(10), 1687 (1979).

SIDERONE

Source: *Sideritis syriaca* L. (syn. *S. Sicula Ucria*)
Mol. formula: $C_{20}H_{30}O_2$
Mol. wt.: 302
Solvent: $CDCl_3$

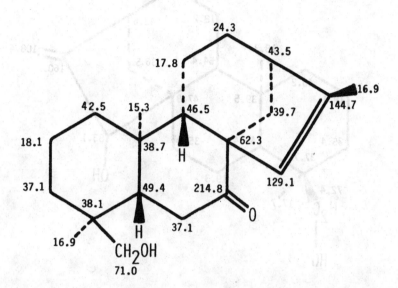

P. Venturella, A. Bellino and M.L. Marino, Phytochemistry, 22(11), 2537 (1983).

||||||||||||||||||||||||||||||||||||||||||||||||||||||||||||||||||||||||||||||||||||||||||||||||||||||||||||||||||||||||||||||||||||||||||||||||

YUCALEXIN B-14

Source: *Manihot esculenta* Crantz
Mol. formula: $C_{20}H_{30}O_2$
Mol. wt.: 302
Solvent: $CDCl_3$

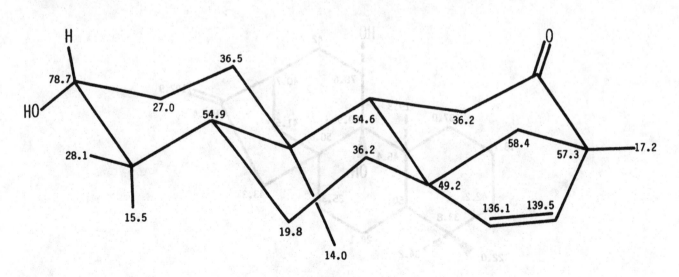

T. Sakai and Y. Nakagawa, Phytochemistry, 27 (12), 3769 (1988).

CANDIDIOL

Source: *Gibberella fujikuroi*
Mol. formula: $C_{20}H_{32}O_2$
Mol. wt.: 304

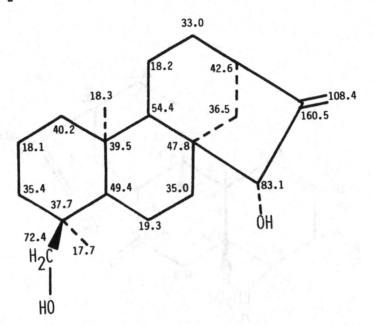

B.M. Fraga, P. Gonzalez, M.G. Hernandez, F.G. Tellado and A. Perales, <u>Phytochemistry</u>, 25 (5), 1235 (1986).

|||||||||||||||||||||||||||||||||||||||||||||||||||||||||||||||||||||||||||||||||||||||||||||||||||||||||||||||||

ENT-9β,11α-DIHYDROXYKAUR-16-ENE

Source: *Vellozia caput-ardeae* L.B. Smith and Ayensu
Mol. formula: $C_{20}H_{32}O_2$
Mol. wt.: 304
Solvent: $CDCl_3$

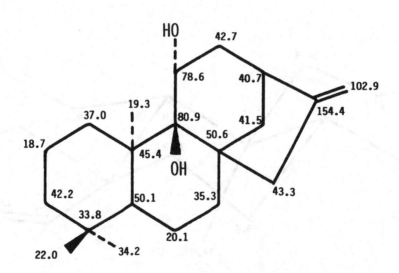

A.C. Pinto, R. Pinchin and S.K. Do Prado, <u>Phytochemistry</u>, 22(9), 2017 (1983).

DITERPENE SL-II

Source: *Stachys lanata*
Mol. formula: $C_{20}H_{32}O_2$
Mol. wt.: 304
Solvent: $CDCl_3$

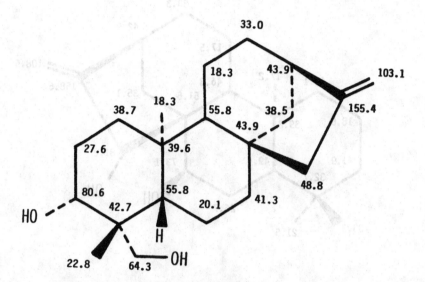

F. Piozzi, G. Savona and J.R. Hanson, <u>Phytochemistry</u>, 19(6), 1237 (1980).

||||||||||||||||||||||||||||||||||||||||||||||||||||||||||||||||||||||||||||||||||||||||||||||||||||||||||||||||||||||

<u>ENT-KAUR-15-EN-3β,17-DIOL</u>

Source: *Croton lacciferus* Linn.
Mol. formula: $C_{20}H_{32}O_2$
Mol. wt.: 304
Solvent: $CDCl_3$

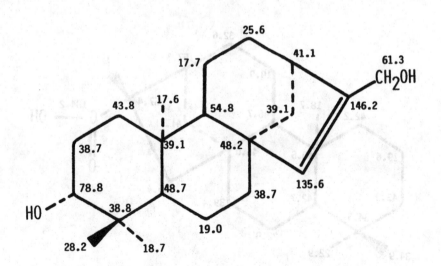

B.M.R. Bandara, W.R. Wimalasiri and J.K. MacLeod, <u>Phytochemistry</u>, 27 (3), 869 (1988).

ENT-KAUR-16-EN-7α,15β-DIOL

Source: *Plagiochila pulcherrima* Hovik

Mol. formula: $C_{20}H_{32}O_2$

Mol. wt.: 304

Solvent: $CDCl_3$

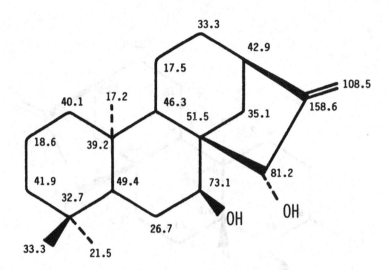

Y. Fukuyama, M. Toyota and Y. Asakawa, Phytochemistry, 27(5), 1425 (1988).

|||||||||||||||||||||||||||||||||||||||||||||||||||||||||||||||||||||||||||||||||||||||||||||||||||||||||||||||||||||||||||||||||||

16α-H-ENT-KAURAN-17-OIC ACID

Source: *Croton lacciferus* Linn.

Mol. formula: $C_{20}H_{32}O_2$

Mol. wt.: 304

Solvent: $CDCl_3$

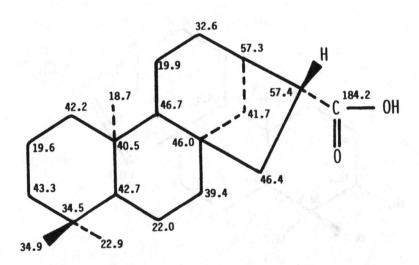

B.M.R. Bandara, W.R. Wimalasiri and J.K. MacLeod, Phytochemistry, 27 (3), 869 (1988).

Source: *Fritillaria thunbergii* Miq.

Mol. formula: $C_{20}H_{34}O_2$

Mol. wt.: 306

Solvent: $CDCl_3$

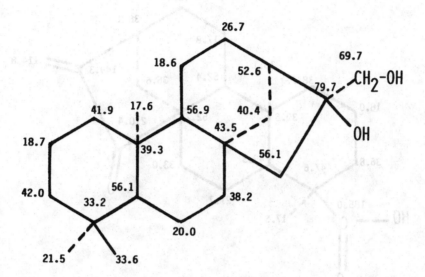

J. Kitajima, T. Komori and T. Kawasaki, Chem. Pharm. Bull., (Tokyo) 30(11), 3912 (1982).

ENT-KAURAN-16β,17-DIOL

Source: *Fritillaria thunbergii* Miq.

Mol. formula: $C_{20}H_{34}O_2$

Mol. wt.: 306

Solvent: $CDCl_3$

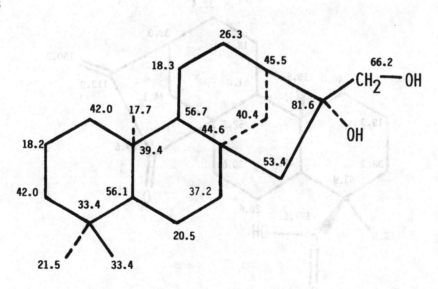

J. Kitajima, T. Komori and T. Kawasaki, Chem. Pharm. Bull., (Tokyo), 30 (11), 3912 (1982).

ENT-KAUR-16-EN-15-OXO-18-OIC ACID

Source: *Croton argyrophylloides* Muell
Mol. formula: $C_{20}H_{28}O_3$
Mol. wt.: 316
Solvent: $CDCl_3$

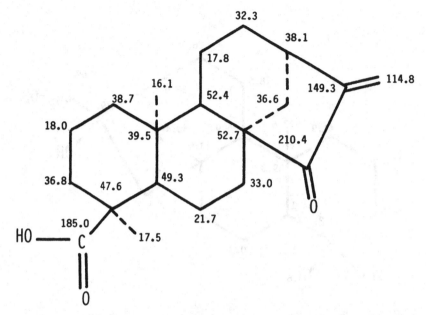

F.J.Q. Monte, E.M.G. Dantas and R. Braz F, Phytochemistry, 27 (10), 3209 (1988).

||||||||||||||||||||||||||||||||||||||||||||||||||||||||||||||||||||||||||||||||||||||||||||||||||||||||||||||||||||||||||||||||

15-OXO(-)-KAUR-16-EN-19-OIC ACID

Source: *Xylopia acutiflora* (Dunal) A. Rich.
Mol. formula: $C_{20}H_{28}O_3$
Mol. wt.: 316
Solvent: $CDCl_3$

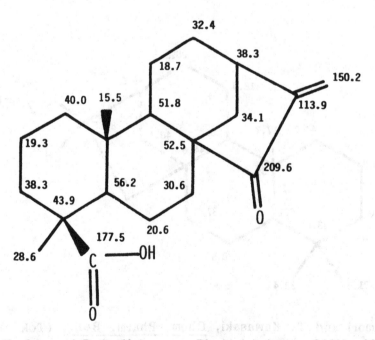

C.M. Hasan, T.M. Healey and P.G. Waterman, Phytochemistry, 24(1), 192 (1985).

ENT-15-OXO-KAUR-16-EN-19-OIC ACID

Source: *Pteris longipes* Don
Mol. formula: $C_{20}H_{28}O_3$
Mol. wt.: 316
Solvent: C_5D_5N

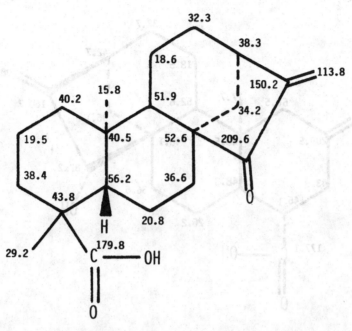

T. Murakami, H. Iida, N. Tanaka, Y. Saiki, C.-M. Chen and Y. Iitaka, Chem. Pharm. Bull., (Tokyo), 29 (3), 657 (1981).

YUCALEXIN B-9

Source: *Manihot esculenta* Crantz
Mol. formula: $C_{20}H_{28}O_3$
Mol. wt.: 316
Solvent: $CDCl_3$

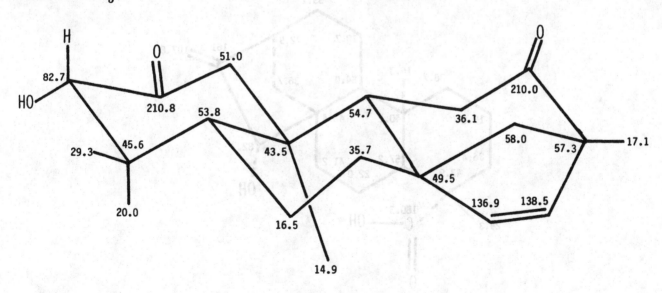

T. Sakai and Y. Nakagawa, Phytochemistry, 27 (12), 3769 (1988).

ATRACTYLIGENONE (2-OXO-15α-HYDROXYATRACTYL-16,17-ENE-4α-CARBOXYLIC ACID)

Source: *Drymaria arenarioides* Willd.

Mol. formula: $C_{19}H_{26}O_4$

Mol. wt.: 318

Solvent: C_5D_5N

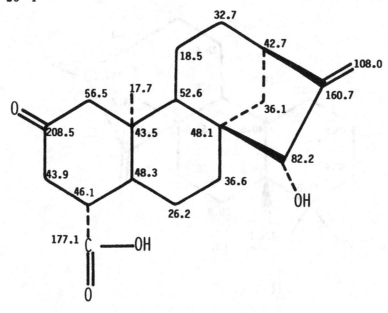

D. Vargas, X.A. Dominguez, K. Acuna-Askar, M. Gutierrez and K. Hostettmann, <u>Phytochemistry</u>, 27(5), 1532 (1988).

|||||||||||||||||||||||||||||||||||||||||||||||||||||||||||||||||||||||||||||||||||||||||||||||||||||||||||||||||||||||||||||

GRANDIFLORIC ACID

Source: *Helianthus niveus* (A.Gray) Heisev

Mol. formula: $C_{20}H_{30}O_3$

Mol. wt.: 318

Solvent: C_5D_5N

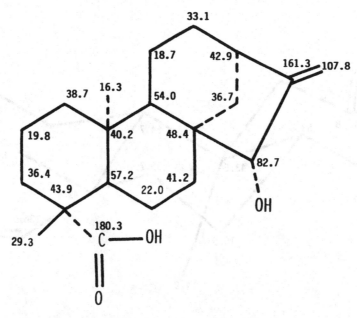

N. Ohno and T.J. Mabry, <u>Phytochemistry</u>, 19(4), 609 (1980).

12α-HYDROXY-ENT-ISOKAUREN-19-OIC ACID

Source: *Helichrysum davenportii* F. Muell.

Mol. formula: $C_{20}H_{30}O_3$

Mol. wt.: 318

Solvent: C_6D_6

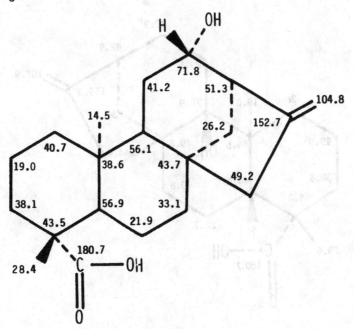

J. Jakupovic, A. Schuster, F. Bohlmann, U. Ganzer, R.M. King and H. Robinson, <u>Phytochemistry</u>, 28(2), 543 (1989).

||||||||||||||||||||||||||||||||||||||||||||||||||||||||||||||||||||||||||||||||||||||||||||||||||||||

12α-HYDROXY-ENT-KAUR-16-EN-19-OIC ACID

Source: *Stevia eupatoria* Will

Mol. formula: $C_{20}H_{30}O_3$

Mol. wt.: 318

Solvent: $CDCl_3$

A. Ortega, F.J. Morales and M. Salmon, <u>Phytochemistry</u>, 24(8), 1850 (1985).

Source: *Helianthus rigidus* (Cas.) Desf.

Mol. formula: $C_{20}H_{30}O_3$

Mol. wt.: 318

Solvent: DMSO-d_6

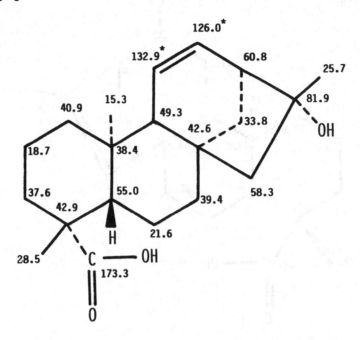

W. Herz, S.V. Govindan and K. Watanabe, Phytochemistry, 21(4), 946 (1982).

|||||||||||||||||||||||||||||||||||||||||||||||||||||||||||||||||||||||||||||||||||||||||||||||||||

PTEROKAURENE L₃

Source: *Pteris longipes* Don

Mol. formula: $C_{20}H_{30}O_3$

Mol. wt.: 318

Solvent: C_5D_5N

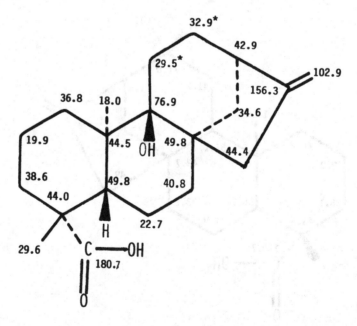

T. Murakami, H. Iida, N. Tanaka, Y. Saiki, C.-M. Chen and Y. Iitaka, Chem.Pharm.Bull. (Tokyo) 29(3), 657 (1981).

PTEROKAURENE L₅

Source: *Pteris longipes* Don
Mol. formula: $C_{20}H_{30}O_3$
Mol. wt.: 318
Solvent: C_5D_5N

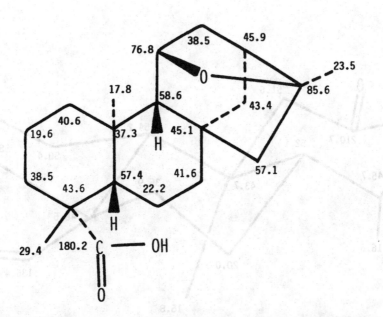

T. Murakami, H. Iida, N. Tanaka, Y. Saiki, C.-M. Chen and Y. Iitaka, <u>Chem. Pharm. Bull.</u>, (Tokyo), 29(3), 657 (1981).

||||||||||||||||||||||||||||||||||||||||||||||||||||||||||||||||||||||||||||||||||||||||||||||||||||||||||||||||||||||||||||||||||||

YUCALEXIN A-19

Source: *Manihot esculenta* Crantz
Mol. formula: $C_{20}H_{30}O_3$
Mol. wt.: 318
Solvent: $CDCl_3$

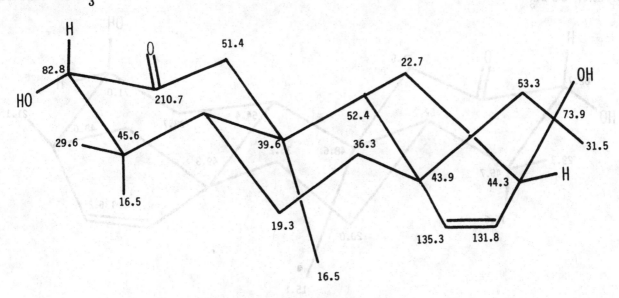

T. Sakai and Y. Nakagawa, <u>Phytochemistry</u>, **27** (12), 3769 (1988).

YUCALEXIN B-18

Source: *Manihot esculenta* Crantz
Mol. formula: $C_{20}H_{30}O_3$
Mol. wt.: 318
Solvent: $CDCl_3$

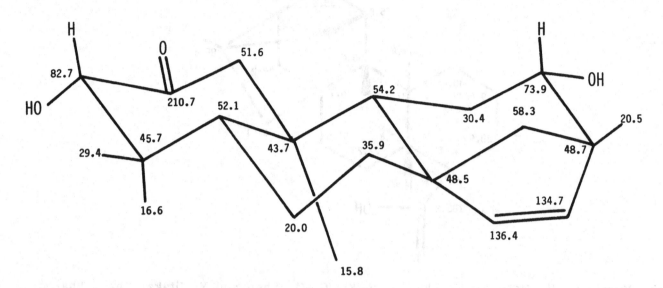

T. Sakai and Y. Nakagawa, Phytochemistry, 27 (12), 3769 (1988).

||||||||||||||||||||||||||||||||||||||||||||||||||||||||||||||||||||||||||||||||

YUCALEXIN B-20

Source: *Manihot esculenta* Crantz
Mol. formula: $C_{20}H_{30}O_3$
Mol. wt.: 318
Solvent: $CDCl_3$

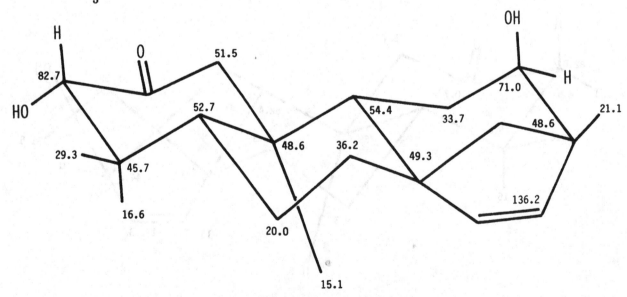

T. Sakai and Y. Nakagawa, Phytochemistry, 27 (12), 3769 (1988).

YUCALEXIN B-22

Source: *Manihot esculenta* Crantz
Mol. formula: $C_{20}H_{30}O_3$
Mol. wt.: 318
Solvent: $CDCl_3$

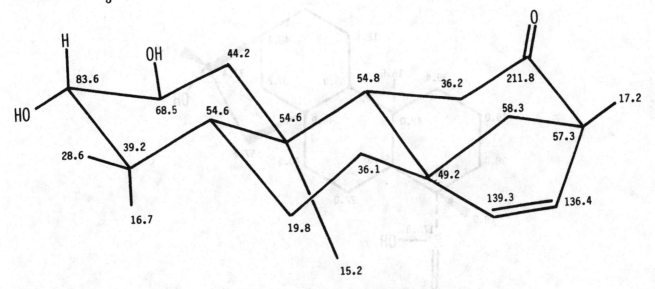

T. Sakai and Y. Nakagawa, <u>Phytochemistry</u>, 27 (12), 3769 (1988).

|||||||||||||||||||||||||||||||||||||||||||||||||||||||||||||||||||||||||||||||||||||||||||||||||||||||||||||||||||||||||||||||||||||||||||||||||||||||||

ATRACTYLIGENIN

Source: *Coffea arabica* L.
Mol. formula: $C_{19}H_{28}O_4$
Mol. wt.: 320
Solvent: $CDCl_3$

H. Richter and G. Spiteller, <u>Chem. Ber.</u>, 111(10), 3506 (1978).

341

16α-HYDROXY-ENT-KAURANOIC ACID

Source: *Viguiera linearis* (Cav.) Sch. Bip. ex Hemsley

Mol. formula: $C_{20}H_{32}O_3$

Mol. wt.: 320

Solvent: $CDCl_3$

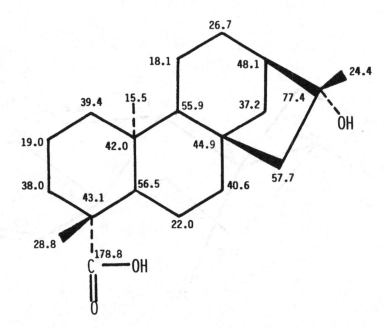

G. Delgado, L. Alvarez and A.R.De Vivar, Phytochemistry, **24**(11), 2736 (1985).

ENT-16α-METHOXY-KAURAN-17-OL

Source: *Fritillaria thunbergii* Miq.

Mol. formula: $C_{21}H_{36}O_2$

Mol. wt.: 320

Solvent: $CDCl_3$

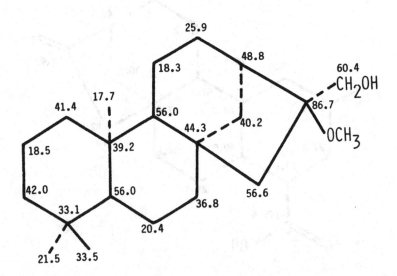

J. Kitajima, T. Komori and T. Kawasaki, Chem. Pharm. Bull., (Tokyo), **30** (11), 3912 (1982).

7,18-DIHYDROXYKAURENOLIDE

Source: *Gibberella fujikuroi*
Mol. formula: $C_{20}H_{28}O_4$
Mol. wt.: 332
Solvent: $CDCl_3$

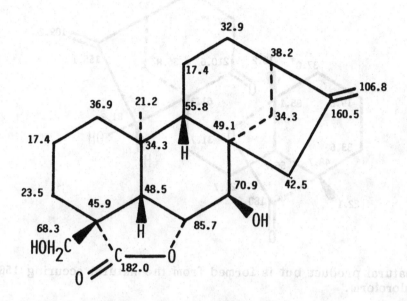

J.R. Hanson and F.Y. Sarah, J. Chem. Soc. Perkin Trans. I, (12), 3151 (1979).

||||||||||||||||||||||||||||||||||||||||||||||||||||||||||||||||||||||||||||||||||||||||||||||||||||||||||||

ENT-11α-HYDROXY-15-OXOKAUR-16-EN-19-OIC ACID

Source: *Eupatorium album* L.
Mol. formula: $C_{20}H_{28}O_4$
Mol. wt.: 332
Solvent: $CDCl_3$

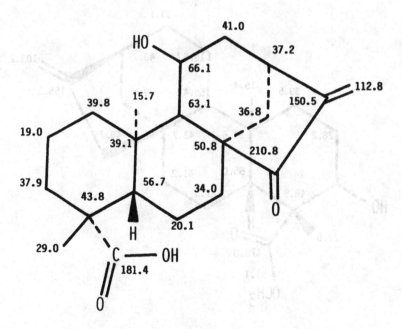

W. Herz and R.P. Sharma, J. Org. Chem., 41(6), 1021 (1976).

15α-HYDROXY-WEDELISECCOKAURENOLIDE[*]

Source: *Gnaphalium undulatum* L.
Mol. formula: $C_{20}H_{28}O_4$
Mol. wt.: 332
Solvent: $CDCl_3$

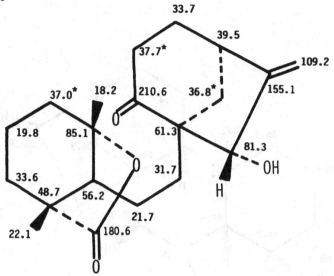

[*]This is not a natural product but is formed from the natural occuring 15β-hydroxy epimer on keeping in chloroform.

F. Bohlmann, J. Ziesche, R.M. King and H. Robinson, <u>Phytochemistry</u>, 20 (4), 751 (1981).

||||||||||||||||||||||||||||||||||||||||||||||||||||||||||||||||||||||||||||||||||||||||||||||||||||||||||||||||||||||||||||||||||||||||||||

DITERPENE SL-V

Source: *Stachys lanata*
Mol. formula: $C_{21}H_{32}O_3$
Mol. wt.: 332
Solvent: $CDCl_3$

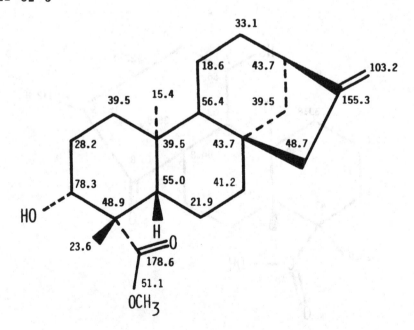

F. Piozzi, G. Savona and J.R. Hanson, <u>Phytochemistry</u>, 19(6), 1237 (1980).

ENT-11α,15α-DIHYDROXY-KAUR-16-EN-19-OIC ACID

Source: *Pteris longipes* Don

Mol. formula: $C_{20}H_{30}O_4$

Mol. wt.: 334

Solvent: C_5D_5N

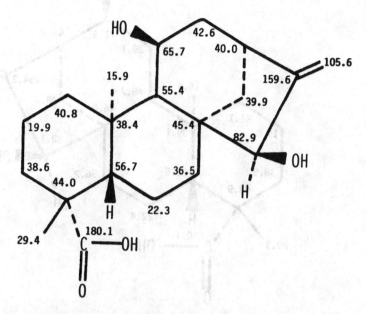

T. Murakami, H. Iida, N. Tanaka, Y. Saiki, C.-M. Chen and Y. Iitaka, <u>Chem. Pharm. Bull.</u>, (Tokyo), **29** (3), 657 (1981).

PTEROKAURENE L₂

Source: *Pteris longipes* Don

Mol. formula: $C_{20}H_{30}O_4$

Mol. wt.: 334

Solvent: C_5D_5N

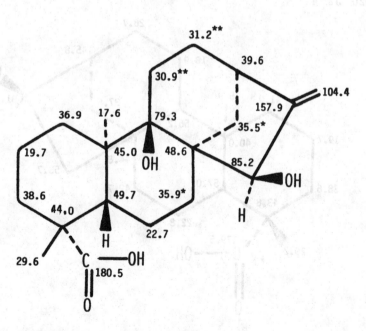

T. Murakami, H. Iida, N. Tanaka, Y. Saiki, C.-M. Chen and Y. Iitaka, <u>Chem. Pharm. Bull.</u>, (Tokyo), **29**(3), 657 (1981).

Source: *Pteris longipes* Don

Mol. formula: $C_{20}H_{30}O_4$

Mol. wt.: 334

Solvent: C_5D_5N

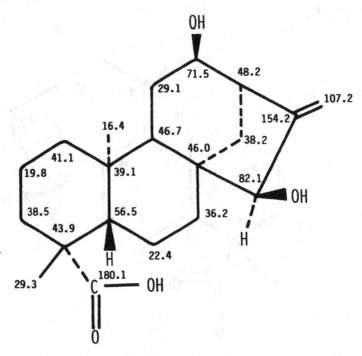

T. Murakami, H. Iida, N. Tanaka, Y. Saiki, C.-M. Chen and Y.Iitaka, Chem. Pharm. Bull. (Tokyo), 29(3), 657 (1981).

|||||||||||||||||||||||||||||||||||||||||||||||||||||||||||||||||||||||||||||||||||||||||||||||||||||

ENT-16β,17-DIHYDROXY-(-)-KAURAN-19-OIC ACID

Source: *Acanthopanax koreanum* Nakai

Mol. formula: $C_{20}H_{32}O_4$

Mol. wt.: 336

Solvent: C_5D_5N

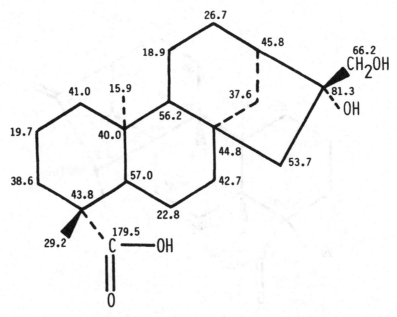

Y.H. Kim, Bo S. Chung and U. Sankawa, J. Nat. Prod., (Lloydia), 51 (6), 1080 (1988).

12β-ETHOXY-ENT-KAUR-9(11),16-DIEN-19-OIC ACID

Source: *Stevia eupatoria* Will

Mol. formula: $C_{22}H_{32}O_3$

Mol. wt.: 344

Solvent: $CDCl_3$

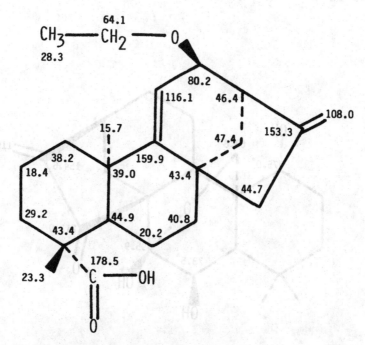

A. Ortega, F.J. Morales and M. Salmon, Phytochemistry, 24(8), 1850 (1985).

||||||||||||||||||||||||||||||||||||||||||||||||||||||||||||||||||||||||||||||||||||||||||||||||||||||||||||||||||||||||||||||||||||||||||||||

DITERPENE RU-IX

Source: *Rabdosia umbrosa* (Maxim.) Hara

Mol. formula: $C_{20}H_{28}O_5$

Mol. wt.: 348

Solvent: C_5D_5N

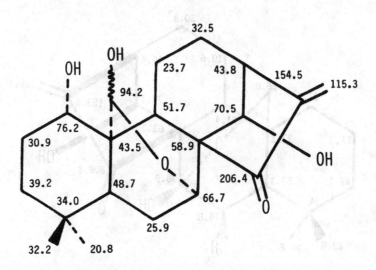

Y. Takeda, I. Ichihara, T. Fujita and A. Ueno, Phytochemistry, 28(6), 1691 (1989).

EFFUSANIN A

Source: *Rabdosia effusa* (Maxim.) Hara

Mol. formula: $C_{20}H_{28}O_5$

Mol. wt.: 348

Solvent: C_5D_5N

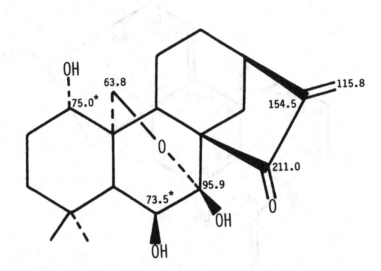

T. Fujita, Y. Takeda, T. Shingu and A. Ueno, <u>Chem. Lett.</u>, 1635 (1980).

||||||||||||||||||||||||||||||||||||||||||||||||||||||||||||||||||||||||||||||||||||||||||||||||||||||||||||||||||||||||

LONGIKAURIN A

Source: *Rabdosia longituba* (Miquel) Hara

Mol. formula: $C_{20}H_{28}O_5$

Mol. wt.: 348

Solvent: C_5D_5N

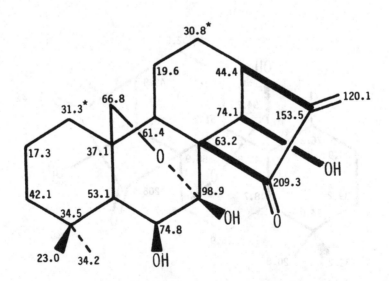

Y. Takeda, T. Fujita and T. Shingu, <u>J. Chem. Soc. Perkin Trans.I</u>, (2), 379 (1988).

COESTINOL

Source: *Plectranthus coesta* Buch-Ham

Mol. formula: $C_{20}H_{30}O_5$

Mol. wt.: 350

Solvent: C_5D_5N

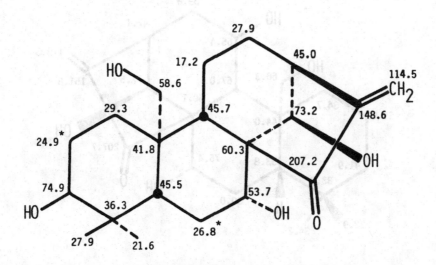

A.P. Phadnis, S.A. Patwardhan and A.S. Gupta, <u>Indian. J. Chem.</u>, **26B(1)**, 15 (1987).

KAMEBAKAURIN

Source: *Rabdosia umbrosa*

Mol. formula: $C_{20}H_{30}O_5$

Mol. wt. 350

Solvent: C_5D_5N

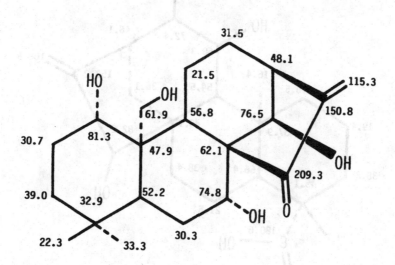

Y. Takeda, T. Ichihara, Y. Takaishi, T. Fujita, T. Shingu and G. Kusano, <u>J. Chem. Soc. Perkin Trans.I</u>, (11), 2403 (1987).

Source: *Rabdosia umbrosa*

Mol. formula: $C_{20}H_{30}O_5$

Mol. wt.: 350

Solvent: C_5D_5N

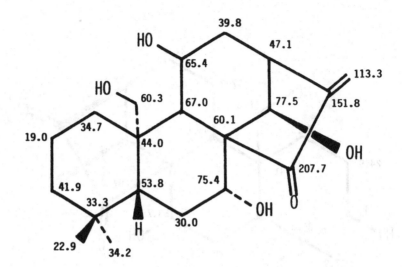

Y. Takeda, T. Ichihara, Y. Takaishi, T. Fujita, T. Shingu and G. Kusano, <u>J. Chem. Soc. Perkin Trans. I</u>, (11), 2403 (1987).

||||||||||||||||||||||||||||||||||||||||||||||||||||||||||||||||||||||||||||||||||||||||||||||||||||||||||||||||||||||||||||||||

<u>ENT</u>-11α,12α,15α-TRIHYDROXYKAUR-16-EN-19-OIC ACID

Source: *Eupatorium album* L.

Mol. formula: $C_{20}H_{30}O_5$

Mol. wt.: 350

Solvent: $CDCl_3$

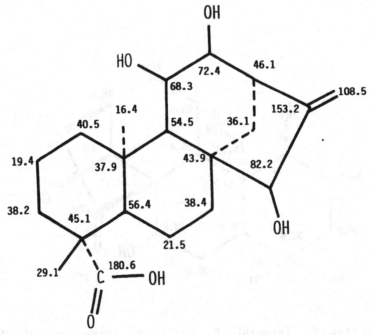

W. Herz and R.P. Sharma, <u>J. Org. Chem.</u>, 41 (6), 1021 (1976).

ENT-3β-ACETOXYKAUR-16-EN-19-OIC ACID

Source: *Stachys lanata*
Mol. formula: $C_{22}H_{32}O_4$
Mol. wt.: 360
Solvent: $CDCl_3$

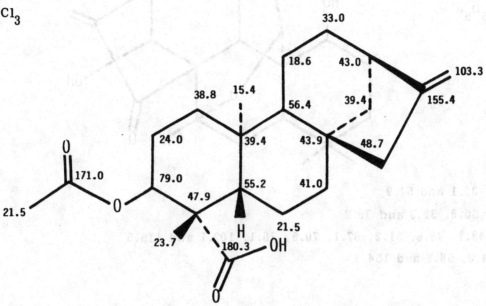

F. Piozzi, G. Savona and J.R. Hanson, <u>Phytochemistry</u>, 19(6), 1237 (1980).

PHYLLOSTACHYSIN A

Source: *Rabdosia phyllostachys* (Diels) Hara
Mol. formula: $C_{20}H_{26}O_6$
Mol. wt.: 362
Solvent: C_5D_5N

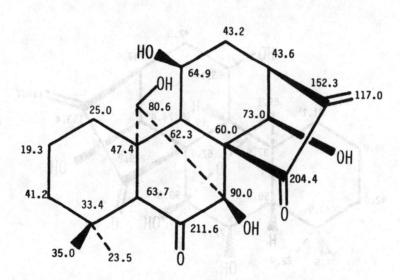

T. Fujita, S. Han-Dong, Y. Takeda, Y. Minami, T. Marunaka, L. Zhon-Wen and Xu Yun-Long, <u>J. Chem. Soc. Chem. Commun.</u>, (23), 1738 (1985).

KAMEBACETAL A

Source: *Rabdosia umbrosa*

Mol. formula: $C_{21}H_{30}O_5$

Mol. wt.: 362

Solvent: C_5D_5N

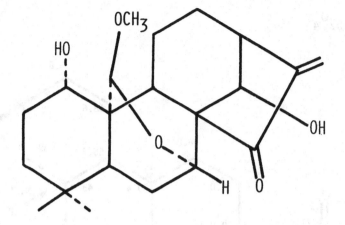

CH_3: 20.7, 32.1 and 54.9

CH_2: 23.5, 30.8, 32.3 and 39.2

CH: 25.7, 43.7, 48.8, 51.2, 67.1, 70.5, 76.1, 102.1 and 115.5

C: 34.1, 44.0, 58.6 and 154.4

C=O: 206.4

Y. Takeda, T. Ichihara, Y. Takaishi, T. Fujita, T. Shingu and G. Kusano, J. Chem. Soc. Perkin Trans. I, (11), 2403 (1987).

||||||||||||||||||||||||||||||||||||||||||||||||||||||||||||||||||||||||||||||||||||||||||||||||||||

LONGIKAURIN G

Source: *Rabdosia longituba* (Miq.) Hara

Mol. formula: $C_{20}H_{28}O_6$

Mol. wt.: 364

Solvent: C_5D_5N

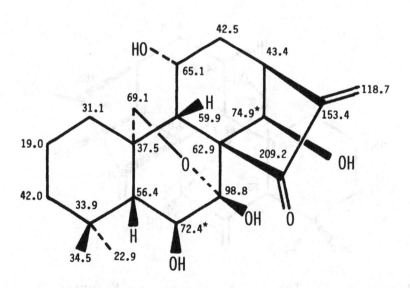

Y. Takeda and T. Fujita, Planta Med., 54(4), 327 (1988).

MAOECRYSTAL K

Source: *Rabdosia eriocalyx* (Dunn) Hara

Mol. formula: $C_{20}H_{30}O_6$

Mol. wt.: 366

Solvent: CD_3OD

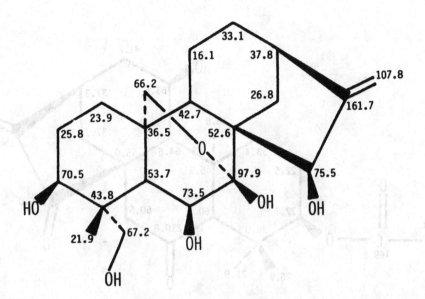

A. Isogai, X. Shen, K. Furihata, H. Kaniwa, H. Sun and A. Suzuki, <u>Phytocheimstry</u>, **28** (9), 2427 (1989).

||||||||||||||||||||||||||||||||||||||||||||||||||||||||||||||||||||||||||||||||||||||||||||||||||||||||||||||||||||||||||||||||||||||||||||||

RUBESCENSIN C

Source: *Radosia rubescens* (Hemsl.) Hara

Mol. formula: $C_{20}H_{30}O_6$

Mol. wt.: 366

Solvent: C_5D_5N

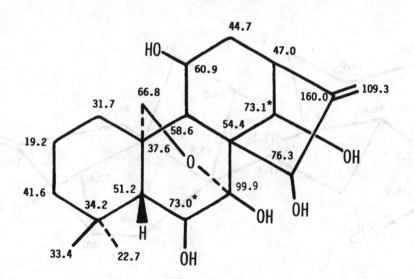

S. Han-Dong, C.Jin-Hau, L. Zhong-Wen, T. Marunaka, Y. Minami and T. Fujita, <u>Chem. Pharm. Bull.</u>, (Tokyo), 30(1), 341 (1982).

Source: *Rabdosia inflexa* (Thunb.) Hara

Mol. formula: $C_{22}H_{30}O_5$

Mol. wt.: 374

Solvent: C_5D_5N

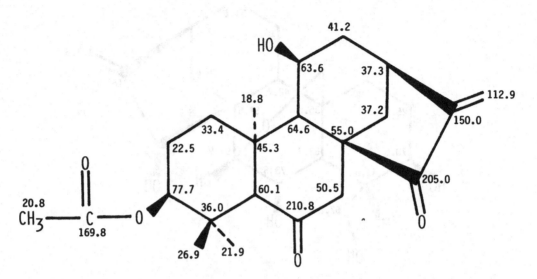

Y. Takeda, T. Ichihara, K. Yamasaki and H. Otsuka, <u>Phytochemistry</u>, **28** (9), 2423 (1989).

ENT-12β-ACETOXYKAUR-16-EN-19-OIC ACID (METHYL ESTER)

Source: *Helianthus decapetalus* Dart

Mol. formula: $C_{23}H_{34}O_4$

Mol. wt.: 374

Solvent; $CDCl_3$

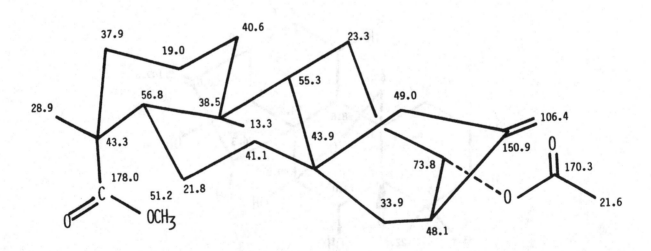

M.H. Beale, J.R. Bearder, J. MacMillan, A. Matsuo and B.O. Phinney, <u>Phytochemistry</u>, 22(4), 875 (1983).

DITERPENE RU-II

Source: *Rabdosia umbrosa*(Maxim.) Hara
Mol. formula: $C_{22}H_{32}O_5$
Mol. wt.: 376
Solvent: $CDCl_3$

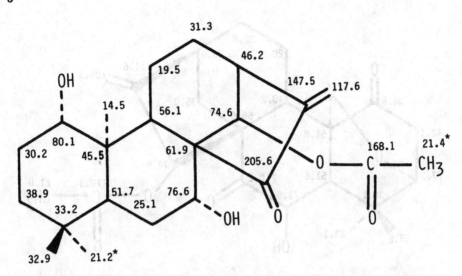

Y. Takeda, T. Ichihara, T. Fujita and A. Ueno, <u>Phytochemistry</u>, 28 (6), 1691 (1989).

||||||||||||||||||||||||||||||||||||||||||||||||||||||||||||||||||||||||||||||||||||||||||||||||||||||||||||||||||||||||||||||||||||||||||||||||||||||

ROSTHORNIN A

Source: *Rabdosia rosthornii* (Diels) Hara
Mol. formula: $C_{22}H_{32}O_5$
Mol. wt.: 376
Solvent: C_5D_5N

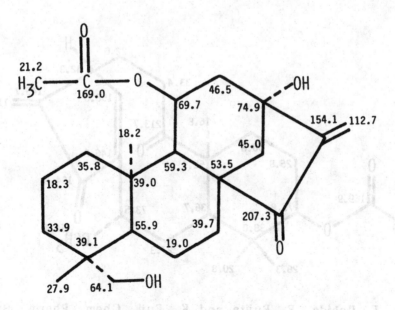

Xu Yunlong and Ma Yunbao, <u>Phytochemistry</u>, 28 (11), 3235 (1989).

MAOECRYSTAL A

Source: *Rabdosia eriocalyx* (Dunn) Hara
Mol. formula: $C_{22}H_{28}O_6$
Mol. wt.: 388
Solvent: C_5D_5N

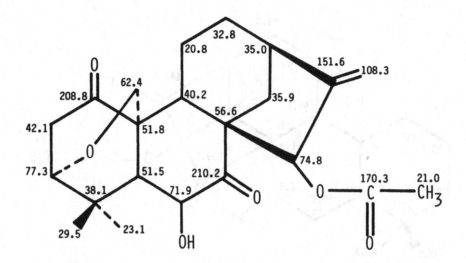

C.-B. Li. H.-D. Sun and J. Zhon, <u>Acta. Chim. Sinica</u> (China), **46** (7), 657 (1988).

O-METHYLSHIKOCCIN

Source: *Rabdosia shikokiana* (Makino) Hara
Mol. formula: $C_{23}H_{32}O_5$
Mol. wt.: 388
Solvent: $CDCl_3$

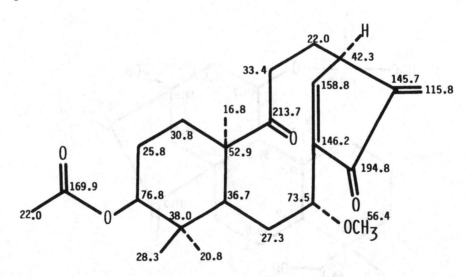

M. Node, N. Ito, I. Uchida, E. Fujita and K. Fuji, <u>Chem. Pharm. Bull</u>., (Tokyo), **33**(3), 1029 (1985).

CANDICANDIOL (DIACETATE)

Source: *Sideritis infernalis*
Mol. formula: $C_{24}H_{36}O_4$
Mol. wt.: 388
Solvent: $CDCl_3$

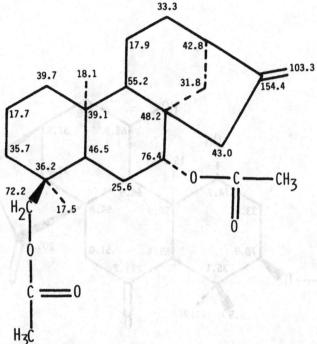

C. Fernandez, B.M. Fraga and M.G. Hernandez, <u>Phytochemistry</u>, 25 (11), 2573 (1986).

||||||||||||||||||||||||||||||||||||||||||||||||||||||||||||||||||||||||||||||||||||||||||||||||||||||||||||||||||

ENT-3β-ANGELOYLOXY-16β-HYDROXYKAURANE

Source: *Robinsonia evenia* Phil
Mol. formula: $C_{25}H_{40}O_3$
Mol. wt.: 388
Solvent: $CDCl_3$

M. Bittner, M. Silva, Z. Rozas, P. Pacheco, T. Stuessy, F. Bohlmann and J. Jakupovic, <u>Phytochemistry</u>, 27(2), 487 (1988).

INFLEXARABDONIN E

Source: *Rabdosia inflexa* (Thunb.) Hara
Mol. formula: $C_{22}H_{30}O_6$
Mol. wt.: 390
Solvent: C_5D_5N

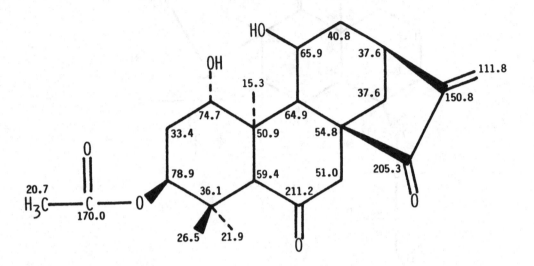

Y. Takeda, T. Ichihara, K. Yamasaki and H. Otsuka, <u>Phytochemistry</u>, **28** (9), 2423 (1989).

||||||||||||||||||||||||||||||||||||||||||||||||||||||||||||||||||||||||||||||||||||||||||||||||||||||||||||||||||||||||||||||||

INFLEXARABDONIN F

Source: *Rabdosia inflexa* (Thunb.) Hara
Mol. formula: $C_{22}H_{34}O_6$
Mol. wt.: 394
Solvent: C_5D_5N

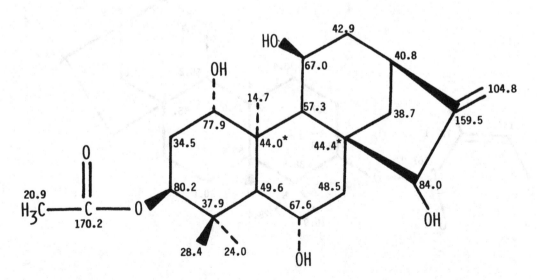

Y. Takeda, T. Ichihara, K. Yamasaki and H. Otsuka, <u>Phytochemistry</u>, **28** (9), 2423 (1989).

ANGELOYLGRANDIFLORIC ACID

Source: *Tetrachyron orizabaensis* Sch. Bip ex Klatt

Mol. formula: $C_{25}H_{36}O_4$

Mol. wt.: 400

Solvent: $CDCl_3$

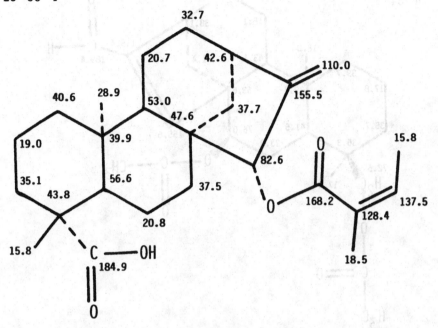

N. Ohno, T.J. Mabry, V. Zabel and W.H. Watson, Phytochemistry, 18(10), 1687 (1979).

|||||||||||||||||||||||||||||||||||||||||||||||||||||||||||||||||||||||||||||||||||||||||||||||||||

15α-TIGLINOYLOXY-KAUR-16-EN-19-OIC ACID

Source: *Wedelia scaberrima* Benth.

Mol. formula: $C_{25}H_{36}O_4$

Mol. wt.: 400

Solvent: $CDCl_3$

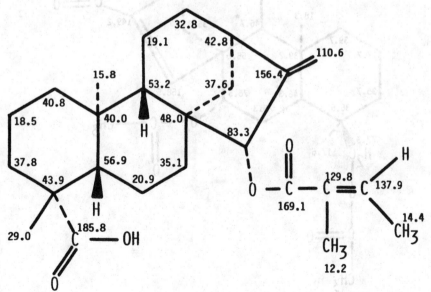

T.C.B. Tomassini and M.E.O. Matos, Phytochemistry, 18(4), 663 (1979).

EPISINFERNAL (DIACETATE)

Source: *Sideritis infernalis*

Mol. formula: $C_{24}H_{34}O_5$

Mol. wt.: 402

Solvent: $CDCl_3$

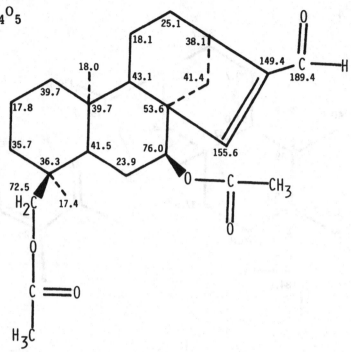

C. Fernandez, B.M. Fraga and M.G. Hernandez, <u>Phytochemistry</u>, 25 (11), 2573 (1986).

SINFERNAL (DIACETATE)

Source: *Sideritis infernalis*

Mol. formula: $C_{24}H_{34}O_5$

Mol. wt.: 402

Solvent: $CDCl_3$

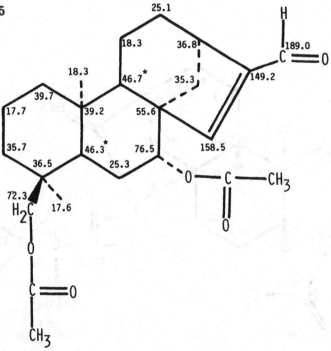

C. Fernandez, B.M. Fraga and M.G. Hernandez, <u>Phytochemistry</u>, 25 (11), 2573 (1986).

LONGIKAURIN B

Source: *Rabdosia longituba* (Miquel) Hara

Mol. formula: $C_{22}H_{30}O_7$

Mol. wt.: 406

Solvent: C_5D_5N

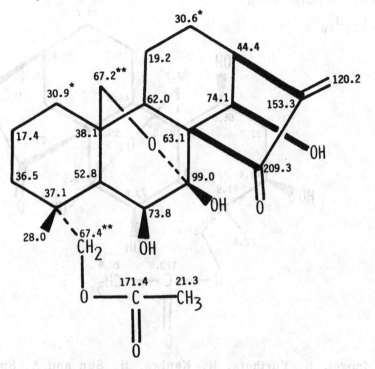

Y. Takeda, T. Fujita and T. Shingu, <u>J. Chem. Soc. Perkin Trans.I</u>, (2), 379 (1988).

||||||||||||||||||||||||||||||||||||||||||||||||||||||||||||||||||||||||||||||||||||||||||||||||||||||||||||||||||||||||||||||||||||||||||||||

ISODOMEDIN

Source: *Isodon skikokianus*

Mol. formula: $C_{23}H_{35}O_6$

Mol. wt.: 407

Solvent: C_5D_5N

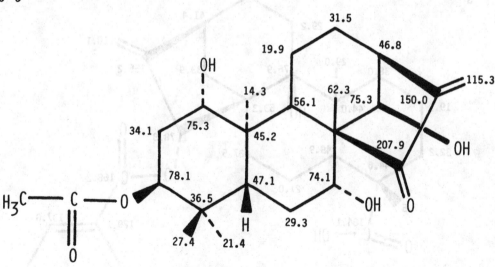

I. Kubo, I. Miura, K. Nakanishi, T. Kamikawa, T. Isobe and T. Kubota, <u>J. Chem. Soc. Chem. Commun.</u>, (16), 555 (1977).

Source: *Rabdosia eriocalyx* Hara

Mol. formula: $C_{22}H_{30}O_8$

Mol. wt.: 422

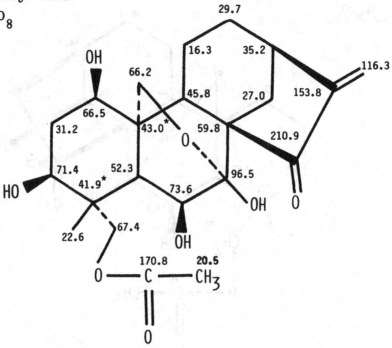

X. Shen, A. Isogai, K. Furihata, H. Kaniwa, H. Sun and A. Suzuki, Phytochemistry, 28 (3), 855 (1989).

15α-TIGLOYLOXY-9β-HYDROXY-ENT-KAUR-16-EN-19-OIC ACID

Source: *Viguiera hypargyrea*

Mol. formula: $C_{25}H_{36}O_5$

Mol. wt.: 416

Solvent: $CDCl_3$

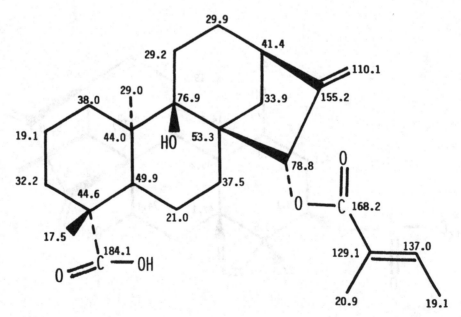

L. Alvarez, R. Mata, G. Delgado and A.R. De Vivar, Phytochemistry, 24(12), 2973 (1985).

MAOECRYSTAL D

Source: *Rabdosia eriocalyx* (Dunn) Hara
Mol. formula: $C_{24}H_{32}O_7$
Mol. wt.: 432
Solvent: C_5D_5N

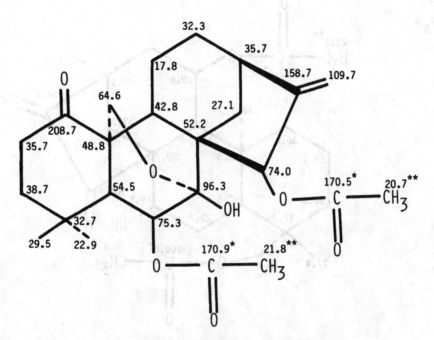

C.-B. Li, H.-D. Sun, and J. Zhon, <u>Acta. Chim. Sinica</u>, **46** (7), 657 (1988).

INFLEXARABDONIN B

Source: *Rabdosia inflexa* (Thunb.) Hara
Mol. formula: $C_{24}H_{34}O_7$
Mol. wt.: 434

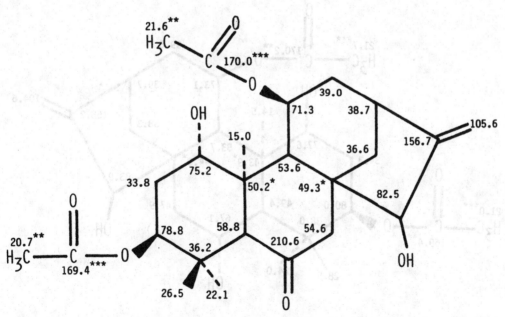

Y. Takeda, T. Ichihara, K. Yamasaki and H. Otsuka, <u>Phytochemistry</u>, **28** (3), 851 (1989).

ROSTHORNIN B

Source: *Rabdosia rosthornii* (Diels) Hara

Mol. formula: $C_{24}H_{34}O_7$

Mol. wt.: 434

Solvent: C_5D_5N

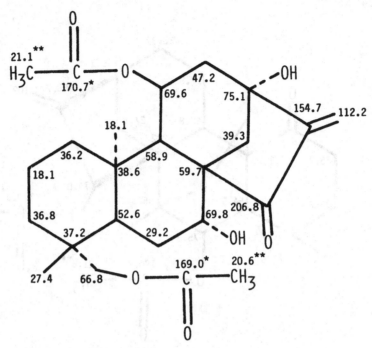

Xu Yunlong and Ma Yunbao, <u>Phytochemistry</u>, **28** (11), 3235 (1989).

||||||||||||||||||||||||||||||||||||||||||||||||||||||||||||||||||||||||||||||||||||||||||||||||||||||||||||||||||||||||||||||||||

INFLEXARABDONIN A

Source: *Rabdosia inflexa* (Thunb.) Hara

Mol. formula: $C_{24}H_{36}O_7$

Mol. wt.: 436

Y. Takeda, T. Ichihara, K. Yamasaki and D. Otsuka, <u>Phytochemistry</u>, **28** (3), 851 (1989).

364

INFLEXARABDONIN D

Source: *Rabdosia inflexa* (Thunb.) Hara

Mol. formula: $C_{24}H_{36}O_7$

Mol. wt.: 436

Solvent: C_5D_5N

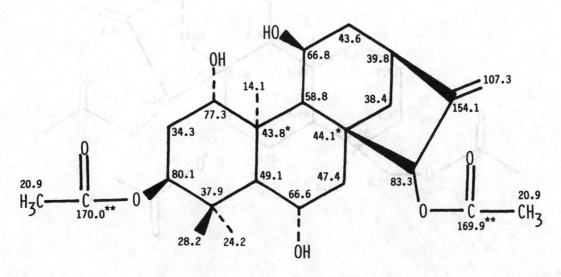

Y. Takeda, T. Ichihara, K. Yamasaki and H. Otsuka, <u>Phytochemistry</u>, **28** (9), 2423 (1989).

||||||||||||||||||||||||||||||||||||||||||||||||||||||||||||||||||||||||||||||||||||||||||||||||||||||

THUJANOL-<u>ENT</u>-KAUR-16-EN-19-OATE

Source: *Helianthus annuus* L.

Mol. formula: $C_{30}H_{46}O_2$

Mol. wt.: 438

Solvent: $CDCl_3$

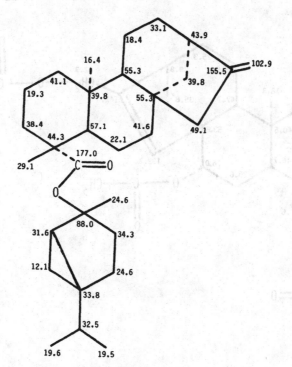

J. St. Pyrek, <u>J. Nat. Prod.</u>, (Lloydia), **47** (5), 822 (1984).

CANDITRIOL (TRIACETATE)

Source: *Sideritis infernalis*
Mol. formula: $C_{26}H_{38}O_6$
Mol. wt.: 446
Solvent: $CDCl_3$

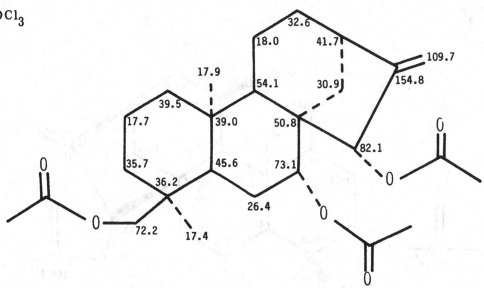

C. Fernandez, B.M. Fraga and M.G. Hernandez, <u>Phytochemistry</u>, 25 (11), 2573 (1986).

SINFERNOL (TRIACETATE)

Source: *Sideritis infernalis*
Mol. formula: $C_{26}H_{38}O_6$
Mol. wt.: 446
Solvent: $CDCl_3$

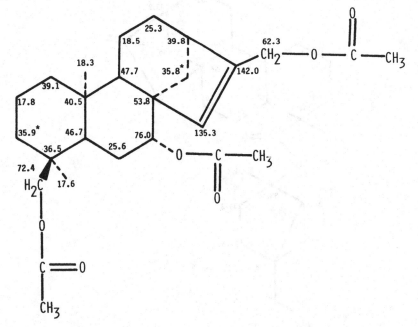

C. Fernandez, B.M. Fraga and M.G. Hernandez, <u>Phytochemistry</u>, 25 (11), 2573 (1986).

EXIDONIN

Source: *Rabdosia henryi* (Hemsl.) Hara
Mol. formula: $C_{24}H_{32}O_8$
Mol. wt.: 448
Solvent: $CDCl_3-C_5D_5N$

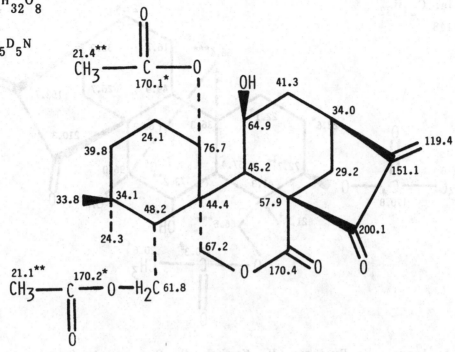

X. Meng, Y. Chen, Y. Cui and J. Cheng, J. Nat. Prod., (Lloydia), 51 (4), 812 (1988).

GANERVOSIN A

Source: *Rabdosia nervosa* Hemsl.
Mol. formula: $C_{24}H_{32}O_8$
Mol. wt.: 448
Solvent: C_5D_5N

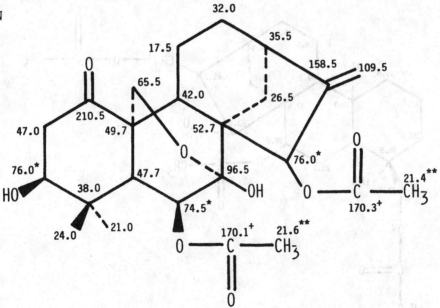

Q.-G. Wang, S.-M. Hua, G. Bai and Y.-Z. Chen, J. Nat. Prod., (Lloydia), 51 (4), 775 (1988).

MAOECRYSTAL J

Source: *Rabdosia eriocalyx* Hara

Mol. formula: $C_{24}H_{32}O_8$

Mol. wt.: 448

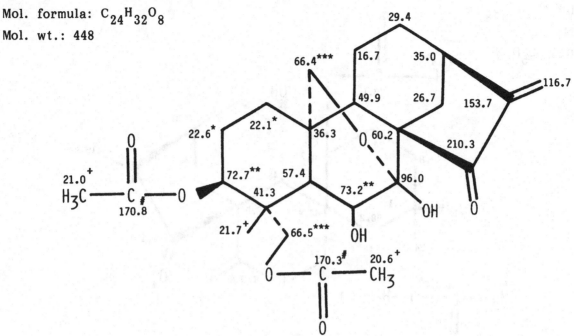

X. Shen, A. Isogai, K. Furihata, H. Kaniwa, H. Sun and A. Suzuki, <u>Phytochemistry</u>, **28** (3), 855 (1989).

EPOXYSINFERNOL (TRIACETATE)

Source: *Sideritis infernalis*

Mol. formula: $C_{26}H_{38}O_7$

Mol. wt.: 462

Solvent: $CDCl_3$

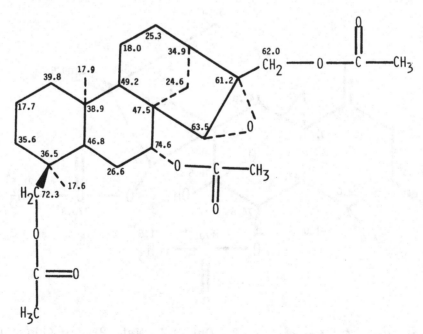

C. Fernandez, B.M. Fraga and M.G. Hernandez, <u>Phytochemistry</u>, **25** (11), 2573 (1986).

LONGIRABDOSIN

Source: *Rabdosia longituba* (Miq.) Hara

Mol. formula: $C_{26}H_{32}O_8$

Mol. wt.: 472

Solvent: C_5D_5N

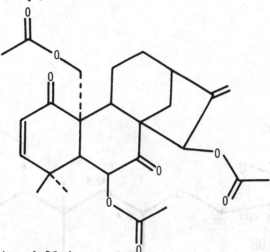

CH_3: 20.7, 20.7, 20.8, 21.4 and 33.4

CH_2: 21.0, 32.5, 34.4, 65.1 and 108.1

CH: 38.1, 39.4, 50.7, 74.4, 75.9, 123.3 and 156.2

C: 37.3, 52.5, 58.1 and 152.3

O=C: 169.5, 170.0, 170.0, 199.6 and 203.1

T. Ichihara, Y. Takeda and H. Otsuka, <u>Phytochemistry</u>, 27 (7), 2261 (1988).

||||||||||||||||||||||||||||||||||||||||||||||||||||||||||||||||||||||||||||||||||||||||||||||||||||||||||||||||||

PANICULOSIDE I

Source: *Stevia paniculata*

Mol. formula: $C_{26}H_{40}O_8$

Mol. wt.: 480

Solvent: C_5D_5N

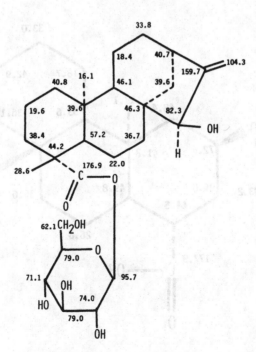

K. Yamasaki, H. Kohda, T. Kobayashi, R. Kasai and O. Tanaka, <u>Tetrahedron Lett.</u>, (13), 1005 (1976).

ATRACTYLIGENIN GLYCOSIDE-II

Source: *Coffea arabica* L.

Mol. formula: $C_{25}H_{38}O_9$

Mol. wt.: 482

Solvent: $CDCl_3$

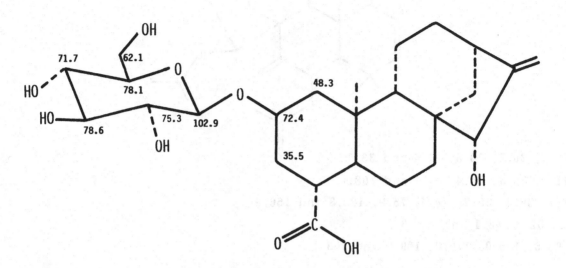

H. Richter and G. Spiteller, <u>Chem. Ber.</u>, 111(10), 3506 (1978).

NORDITERPENE DA-II

Source: *Drymaria arenarioides* Willd.

Mol. formula: $C_{25}H_{38}O_9$

Mol. wt.: 482

Solvent: C_5D_5N

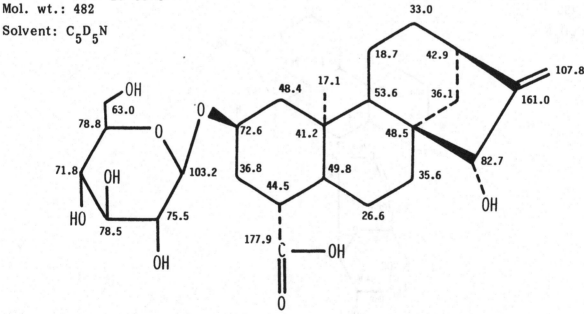

D. Vargas, X.A. Dominguez, K. Acuna-Askar, M. Gutierrez and K. Hostettmann, <u>Phytochemistry</u>, 27(5), 1532 (1988).

DITERPENE LJ-IX

Source: *Lindsaea javanensis* BL.

Mol. formula: $C_{26}H_{44}O_8$

Mol. wt.: 484

Solvent: C_5D_5N

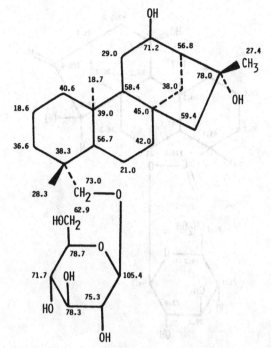

T. Satake, T. Murakami, Y. Saiki and C.-M. Chen, <u>Chem. Pharm. Bull.</u>, (Tokyo), 31 (11), 3865 (1983).

PANICULOSIDE-III

Source: *Stevia paniculata*

Mol. formula: $C_{26}H_{38}O_9$

Mol. wt.: 494

Solvent: C_5D_5N

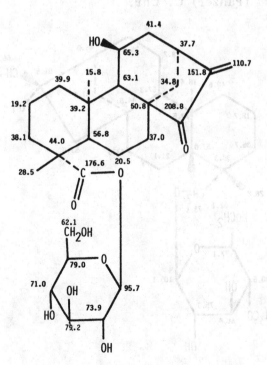

K. Yamasaki, H. Kohda, T. Kobayashi, R. Kasai and O. Tanaka, <u>Tetrahedron Lett.</u>, (13), 1005 (1976).

PANICULOSIDE-II

Source: *Stevia paniculata*

Mol. formula: $C_{26}H_{40}O_9$

Mol. wt.: 496

Solvent: C_5D_5N

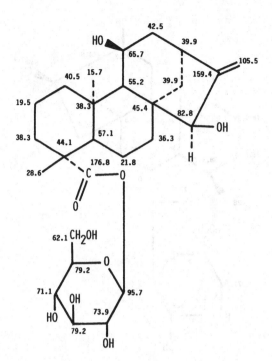

K. Yamasaki, H. Kohda, T. Kobayashi, R. Kasai and O. Tanaka, <u>Tetrahedron Lett.</u>, (13), 1005 (1976).

MICROLEPIN

Source: *Microlepia marginata* (Panzer) C. Chr.

Mol. formula: $C_{27}H_{46}O_8$

Mol. wt.: 498

Solvent: C_5D_5N

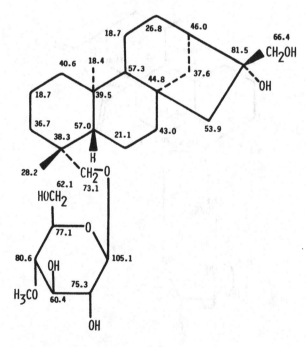

T. Kuraishi, T. Taniguchi, T. Murakami, N. Tanaka, Y. Saiki and C.-M. Chen, <u>Chem. Pharm. Bull.</u>, (Tokyo), **31** (5), 1494 (1983).

DITERPENE LJ-VII

Source: *Lindsaea javanensis* BL.

Mol. formula: $C_{26}H_{44}O_9$

Mol. wt.: 500

Solvent: C_5D_5N

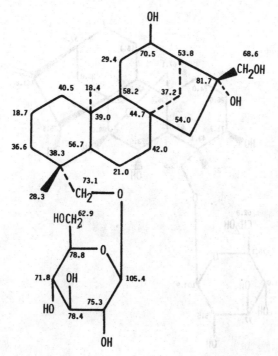

T. Satake, T. Murakami, Y. Saiki and C.-M. Chen, <u>Chem. Pharm. Bull.</u> (Tokyo), 31(11), 3865 (1983).

||||||||||||||||||||||||||||||||||||||||||||||||||||||||||||||||||||||||||||||||||||||||||||||||||||||

SHIKOKIASIDE B

Source: *Rabdosia shikokiana* (Makino) Hara

Mol. formula: $C_{26}H_{40}O_{10}$

Mol. wt.: 512

Solvent: C_5D_5N

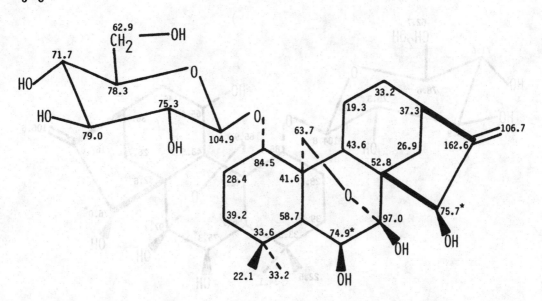

T. Isobe, Y. Noda, K. Shibata and T. Kubota, <u>Chem. Lett.</u>, 1225 (1981).

RABDOSIDE-I

Source: *Rabdosia eriocalyx* (Dunn) Hara

Mol. formula: $C_{26}H_{40}O_{11}$

Mol. wt.: 528

Solvent: CD_3OD-D_2O (1:1)

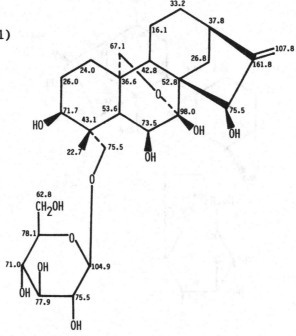

A. Isogai, X. Shen, K. Furihata, H. Kaniwa, H. Sun and A. Suzuki, <u>Phytochemistry</u>, 28 (9), 2427 (1989).

SHIKOKIASIDE A

Source: *Rabdosia shikokiana* (Makino) Hara

Mol. formula: $C_{26}H_{40}O_{11}$

Mol. wt.: 528

Solvent: C_5D_5N

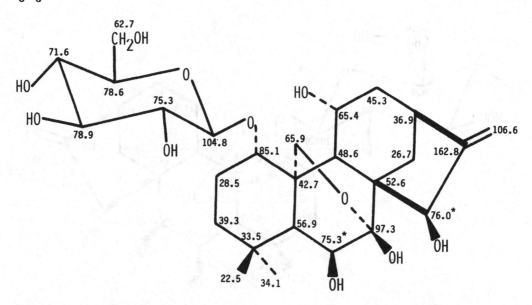

T. Isobe, Y. Noda, K. Shibata and T. Kubota, <u>Chem. Lett.</u>, 1225 (1981).

17-O-ACETYLMICROLEPIN

Source: *Microlepia marginata* (Panzer) C. Chr.

Mol. formula: $C_{29}H_{48}O_9$

Mol. wt.: 540

Solvent: C_5D_5N

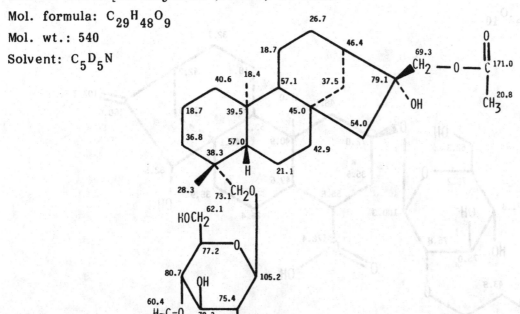

T. Kuraishi, T. Taniguchi, T. Murakami, N. Tanaka, Y. Saiki and C.-M. Chen, <u>Chem. Pharm. Bull.</u>, (Tokyo), **31** (5), 1494 (1983).

RABDOSIDE-II

Source: *Rabdosia eriocalyx* (Dunn) Hara

Mol. formula: $C_{26}H_{40}O_{12}$

Mol. wt.: 544

Solvent: CD_3OD-D_2O (1:1)

A. Isogai, X. Shen, K. Furihata, H. Kaniwa, H. Sun and A. Suzuki, <u>Phytochemistry</u>, **28** (9), 2427 (1989).

2-O-(2-O-ISOVALERYL-β-D-GLUCOPYRANOSYL) ATRACTYLIGENIN

Source: *Coffea arabica*

Mol. formula: $C_{30}H_{46}O_{10}$

Mol. wt.: 566

Solvent: C_5D_5N

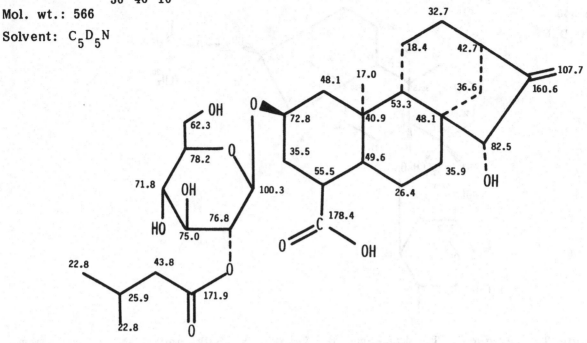

H. Richter and G. Spiteller, <u>Chem. Ber.</u>, 111(10), 3506 (1978).

STEVIOLBIOSIDE

Source: *Stevia rebaudiana* Bertoni

Mol. formula: $C_{32}H_{50}O_{13}$

Mol. wt.: 642

Solvent: C_5D_5N

H. Kohda, R. Kasai, K. Yamasaki, K. Murakami and O. Tanaka, <u>Phytochemistry</u>, 15(6), 981 (1976).

376

ATRACTYLIGENIN GLYCOSIDE-I

Source: *Coffea arabica* L.
Mol. formula: $C_{36}H_{56}O_{15}$
Mol. wt.: 728
Solvent: $CDCl_3$

H. Richter and G. Spiteller, <u>Chem. Ber.</u>, 111(10), 3506 (1978).

||||||||||||||||||||||||||||||||||||||||||||||||||||||||||||||||||||||||||||||||||||||||||||||||||||||||||

REBAUDIOSIDE B

Source: *Stevia rebaudiana* Bertoni
Mol. formula: $C_{38}H_{60}O_{18}$
Mol. wt.: 804
Solvent: C_5D_5N

H. Kohda, R. Kasai, K. Yamasaki, K. Murakami and O. Tanaka, <u>Phytochemistry</u>, 15(6), 981 (1976).

Source: *Stevia rebaudiana* Bertoni

Mol. formula: $C_{38}H_{60}O_{18}$

Mol. wt.: 804

Solvent: C_5D_5N

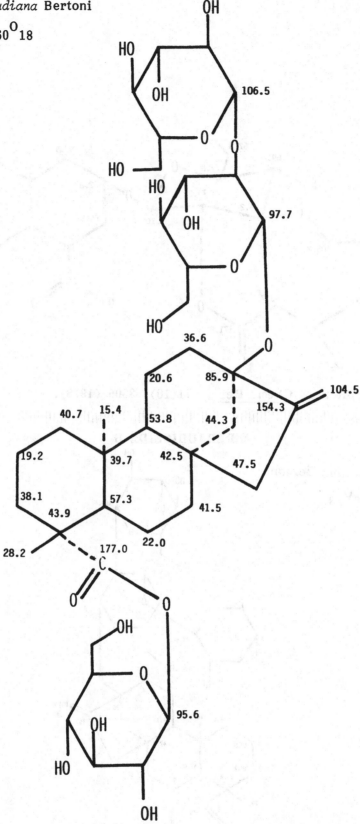

H. Kohda, R. Kasai, K. Yamasaki, K. Murakami and O. Tanaka, <u>Phytochemistry</u>, 15(6), 981 (1976).

REBAUDIOSIDE A

Source: *Stevia rebaudiana* Bertoni

Mol. formula: $C_{44}H_{70}O_{23}$

Mol. wt.: 966

Solvent: C_5D_5N

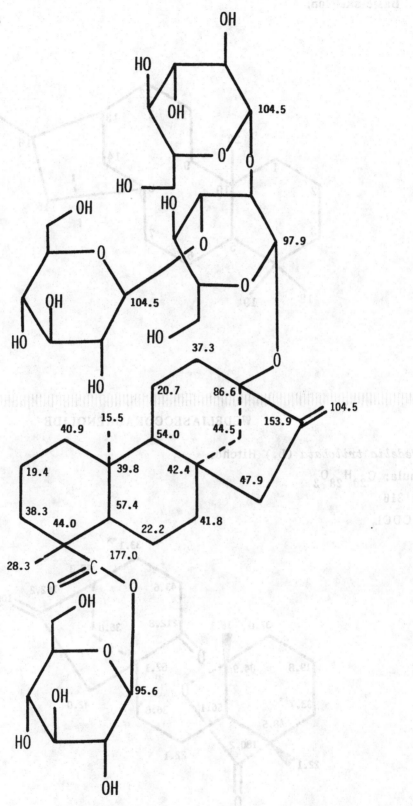

H. Kohda, R. Kasai, K. Yamasaki, K. Murakami and O. Tanaka, <u>Phytochemistry</u>, 15(6), 981 (1976).

Basic skeleton:

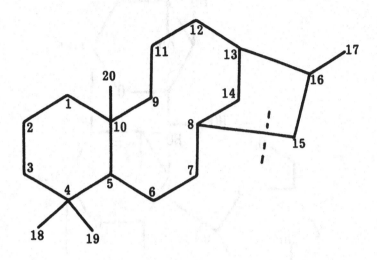

WEDELIASECCOKAURENOLIDE

Source: *Wedelia trilobata* (L.) Hitchc.
Mol. formula: $C_{20}H_{28}O_3$
Mol. wt.: 316
Solvent: $CDCl_3$

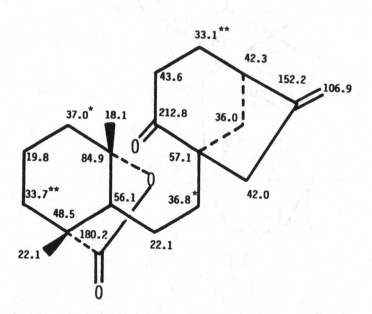

F. Bohlmann, J. Ziesche, R.M. King and H. Robinson, <u>Phytochemistry</u>, **20** (4), 751 (1981).

RABDOLATIFOLIN

Source: *Rabdosia umbrosa* (Maxim.) Hara

Mol. formula: $C_{20}H_{28}O_4$

Mol. wt.: 332

Solvent: C_5D_5N

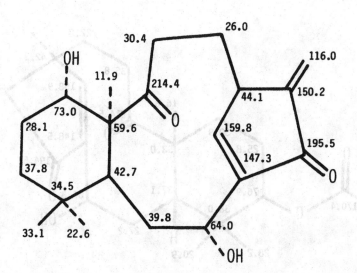

Y. Takeda, T. Fujita and A. Ueno, <u>Phytochemistry</u>, **22** (11) 2531 (1983).

||||||||||||||||||||||||||||||||||||||||||||||||||||||||||||||||||||||||||||||||||||||||||||||||||||||||||||||||

Aa-7CM

Source: *Croton argyrophylloides* Muell. Arg.

Mol. formula: $C_{21}H_{30}O_4$

Mol. wt.: 346

Solvent: $CDCl_3$

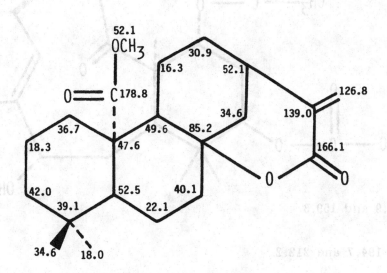

F.J.Q. Monte, C.H.S. Andrade, A.A. Craveiro and R.B. Filho, <u>J. Nat. Prod.</u>, (Lloydia), **47** (1), 55 (1984).

SHIKOCCIN

Source: *Rabdosia shikokiana* (Makino) Hara

Mol. formula: $C_{22}H_{30}O_5$

Mol. wt.: 374

Solvent: $CDCl_3$

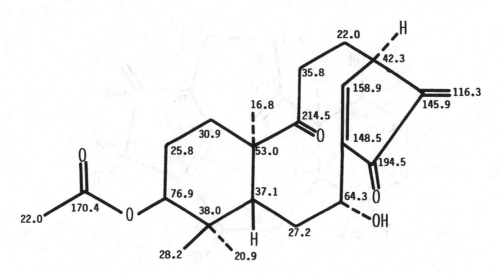

M. Node, N. Ito, I. Uchida, E. Fujita and K. Fuji, <u>Chem. Pharm. Bull.</u>, (Tokyo), 33(3), 1029 (1985).

SHIKODOMEDIN

Source: *Rabdosia shikokiana* (Makino) Hara.

Mol. formula: $C_{24}H_{32}O_7$

Mol. wt.: 432

Solvent: $CDCl_3$

CH_2: 116.8

CH: 64.0, 71.0, 76.9 and 159.3

C: 145.7 and 148.6

C=O: 170.3, 170.7, 194.7 and 212.2

T. Fujita, Y. Takeda, T. Shingu, M. Kido and Z. Taira, <u>J. Chem. Soc. Chem. Commun</u>; (3) 162 (1982).

BEYRANE

Basic skeleton:

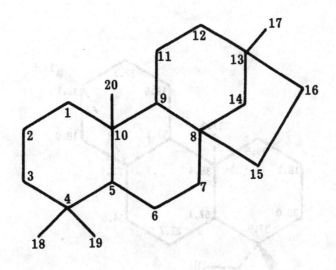

19-HYDROXY-15,16-BISEPI-MYRIOPHAN

Source: *Myriocephalus stuartii* (F. Mull. et Sonder ex Sonder) Benth.
Mol. formula: $C_{18}H_{30}O$
Mol. wt.: 262
Solvent: $CDCl_3$

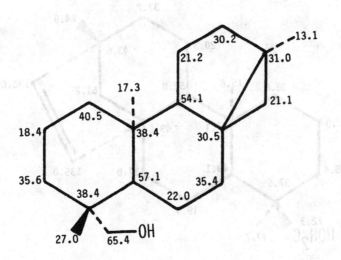

C. Zdero, F. Bohlmann, L. Haegi and R.M. King, <u>Liebigs Ann. Chem.</u>, 665 (1987).

19-HYDROXYMYRIOPHAN

Source: *Myriocephalus stuartii* (F. Mull. et Sonder ex Sonder) Benth.

Mol. formula: $C_{18}H_{30}O$

Mol. wt.: 262

Solvent: $CDCl_3$

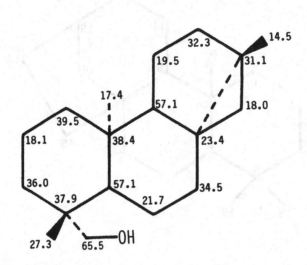

C. Zdero, F. Bohlmann, L. Haegi and R.M. King, <u>Liebigs Ann. Chem.</u>, 665 (1987).

||||||||||||||||||||||||||||||||||||||||||||||||||||||||||||||||||||||||||||||||||||||||||||||||||||||||||||||||||||||||||||||

<u>ENT-BEYER-15-EN-18-OL</u>

Source: *Baccharis tola*

Mol. formula: $C_{20}H_{32}O$

Mol. wt.: 288

Solvent: $CDCl_3$

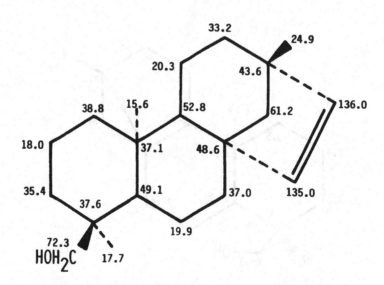

A.S. Martin, J. Rovirosa, R. Becker and M. Castillo, <u>Phytochemistry</u>, 19(9), 1985 (1980).

ERYTHROXYLOL A

Source: *Baccharis tola*
Mol. formula: $C_{20}H_{32}O$
Mol. wt.: 288
Solvent: $CDCl_3$

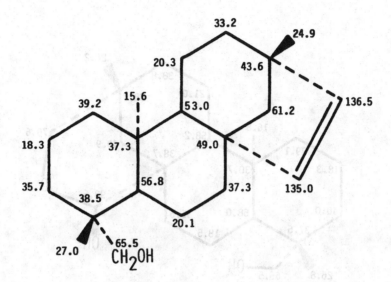

A.S. Martin, J. Rovirosa, R. Becker and M. Castillo, <u>Phytochemistry</u>, 19(9), 1985 (1980).

<u>ENT-BEYER-15-EN-17,19-DIOL</u>

Source: *Baccharis tola*
Mol. formula: $C_{20}H_{32}O_2$
Mol. wt.: 304
Solvent: $CDCl_3$

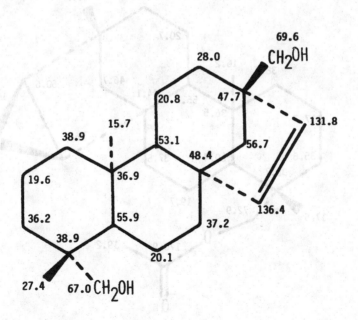

A.S. Martin, J. Rovirosa and M. Castillo, <u>Phytochemistry</u>, 22(6), 1461 (1983).

385

15β,19-DIHYDROXY-15,16-EPOXY-SECO-BEYERAN

Source: *Myriocephalus stuartii* (F. Mull et Sonder ex Sonder) Benth.
Mol. formula: $C_{20}H_{34}O_3$
Mol. wt.: 322
Solvent: $CDCl_3$

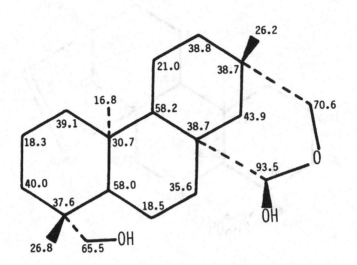

C. Zdero, F. Bohlmann, L. Haegi and R.M. King, <u>Liebigs Ann. Chem.</u>, 665 (1987).

||||||||||||||||||||||||||||||||||||||||||||||||||||||||||||||||||||||||||||||||||||||||||

15β,16β-EPOXIDE-<u>ENT</u>-BEYERAN-18-OL (ACETATE)

Source: *Baccharis tola* Phil
Mol. formula: $C_{22}H_{34}O_3$
Mol. wt.: 346
Solvent: $CDCl_3$

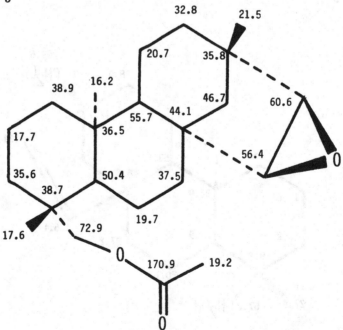

A.S. Martin, J. Rovirosa and M. Castillo, <u>Phytochemistry</u>, 22(6), 1461 (1983).

ENT-BEYER-15-EN-18,19-DIOL (DIACETATE)

Source: *Baccharis tola* Phil

Mol. formula: $C_{24}H_{36}O_4$

Mol. wt.: 388

Solvent: $CDCl_3$

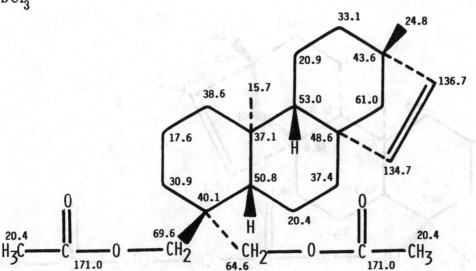

S.A. Martin, J. Rovirosa and M. Castillo, <u>Phytochemistry</u>, 22 (6), 1461 (1983).

18-HYDROXY-ENT-BEYER-15-EN-MALONIC ACID (METHYL ESTER)

Source: *Helipterum floribundum* DC.

Mol. formula: $C_{24}H_{36}O_4$

Mol. wt.: 388

Solvent: $CDCl_3$

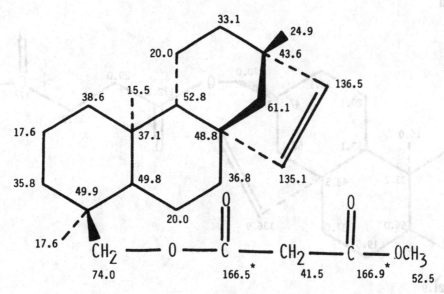

C. Zdero, F. Bohlmann, R.M. King and H. Robinson, <u>Phytochemistry</u>, 28(2), 517 (1989).

18-HYDROXY-ENT-BEYER-15-EN-SUCCINIC ACID (METHYL ESTER)

Source: *Helipterum floribundum* DC.

Mol. formula: $C_{25}H_{38}O_4$

Mol. wt.: 402

Solvent: $CDCl_3$

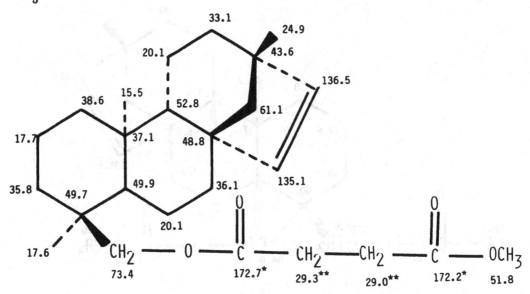

C. Zdero, F. Bohlmann, R.M. King and H. Robinson, <u>Phytochemistry</u>, **28**(2), 517 (1989).

||||||||||||||||||||||||||||||||||||||||||||||||||||||||||||||||||||||||||||||||||||||||||||||||||||||||

ENT-17-SUCCINYLOXYBEYER-15-ENE (METHYL ESTER)

Source: *Myriocephalus stuartii* (F. Mull. et Sonder ex Sonder) Benth.

Mol. formula: $C_{25}H_{38}O_4$

Mol. wt.: 402

Solvent: $CDCl_3$

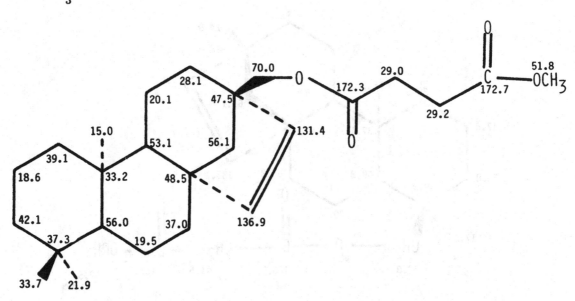

C. Zdero, F. Bohlmann, L. Haegi and R.M. King, <u>Liebigs Ann. Chem.</u>, 665 (1987).

388

ENT-19-SUCCINYLOXYBEYER-15-ENE (METHYL ESTER)

Source: *Myriocephalus stuartii* (F. Mull. et Sonder ex Sonder) Benth.

Mol. formula: $C_{25}H_{38}O_4$

Mol. wt.: 402

Solvent: $CDCl_3$

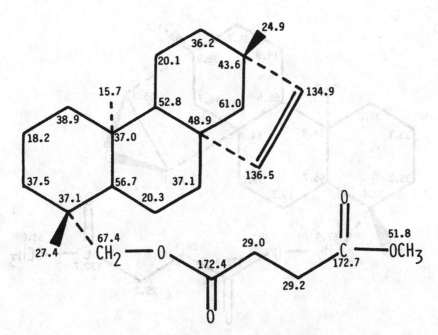

C. Zdero, F. Bohlmann, L. Haegi and R.M. King, <u>Liebigs Ann. Chem.</u>, 665 (1987).

7β,18-DIHYDROXY-ENT-BEYER-15-EN-18-O-SUCCINIC ACID (METHYL ESTER)

Source: *Helipterum floribundum* DC.

Mol. formula: $C_{25}H_{38}O_5$

Mol. wt.: 418

Solvent: $CDCl_3$

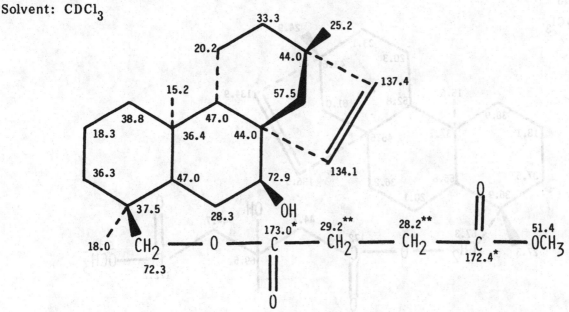

C. Zdero, F. Bohlmann, R.M. King and H. Robinson, <u>Phytochemistry</u>, 28(2), 517 (1989).

ENT-19-SUCCINYLOXY-15α,16α-EPOXYBEYERAN (METHYL ESTER)

Source: *Myriocephalus stuartii* (F. Mull. et Sonder ex Sonder) Benth.

Mol. formula: $C_{25}H_{38}O_5$

Mol. wt.: 418

Solvent: $CDCl_3$

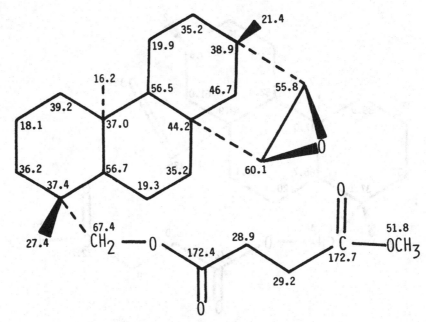

C. Zdero, F. Bohlmann, L. Haegi and R.M. King, <u>Liebigs Ann. Chem.</u>, 665 (1987).

ENT-19-(3-HYDROXY-3-METHYLGLUTAROYLOXY)BEYER-15-ENE(METHYL ESTER)

Source: *Myriocephalus stuartii* (F. Mull. et Sonder ex Sonder) Benth.

Mol. formula: $C_{27}H_{42}O_5$

Mol. wt.: 446

Solvent: $CDCl_3$

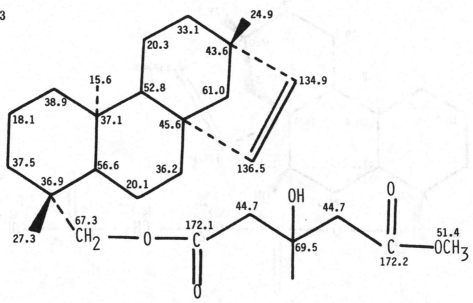

C. Zdero, F. Bohlmann, L. Haegi and R.M. King, <u>Liebigs Ann. Chem.</u>, 665 (1987).

390

Basic skeleton:

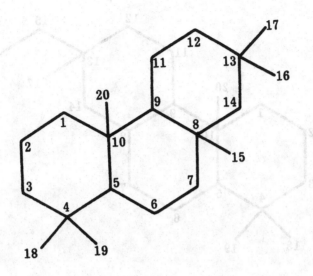

19-HYDROXY-15-OXO-SECO-BEYERAN-16-OIC ACID (METHYL ESTER)

Source: *Myriocephalus stuartii* (F. Mull. et Sonder ex Sonder) Benth.

Mol. formula: $C_{21}H_{34}O_4$

Mol. wt.: 350

Solvent: $CDCl_3$

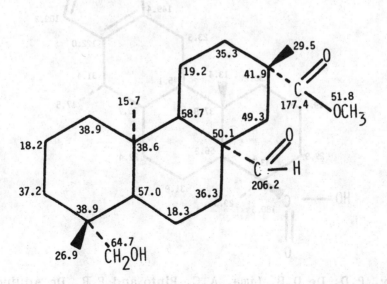

C. Zdero, F. Bohlmann, L. Haegi and R.M. King, <u>Liebigs Ann. Chem.</u>, 665 (1987).

CASSANE

Basic skeleton:

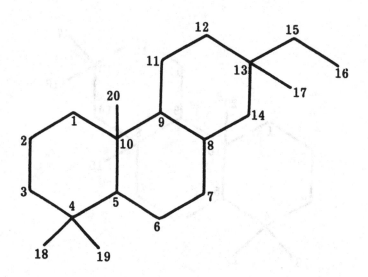

VOUACAPENIC ACID

Source: *Dipterix odorata* Willd (Aubl.) (syn. *coumarouna odorata* Aubl.)

Mol. formula: $C_{20}H_{28}O_3$

Mol. wt.: 316

Solvent: $CDCl_3$

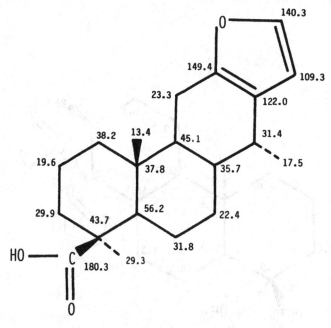

R.L. De O. Godoy, P.D. De D.B. Lima, A.C. Pinto and F.R. De Aquino Neto, <u>Phytochemistry</u>, **28**(2), 642 (1989).

3β-ACETOXYVOUACAPENOL

Source: *Dipterix odorata* Willd (Aubl.) (syn. *coumarouna odorata* Aubl.)
Mol. formula: $C_{22}H_{32}O_4$
Mol. wt.: 360
Solvent: $CDCl_3$

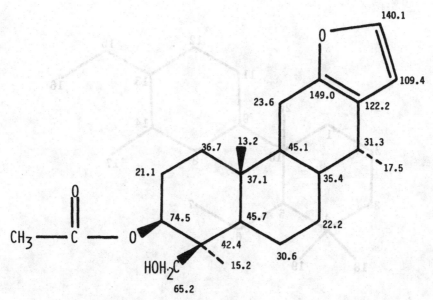

R.L. De O. Godoy, P.D. De D.B. Lima, A.C. Pinto and F.R. De Aquino Neto, <u>Phytochemistry</u>, **28**(2), 642 (1989).

6β-CINNAMOYL-7β-HYDROXY-VOUACAPEN-5α-OL

Source: *Caesalpinia pulcherrima* Swartz
Mol. formula: $C_{29}H_{36}O_5$
Mol. wt.: 464
Solvent: $CDCl_3$

D.D. Mc Pherson, C.-T. Che, G.A. Cordell, D.D. Soejarto, J.M. Pezzuto and H.H.S. Fong, <u>Phytochemistry</u>, **25** (1), 167 (1986).

Basic skeleton:

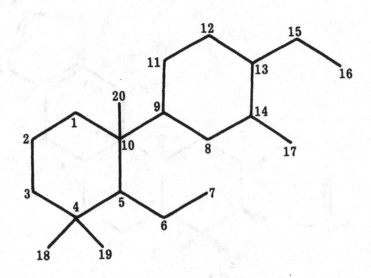

||||||||||||||||||||||||||||||||||||||||||||||||||||||||||||||||||||||||||||||||||||||||

DITERPENOID A

Source: *Acacia jacquemontii*
Mol. formula: $C_{20}H_{32}O_2$
Mol. wt.: 304
Solvent: $CDCl_3$

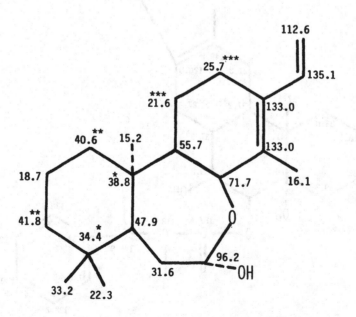

K.C. Joshi, R.K. Bansal, T. Sharma, R.D.H. Murray, I.T. Forbes, A.F. Cameron and A. Maltz, <u>Tetrahedron</u>, **35**(11), 1449 (1979).

Source: *Acacia jacquemontii*

Mol. formula: $C_{20}H_{32}O_2$

Mol. wt.: 304

Solvent: $CDCl_3$

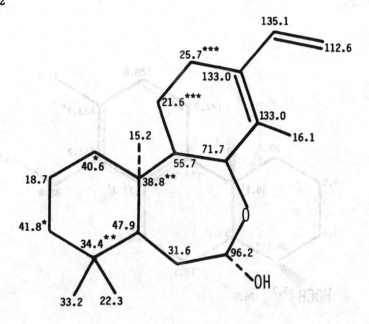

K.C. Joshi, R.K. Bansal, T. Sharma, R.D.H. Murray, I.T. Forbes, A.F. Cameron and A. Maltz, <u>Tetrahedron</u>, **35** (11), 1449 (1979).

||||||||||||||||||||||||||||||||||||||||||||||||||||||||||||||||||||||||||||||||||||||||||||||||||||||||||||

CLEISTANTHANE

Basic skeleton:

(4R,5S,10S)-CLEISTANTHA-8,11,13-TRIEN-19-OL

Source: *Vellozia flavicans*

Mol. formula: $C_{20}H_{30}O$

Mol. wt.: 286

Solvent: $CDCl_3$

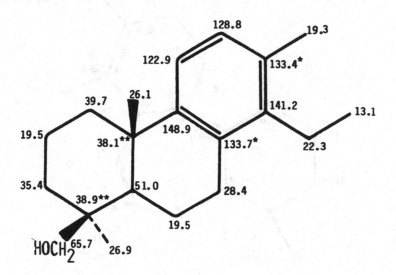

A.C. Pinto, D.H.T. Zocher, P.P.S. Queiroz and A. Kelecom, <u>Phytochemistry</u>, 26 (8), 2409 (1987).

|||||||||||||||||||||||||||||||||||||||||||||||||||||||||||||||||||||||||||||||||||||||||||||||||||||||||||||||||||||

SONDERIANOL

Source: *Croton sonderianus* Muell.Arg.

Mol. formula: $C_{20}H_{26}O_2$

Mol. wt.: 298

Solvent: $CDCl_3$

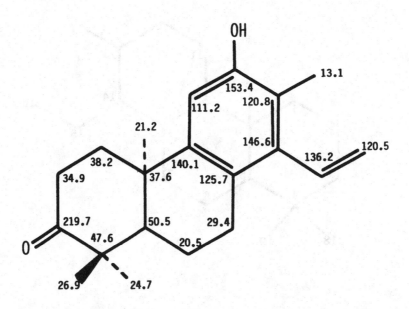

A.A. Craveiro and E.R. Silveira, <u>Phytochemistry</u>, 21(10), 2571 (1982).

11-HYDROXYCLEISTANTHA-8,11,13-TRIEN-7-ONE

Source: *Vellozia nivea* L.B. Smith and Ayensu
Mol. formula: $C_{20}H_{28}O_2$
Mol. wt.: 300
Solvent: $CDCl_3$

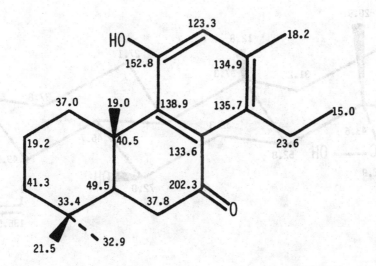

A.C. Pinto and A.M.P.Macaira, <u>Phytochemistry</u>, 27 (12), 3973 (1988).

7,11-DIKETO-14α-HYDROXYCLEISTANTHA-8,12-DIENE

Source: *Vellozia nivea* L.B. Smith and Ayensu
Mol. formula: $C_{20}H_{28}O_3$
Mol. wt.: 316
Solvent: $CDCl_3$

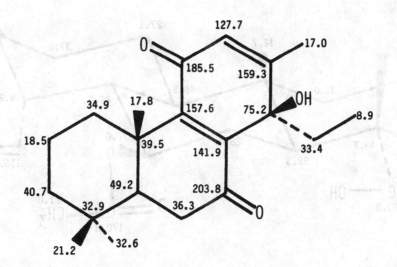

A.C. Pinto and A.M.P. Macaira, <u>Phytochemistry</u>, 27 (12), 3973 (1988).

7-HYDROXYCLEISTANTH-13,15-DIEN-18-OIC ACID

Source: *Pogostemon auricularis* Hassk

Mol. formula: $C_{20}H_{30}O_3$

Mol. wt.: 318

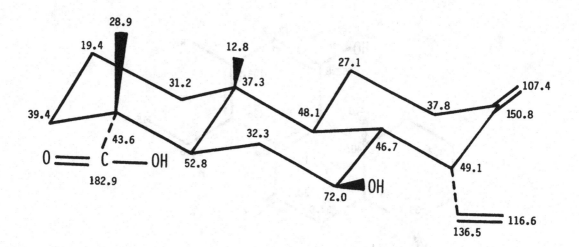

F.A. Hussaini, S. Agarwal, R. Roy, O. Prakash and A. Shoeb, J. Nat. Prod., (Lloydia), 51 (2), 212 (1988).

7-ACETOXYCLEISTANTH-13,15-DIEN-18-OIC ACID

Source: *Pogostemon auricularis* Hassk

Mol. formula: $C_{22}H_{32}O_4$

Mol. wt.: 360

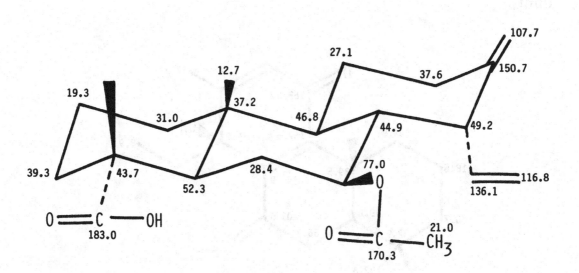

F.A. Hussaini, S. Agarwal, R. Roy, O. Prakash and A. Shoeb, J. Nat. Prod., (Lloydia), 51 (2), 212 (1988).

Seco-CLEISTANTHANE

Basic skeleton:

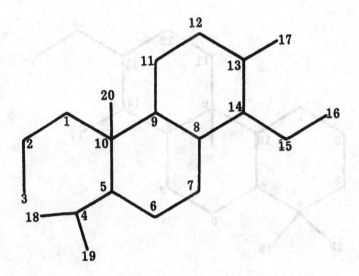

|||||||||||||||||||||||||||||||||||||||||||||||||||||||||||||||||||||||||||||||||||||||||||||||||||||||||||||||||||||||||||||||||||||||||||||||||||||||

3,4-SECOSONDERIANOL

Source: *Croton sonderianus* Muell.Arg.

Mol. formula: $C_{21}H_{28}O_3$

Mol. wt.: 328

Solvent: $CDCl_3$

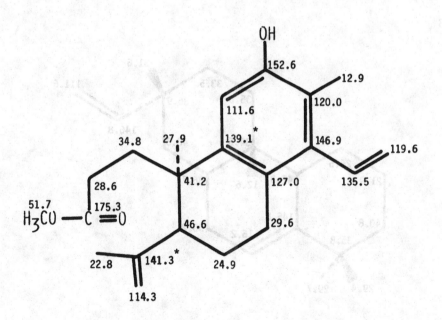

A.A. Craveiro and E.R. Silveira, <u>Phytochemistry</u>, **21**(10), 2571 (1982).

ROSANE

Basic skeleton:

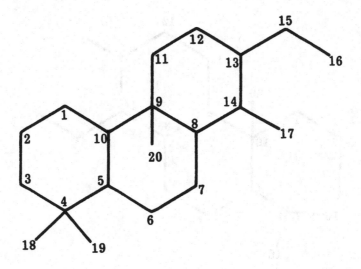

||||||||||||||||||||||||||||||||||||||||||||||||||||||||||||||||||||||||||||||||||||||||||||||||||||||

RIMUENE

Source: *Dacrydium cupressinum* Lamb.

Mol. formula: $C_{20}H_{32}$

Mol. wt.: 272

Solvent: $CDCl_3$

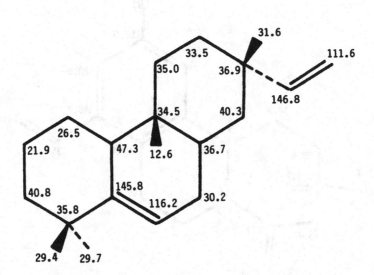

I. Salasoo, <u>Phytochemistry</u>, 23 (1), 192 (1984).

Source: *Sideritis serrata* Lag.

Mol. formula: $C_{20}H_{32}$

Mol. wt.: 272

Solvent: $CDCl_3$

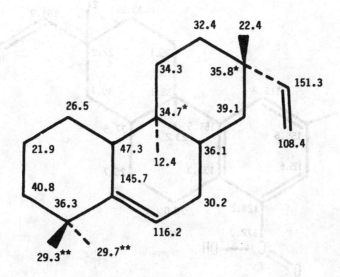

M.C. Garcia-Alvarez, B. Rodriguez, S. Valverde, B.M. Fraga and A.G. Gonzalez, Phyto-chemistry, 20(1), 167 (1981).

IIIIIIIIIIIIIIIIIIIIIIIIIIIIIIIIIIIIIIIIIIIIIIIIIIIIIIIIIIIIIIIIIIIIIIIIIIIIIIIIIIIIIIIIIIIIIIIIIIIIIIIII

Ar-MAXIMOL

Source: *Arachniodes maximowiczii* Ohwl

Mol. formula: $C_{19}H_{26}O_2$

Mol. wt.: 286

Solvent: CD_3OD

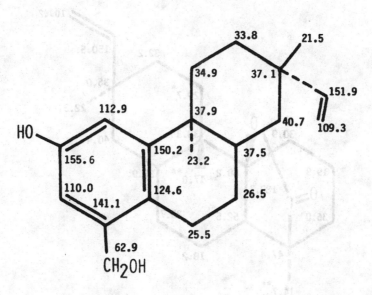

N. Tanaka, H. Sakai, T. Murakami, Y. Saiki, C.-M. Chen and Y. Iitaka, Chem. Pharm. Bull. (Tokyo), 34 (3), 1015 (1986).

Ar-MAXIMIC ACID

Source: *Arachniodes maximowiczii* Ohwi
Mol. formula: $C_{19}H_{24}O_3$
Mol. wt.: 300
Solvent: CD_3OD

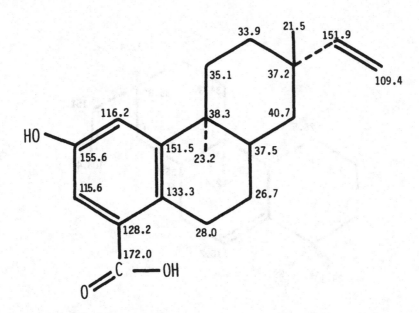

N. Tanaka, H. Sakai, T. Murakami, Y. Saiki, C.-M. Chen and Y. Iitaka, Chem. Pharm. Bull., (Tokyo), 34(3), 1015 (1986).

||||||||||||||||||||||||||||||||||||||||||||||||||||||||||||||||||||||||||||||||||||||||||||||||||||||||||||||||||||||||||

DESOXYROSENONOLACTONE

Source: *Trichothecium roseum*
Mol. formula: $C_{20}H_{30}O_2$
Mol. wt.: 302
Solvent: $CDCl_3$

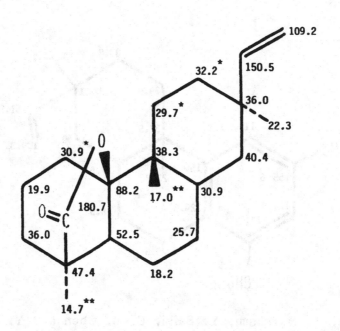

B. Dockerill, J.R. Hanson and M. Siverns, Phytochemistry, 17(3), 572 (1978).

ROSENONOLACTONE

Source: *Trichothecium roseum*
Mol. formula: $C_{20}H_{28}O_3$
Mol. wt.: 316
Solvent: $CDCl_3$

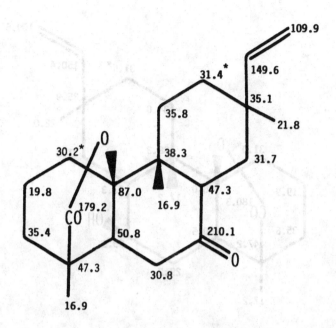

B. Dockerill, J.R. Hanson and M. Siverns, <u>Phytochemistry</u>, 17(3), 572 (1978).

||||||||||||||||||||||||||||||||||||||||||||||||||||||||||||||||||||||||||||||||||||||||||||||||||||||||||||||||||||||||||||||||||||||||||||||

INEKETONE

Source: *Oryza sativa* L.cv Koshihikari
Mol. formula: $C_{20}H_{30}O_3$
Mol. wt.: 318
Solvent: $CDCl_3$

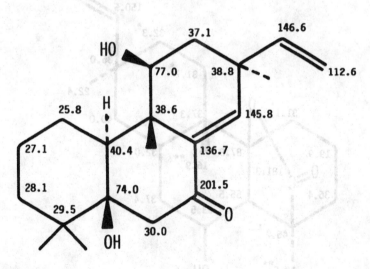

CH_3: 14.9, 16.8, 23.8 and 27.7

T. Kato, M. Tsunakawa, N. Sasaki, H. Aizawa, K. Fujita, Y. Kitahara and N. Takahashi, <u>Phytochemistry</u>, 16(1), 45 (1977).

ROSENOLOLACTONE

Source: *Trichothecium roseum*

Mol. formula: $C_{20}H_{30}O_3$

Mol. wt.: 318

Solvent: $CDCl_3$

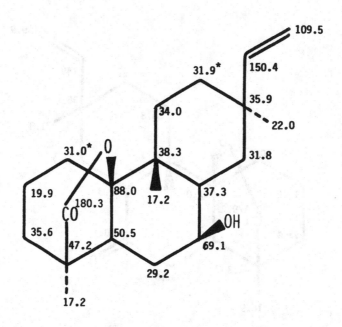

B. Dockerill, J.R. Hanson and M. Siverns, <u>Phytochemistry</u>, 17(3), 572 (1978).

ROSOLOLACTONE

Source: *Trichothecium roseum*

Mol. formula: $C_{20}H_{30}O_3$

Mol. wt.: 318

Solvent: $CDCl_3$

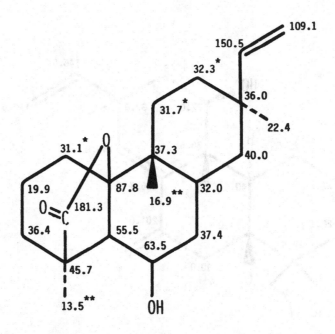

B. Dockerill, J.R. Hanson and M. Siverns, <u>Phytochemistry</u>, 17(3), 572 (1978).

EPOXIVELLOZIN

Source: *Vellozia candida* Mikan
Mol. formula: $C_{20}H_{28}O_4$
Mol. wt.: 332
Solvent: $CDCl_3$

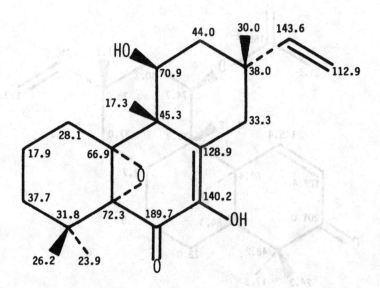

A.C. Pinto, T.C.V. Scofield and R.B. Filho, Tetrahedron Lett., 24 (46), 5043 (1983).

JESROMOTETROL

Source: *Palafoxia rosea* Dominguez
Mol. formula: $C_{20}H_{34}O_4$
Mol. wt.: 338
Solvent: $CDCl_3$-C_5D_5N (9:1)

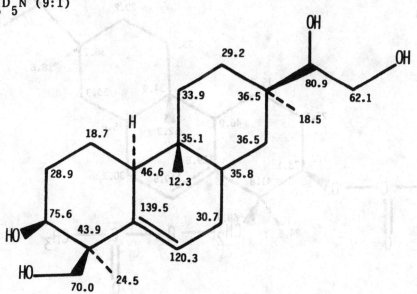

W.B. Smith, J. Org. Chem., 44 (10), 1631 (1979).

EPOXICORCOVADIN (ACETATE)

Source: *Vellozia candida* Mikan
Mol. formula: $C_{22}H_{30}O_4$
Mol. wt.: 358
Solvent: $CDCl_3$

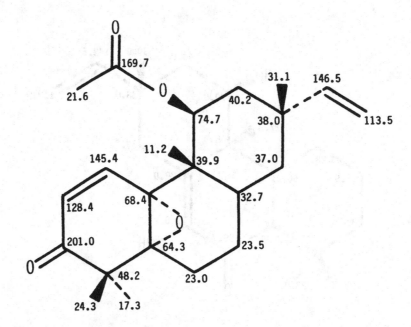

A.C. Pinto, T.C.V. Scofield and R.B. Filho, <u>Tetrahedron Lett.</u>, **24** (46), 5043 (1983).

||||||||||||||||||||||||||||||||||||||||||||||||||||||||||||||||||||||||||||||||||||||||||

3β,19-DIACETOXYJESROMOTETROL

Source: *Palafoxia texana* DC.
Mol. formula: $C_{24}H_{38}O_6$
Mol. wt.: 422
Solvent: $CDCl_3$

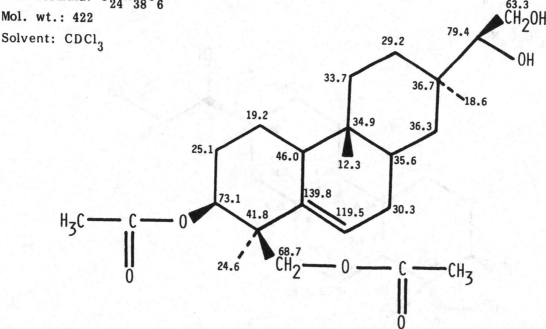

A.G. Gonzalez, J.J. Mendoza, J.G. Luis, A.G. Ravelo, X.A. Dominguez and G. Cano, <u>Phytochemistry</u>, **24** (12), 3056 (1985).

LANUGON

Basic skeleton:

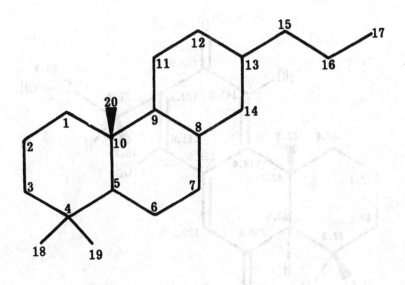

||||||||||||||||||||||||||||||||||||||||||||||||||||||||||||||||||||||||||||||||||||||||||||||||||||||||||||||||||||||||||||||||||||||||||||||||||

LANUGON M

Source: *Plectranthus laguinosus*
Mol. formula: $C_{20}H_{24}O_3$
Mol. wt.: 312
Solvent: $CDCl_3$

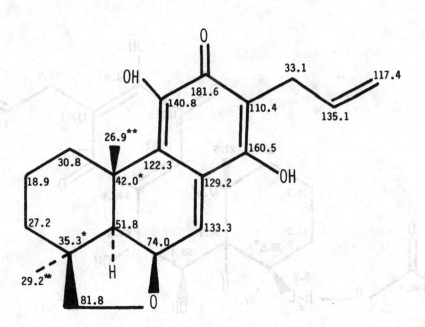

J.M. Schmid, P. Ruedi and C.H. Eugster, <u>Helv. Chim. Acta</u>, 65 (7), 2136 (1982).

LANUGON O

Source: *Plectranthus lanuginosus*
Mol. formula: $C_{20}H_{26}O_5$
Mol. wt.: 346
Solvent: $(CD_3)_2CO$

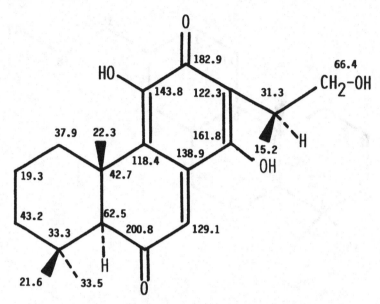

J.M. Schmid, P. Ruedi and C.H. Eugster, <u>Helv. Chim. Acta</u>, 65 (7), 2136 (1982).

LANUGON D

Source: *Plectranthus lanuginosus*
Mol. formula: $C_{21}H_{26}O_7$
Mol. wt.: 390
Solvent: $CDCl_3$

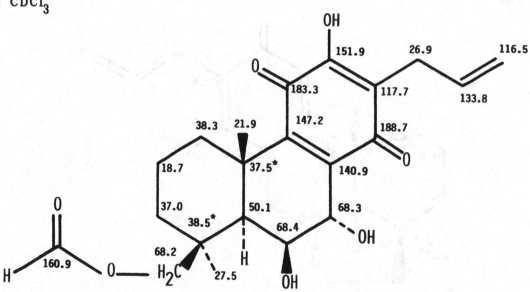

J.M. Schmid, P. Ruedi and C.H. Eugster, <u>Helv. Chim. Acta</u>, 65(7), 2136 (1982).

LANUGON E

Source: *Plectranthus lanuginosus*
Mol. formula: $C_{22}H_{30}O_6$
Mol. wt.: 390
Solvent: $CDCl_3$

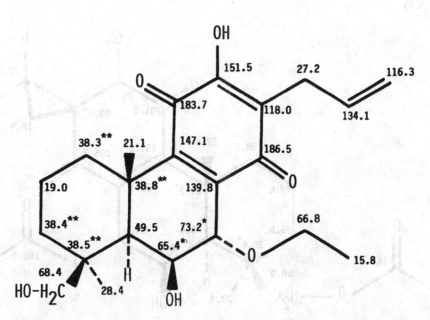

J.M. Schmid, P. Ruedi and C.H. Eugster, <u>Helv. Chim. Acta</u>, 65(7), 2136 (1982).

LANUGON P

Source: *Plectranthus lanuginosus*
Mol. formula: $C_{21}H_{26}O_7$
Mol. wt.: 390
Solvent: $CDCl_3$

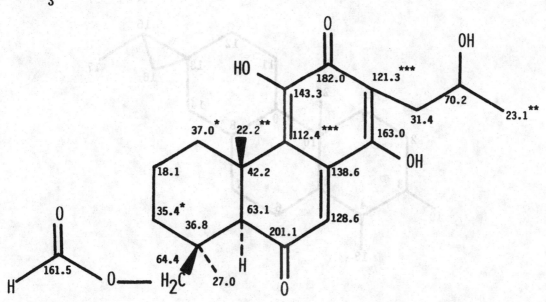

J.M. Schmid, P.Ruedi and C.H. Eugster, <u>Helv. Chim. Acta</u>, 65(7), 2136 (1982).

Source: *Plectranthus lanuginosus*
Mol. formula: $C_{22}H_{28}O_8$
Mol. wt.: 420
Solvent: $(CD_3)_2CO$

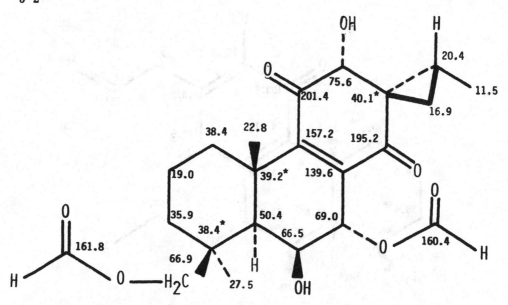

J.M. Schmid, P. Ruedi and C.H. Eugster, <u>Helv. Chim. Acta</u>, 65(7), 2136 (1982).

COLEON

Basic skeleton:

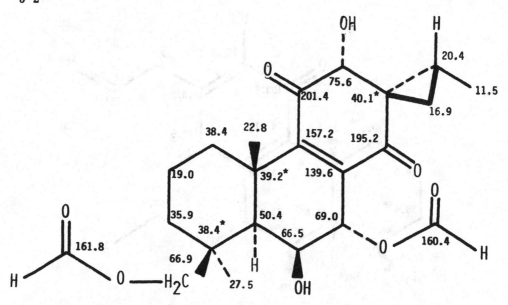

Source: *Coleus fredericii* G. Tayl.

Mol. formula: $C_{20}H_{22}O_5$

Mol. wt.: 342

Solvent: $CDCl_3$

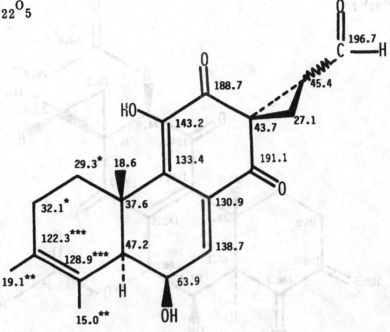

Z.-Y. Zhu, H. Nayeshiro, R. Prewo, P. Ruedi and C.H. Eugster, <u>Helv. Chim. Acta</u>, 71(3), 577 (1988).

||||||||||||||||||||||||||||||||||||||||||||||||||||||||||||||||||||||||||||||||||||||||||||||||||||||||||||||||||||||||||||||||||||

12β-O-ACETYL-7-O-FORMYL-7-O-DESACETYLCOLEON Z

Source: *Solenostemon monostachys* (P.Beauv.) Briq.

Mol. formula: $C_{23}H_{26}O_7$

Mol. wt.: 414

Solvent: $CDCl_3$

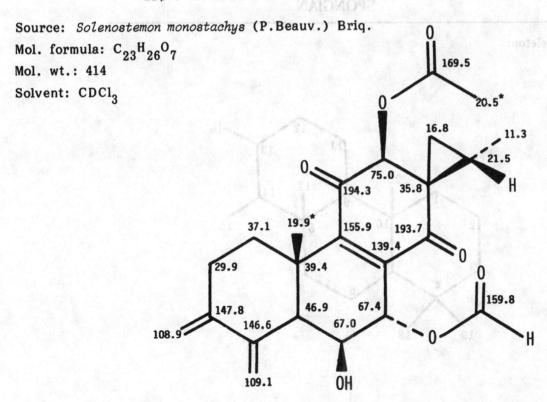

T. Miyase, P. Ruedi and C.H. Eugster, <u>Helv. Chim. Acta</u>, 63(1), 95 (1980).

411

Source: *Solenostemon monostachys* (P. Beauv.) Briq.

Mol. formula: $C_{24}H_{28}O_7$

Mol. wt.: 428

Solvent: $CDCl_3$

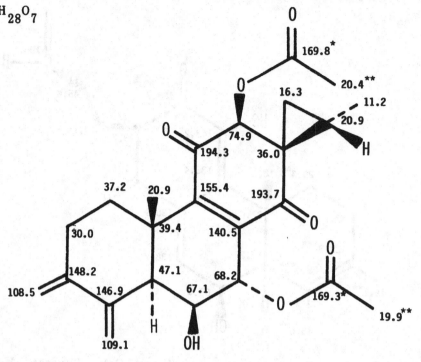

T. Miyase, P. Ruedi and C.H. Eugster, Helv. Chim. Acta , 63(1), 95 (1980).

||||||||||||||||||||||||||||||||||||||||||||||||||||||||||||||||||||||||||||||||||||||||||||||||||

SPONGIAN

Basic skeleton:

SPONGIA-13(16),14-DIENE

Source: *Hyatella intestinalis* (Lamarck)

Mol. formula: $C_{20}H_{30}O$

Mol. wt.: 286

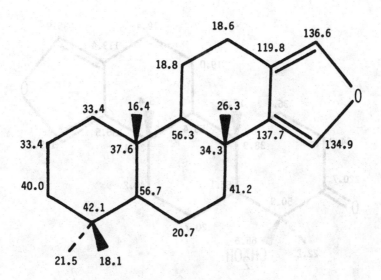

R.C. Cambie, P.A. Craw, M.J. Stone and P.R. Bergquist, <u>J. Nat. Prod.</u>, (Lloydia), 51 (2), 293 (1988).

||||||||||||||||||||||||||||||||||||||||||||||||||||||||||||||||||||||||||||||||||||||||||||||||||||||||||||||||||||||

POLYRHAPHIN D

Source: *Aplysilla polyrhaphis*

Mol. formula: $C_{20}H_{30}O_2$

Mol. wt.: 302

Solvent: $CDCl_3$

CH_3: 12.2, 16.2, 21.5 and 33.3

CH_2: 18.2, 18.6, 21.0, 27.2, 40.4, 40.5 and 41.9

CH: 54.7, 56.6, 91.8 and 112.8

C: 33.4, 37.8, 42.0 and 169.3

C=O: 169.3

S.C. Bobzin and D.J. Faulkner, <u>J. Org. Chem.</u>, 54 (16), 3902 (1989).

19-HYDROXYSPONGIA-13(16),14-DIEN-3-ONE

Source: *Hyatella intestinalis* (Lamarck)

Mol. formula: $C_{20}H_{28}O_3$

Mol. wt.: 316

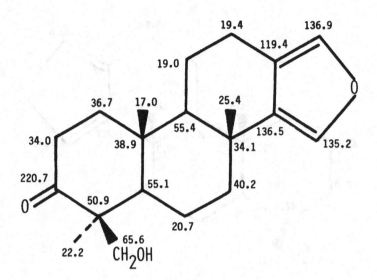

R.C. Cambie, P.A. Craw, M.J. Stone and P.R. Bergquist, J. Nat. Prod., (Lloydia), 51 (2), 293 (1988).

||||||||||||||||||||||||||||||||||||||||||||||||||||||||||||||||||||||||||||||||||||||||||||||||||||||||||||||||

DITERPENE CB-VII

Source: *Ceratosoma brevicaudatum* (Abraham)

Mol. formula: $C_{20}H_{30}O_3$

Mol. wt.: 318

Solvent: $CDCl_3$

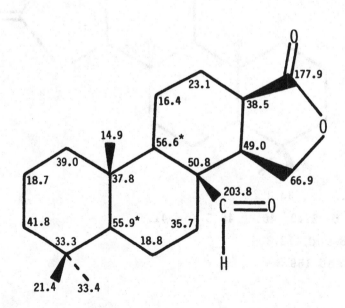

M.B. Ksebati and F.J. Schmitz, J. Org. Chem., 52 (17), 3766 (1987).

2α,19-DIHYDROXYSPONGIA-13(16),14-DIEN-3-ONE

Source: *Hyatella intestinalis* (Lamarck)

Mol. formula: $C_{20}H_{28}O_4$

Mol. wt.: 332

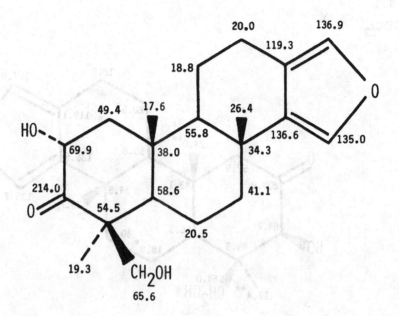

R.C. Cambie, P.A. Craw, M.J. Stone and P.R. Bergquist, <u>J. Nat. Prod.</u>, (Lloydia), 51 (2), 293 (1988).

||||||||||||||||||||||||||||||||||||||||||||||||||||||||||||||||||||||||||||||||||||||||||||||||||||||||||||||||||||||||||||||||||||||

3β,19-DIHYDROXYSPONGIA-13(16),14-DIEN-2-ONE

Source: *Hyatella intestinalis* (Lamarck)

Mol. formula: $C_{20}H_{28}O_4$

Mol. wt.: 332

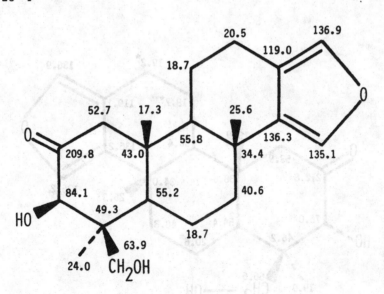

R.C. Cambie, P.A. Craw, M.J. Stone and P.R. Bergquist, <u>J. Nat. Prod.</u>, (Lloydia), 51 (2), 293 (1988).

EPISPONGIADIOL

Source: *Spongia* sp.

Mol. formula: $C_{20}H_{28}O_4$

Mol. wt.: 332

Solvent: $CDCl_3$

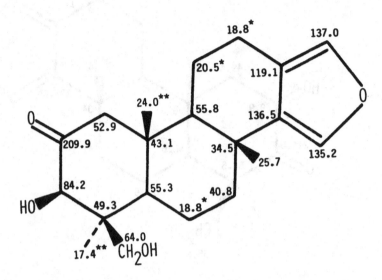

R. Kazlauskas, P.T. Murphy, R.J. Wells, K. Noack, W.E. Oberhansli and P. Schonholzer, Aust. J. Chem., 32, 867 (1979).

SPONGIADIOL

Source: *Spongia* sp.

Mol. formula: $C_{20}H_{28}O_4$

Mol. wt.: 332

Solvent: $CDCl_3$

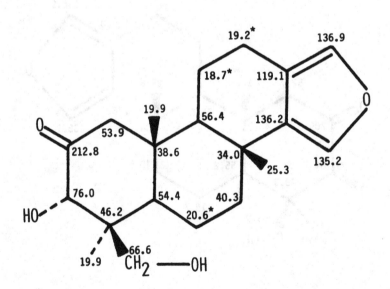

R. Kazlauskas, P.T. Murphy, R.J. Wells, K. Noack, W.E. Oberhansli and P. Schonholzer, Aust. J. Chem., 32, 867 (1979).

DENDRILLOL-1

Source: *Dendrilla rosea*

Mol. formula: $C_{20}H_{30}O_4$

Mol. wt.: 334

Solvent: $CDCl_3$

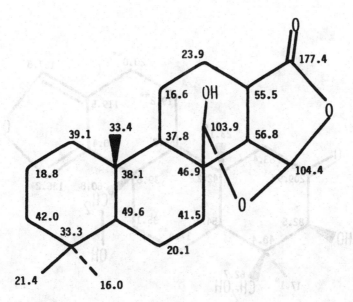

P. Karuso, P.R. Bergquist, R.C. Cambie, J.S. Buckleton, G.R. Clark and C.E.F. Rickard, Aust. J. Chem., **39**(10), 1643 (1986).

||||||||||||||||||||||||||||||||||||||||||||||||||||||||||||||||||||||||||||||||||||||||||||||||||||||

DITERPENE CB-VI

Source: *Ceratosoma brevicaudatum* (Abraham)

Mol. formula: $C_{20}H_{30}O_4$

Mol. wt.: 334

Solvent: $CDCl_3$

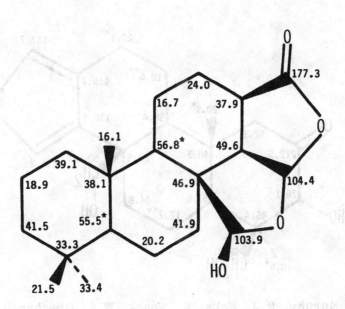

M.B. Ksebat and F.J. Schmitz, J. Org. Chem., **52** (17), 3766 (1987).

EPISPONGIATRIOL

Source: *Spongia* sp.

Mol. formula: $C_{20}H_{28}O_5$

Mol. wt.: 348

Solvent: CD_3SOCD_3

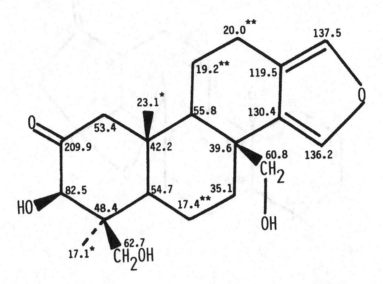

R. Kazlauskas, P.T. Murphy, R.J. Wells, K. Noack, W.E. Oberhansli and P. Schonholzer,
Aust. J. Chem., 32, 867 (1979).

||||||||||||||||||||||||||||||||||||||||||||||||||||||||||||||||||||||||||||||||||||||||

SPONGIATRIOL

Source: *Spongia* sp.

Mol. formula: $C_{20}H_{28}O_5$

Mol. wt.: 348

Solvent: CD_3SOCD_3

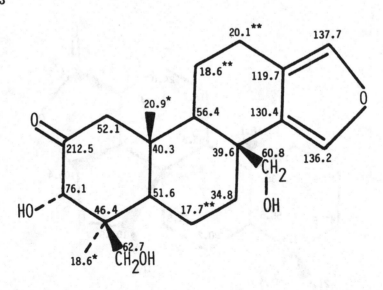

R. Kazlauskas, P.T. Murphy, R.J. Wells, K. Noack, W.E. Oberhansli and P. Schonholzer,
Aust. J. Chem., 32, 867 (1979).

DENDRILLOL-III

Source: *Dendrilla rosea*
Mol. formula: $C_{21}H_{32}O_4$
Mol. wt.: 348
Solvent: $CDCl_3$

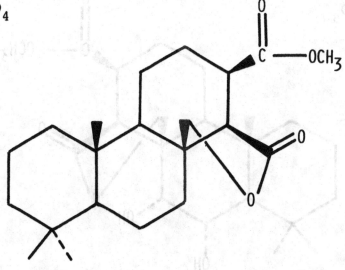

14.6, 17.3, 18.3, 18.9, 21.5, 22.0, 33.1, 33.4, 37.8, 38.4, 39.1, 41.0, 41.8, 42.6, 49.9, 51.1, 51.9, 56.2, 72.9, 173.3 and 177.3

P. Karuso, P.R. Bergquist, R.C. Cambie, J.S. Buckleton, G.R. Clark and C.E.F. Rickard, Aust. J. Chem., 39(10), 1643 (1986).

|||||||||||||||||||||||||||||||||||||||||||||||||||||||||||||||||||||||||||||||||||||||||||||||||||||||||||||||

3β-ACETOXY-19-HYDROXYSPONGIA-13(16),14-DIEN-2-ONE

Source: *Hyatella intestinalis* (Lamarck)
Mol. formula: $C_{22}H_{30}O_5$
Mol. wt.: 374

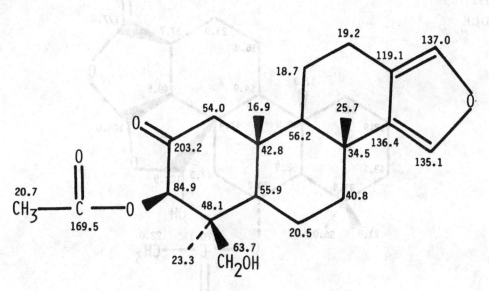

R.C. Cambie, P.A. Craw, M.J. Stone and P.R. Bergquist, J. Nat. Prod., (Lloydia), 51 (2), 293 (1988).

DENDRILLOL-IV

Source: *Dendrilla rosea*

Mol. formula: $C_{21}H_{32}O_6$

Mol. wt.: 380

Solvent: $CDCl_3$

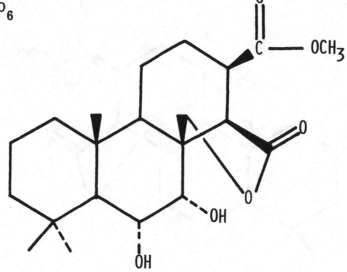

15.3, 16.9, 18.2, 21.6, 22.0, 33.0, 36.5, 38.4, 38.6, 39.1, 42.3, 43.4, 45.0, 47.3, 51.3, 52.1, 69.6, 72.2, 75.3, 173.5 and 177.7

P. Karuso, P.R. Bergquist, R.C. Cambie, J.S. Buckleton, G.R. Clark and C.E.F. Rickard, Aust. J. Chem., **39**(10), 1643 (1986).

||||||||||||||||||||||||||||||||||||||||||||||||||||||||||||||||||||||||||||||||||||||||||||||||||||||||||||||||||||||||||||||||

DITERPENE CB-III

Source: *Ceratosoma brevicaudatum* (Abraham)

Mol. formula: $C_{22}H_{32}O_6$

Mol. wt.: 392

Solvent: $CDCl_3$

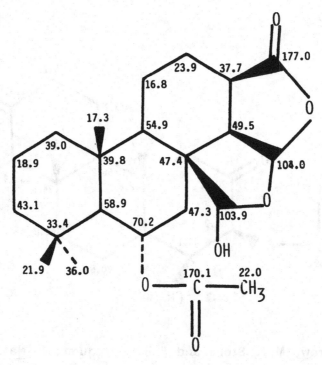

M.B. Ksebati and F.J. Schmitz, J. Org. Chem., 52 (17), 3766 (1987).

3β,19-DIACETOXYSPONGIA-13(16),14-DIEN-2-ONE

Source: *Hyatella intestinalis* (Lamarck)

Mol. formula: $C_{24}H_{32}O_6$

Mol. wt.: 416

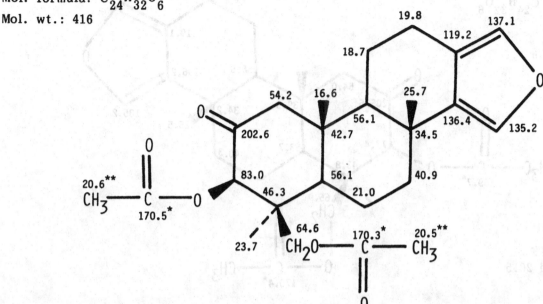

R.C. Cambie, P.A. Craw, M.J. Stone and P.R. Bergquist, <u>J. Nat. Prod.</u>, (Lloydia), 51 (2), 293 (1988).

||||||||||||||||||||||||||||||||||||||||||||||||||||||||||||||||||||||||||||||||||||||||||||||||||||||||

EPISPONGIADIOL DIACETATE

Source: *Spongia* sp.

Mol. formula: $C_{24}H_{32}O_6$

Mol. wt.: 416

Solvent: $CDCl_3$

CH_3: 16.6, 20.5, 20.5 and 23.8

R. Kazlauskas, P.T. Murphy, R.J. Wells, K. Noack, W.E. Oberhansli and P. Schonholzer, <u>Aust. J. Chem.</u>, 32, 867 (1979).

SPONGIADIOL DIACETATE

Source: *Spongia* sp.

Mol. formula: $C_{24}H_{32}O_6$

Mol. wt.: 416

Solvent: $CDCl_3$

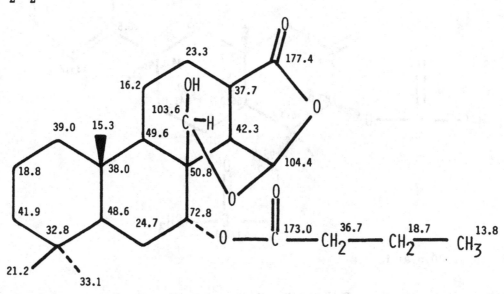

CH$_3$: 19.3 and 20.5
CH$_2$: 19.2
19.8, 20.7, 20.8 and 21.2

R. Kazlauskas, P.T. Murphy, R.J. Wells, K. Noack, W.E. Oberhansli and P. Schonholzer, Aust. J. Chem., 32, 867 (1979).

|||||||||||||||||||||||||||||||||||||||||||||||||||||||||||||||||||||||||||||||||||||||||||||||||||||||||||||

APLYROSEOL-I

Source: *Aplysilla rosea* Barrois

Mol. formula: $C_{24}H_{36}O_6$

Mol. wt.: 420

Solvent: CD_2Cl_2

P. Karuso and W.C. Taylor, Aust. J. Chem., 39(10), 1629 (1986).

7α-17β-DIHYDROXY-15,17-OXIDOSPONGIAN-16-ONE-7-BUTYRATE

Source: *Igernella notabilis* (Duch and Mich.)

Mol. formula: $C_{24}H_{36}O_6$

Mol. wt.: 420

Solvent: $CDCl_3$

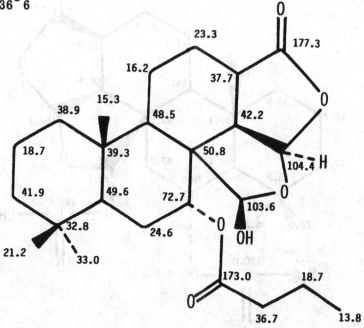

F.J. Schmitz, J.S. Chang, M.B. Hossain and D.V.Der Helm, <u>J. Org. Chem.</u>, **50** (16), 2862 (1985)

||||||||||||||||||||||||||||||||||||||||||||||||||||||||||||||||||||||||||||||||||||||||||||||||||||||||||||||||||

(5R*,7S*,8S*,9S*,10R*,13S*,14S*)-16-OXOSPONGIAN-7,17-DIYL DIACETATE

Source: *Aplysilla rosea* Barrois

Mol. formula: $C_{24}H_{36}O_6$

Mol. wt.: 420

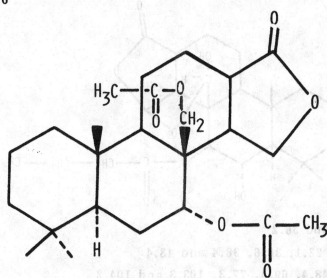

16.3, 17.0, 18.4, 21.0, 21.3, 22.0, 23.3, 29.7, 32.7, 32.9, 37.5, 37.6, 39.9, 40.8, 41.7, 47.5, 51.9, 62.8, 67.7, 70.1, 170.0, 170.3 and 178.8

P. Karuso and W.C. Taylor, <u>Aust. J. Chem.</u>, **39** (10), 1629 (1986).

423

DITERPENE CB-IV

Source: *Ceratosoma brevicaudatum* (Abraham)

Mol. formula: $C_{24}H_{34}O_7$

Mol. wt.: 434

Solvent: $CDCl_3$

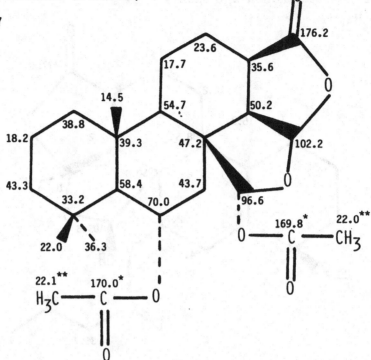

M.B. Ksebati and F.J. Schmitz, J. Org. Chem., 52 (17), 3766 (1987).

|||||||||||||||||||||||||||||||||||||||||||||||||||||||||||||||||||||||||||||||||||||||||||||||||||||

APLYROSEOL-III

Source: *Aplysilla rosea* Barrois

Mol. formula: $C_{24}H_{36}O_7$

Mol. wt.: 436

CH_3: 13.6, 16.3, 21.6 and 36.2

CH_2: 16.3, 16.7, 18.6, 23.1, 36.6, 38.4 and 43.4

CH: 37.5, 42.6, 43.3, 48.4, 69.4, 77.3, 103.3 and 104.2

C: 33.1, 38.9 and 51.2

C=O: 174.7 and 177.6

P. Karuso and W.C. Tylor, Aust. J. Chem., 39 (10), 1629 (1986).

APLYROSEOL-V

Source: *Aplysilla rosea* Barrois
Mol. formula: $C_{24}H_{36}O_7$
Mol. wt.: 436
Solvent: CD_2Cl_2

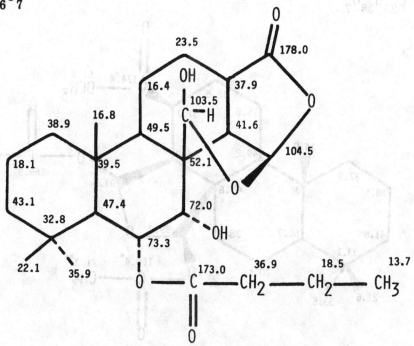

P. Karuso and W.C. Taylor, <u>Aust. J. Chem.</u>, 39(10), 1629 (1986).

||||||||||||||||||||||||||||||||||||||||||||||||||||||||||||||||||||||||||||||||||||||||||||||||||||||||||||||||||||||

DENDRILLOL-II

Source: *Dendrilla rosea*
Mol. formula: $C_{24}H_{34}O_8$
Mol. wt.: 450
Solvent: $CDCl_3$

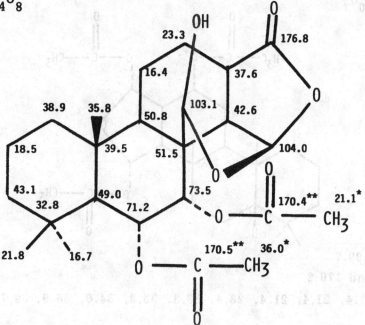

P. Karuso, P.R. Bergquist, R.C. Cambie, J.S. Buckleton, G.R. Clark and C.E.F. Rickard, <u>Aust. J. Chem.</u>, 39(10), 1643 (1986).

DITERPENE CB-X

Source: *Ceratosoma brevicaudatum* (Abraham)

Mol. formula: $C_{25}H_{38}O_7$

Mol. wt.: 450

Solvent: $CDCl_3$

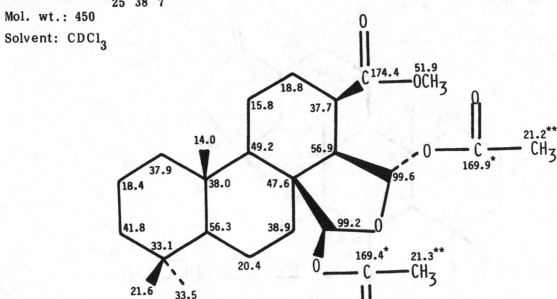

M.B. Ksebati and F.J. Schmitz, <u>J. Org. Chem.</u>, **52** (17), 3766 (1987).

||||||||||||||||||||||||||||||||||||||||||||||||||||||||||||||||||||||||||||||||||||||||||||||||||||||||||||||||||||||||||||||||||||||

APLYSILLIN

Source: *Aplysilla rosea*

Mol. formula: $C_{26}H_{40}O_7$

Mol. wt.: 464

Solvent: $CDCl_3$

CH: 67.8, 99.7 and 99.7

C=O: 169.7, 170.0 and 170.2

16.2, 16.9, 18.2, 21.4, 21.4, 21.4, 23.4, 33.3, 33.3, 34.6, 36.9, 39.7, 41.9, 42.6, 45.5, 49.9, 56.7 and 57.1

R. Kazlauskas, P.T. Murphy, R.J. Wells and J.J. Daly, <u>Tetrahedron Lett.</u>, (10), 903 (1979).

EPISPONGIATRIOL TRIACETATE

Source: *Spongia* sp.

Mol. formula: $C_{26}H_{34}O_8$

Mol. wt.: 474

Solvent: $CDCl_3$

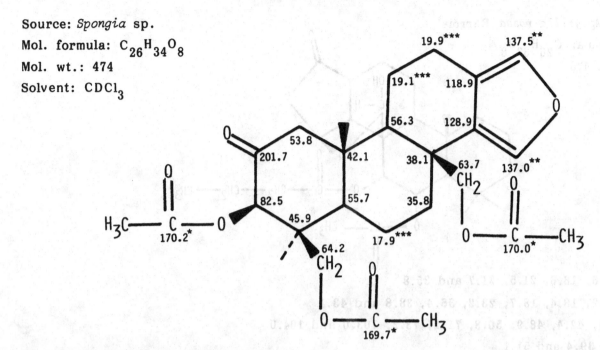

CH_3: 16.6, 20.5, 20.5 and 23.4

R. Kazlauskas, P.T. Murphy, R.J. Wells, K. Noack, W.E. Oberhansli and P. Schonholzer, Aust. J. Chem., 32, 867 (1979).

||||||||||||||||||||||||||||||||||||||||||||||||||||||||||||||||||||||||||||||||||||||||||||||||||||||||||||||

SPONGIATRIOL TRIACETATE

Source: *Spongia* sp.

Mol. formula: $C_{26}H_{34}O_8$

Mol. wt.: 474

Solvent: $CDCl_3$

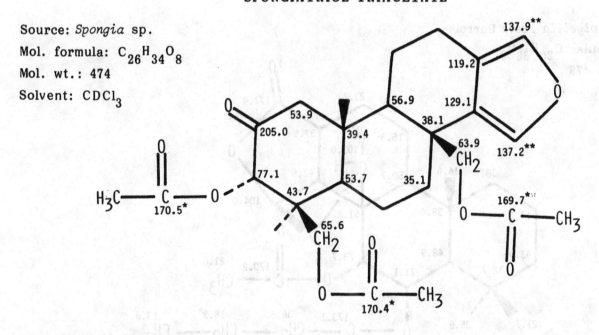

CH_3: 20.5, 20.9 and 21.2

18.8, 19.3 and 19.9

R. Kazlauskas, P.T. Murphy, R.J. Wells, K. Noack, W.E. Oberhansli and P. Schonholzer, Aust. J. Chem., 32, 867 (1979).

APLYROSEOL-IV

Source: *Aplysilla rosea* Barrois
Mol. formula: $C_{26}H_{38}O_8$
Mol. wt.: 478

CH$_3$: 13.6, 16.6, 21.5, 21.7 and 35.8
CH$_2$: 16.2, 18.4, 18.7, 23.2, 36.4, 38.8 and 43.1
CH: 37.5, 42.4, 48.9, 50.8, 71.2, 73.2, 103.0 and 104.0
C: 32.7, 39.4 and 51.1
C=O: 170.8, 173.2 and 177.6

P. Karuso and W.C. Taylor, <u>Aust. J. Chem.</u>, 39(10), 1629 (1986).

||||||||||||||||||||||||||||||||||||||||||||||||||||||||||||||||||||||||||||||||||||||||||||||||||||||||||||||||||||||||||

APLYROSEOL-VI

Source: *Aplysilla rosea* Barrois
Mol. formula: $C_{26}H_{38}O_8$
Mol. wt.: 478

P. Karuso and W.C. Taylor, <u>Aust. J. Chem.</u>, 39(10), 1629 (1986).

DITERPENE CB-VIII

Source: *Ceratosoma brevicaudatum* (Abraham)
Mol. formula: $C_{27}H_{40}O_9$
Mol. wt.: 508
Solvent: $CDCl_3$

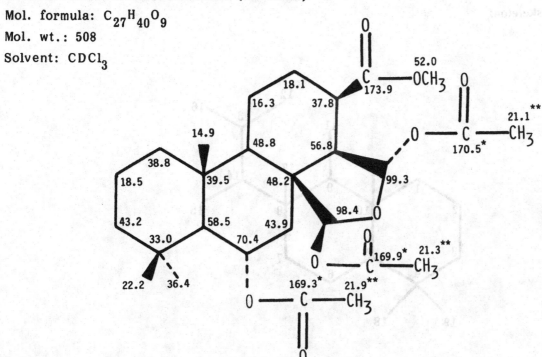

M.B. Ksebati and F.J. Schmitz, <u>J. Org. Chem.</u>, 52 (17), 3766 (1987).

||||||||||||||||||||||||||||||||||||||||||||||||||||||||||||||||||||||||||||||||||||||||||||

DITERPENE CB-IX

Source: *Ceratosoma brevicaudatum* (Abraham)
Mol. formula: $C_{29}H_{44}O_9$
Mol. wt.: 536
Solvent: $CDCl_3$

M.B. Ksebati and F.J. Schmitz, <u>J. Org. Chem.</u>, 52 (17), 3766 (1987).

Basic skeleton:

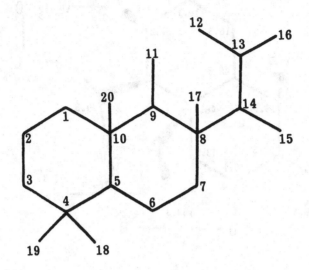

||||||||||||||||||||||||||||||||||||||||||||||||||||||||||||||||||||||||||||||||||||||||||||||||||||||||||||||||||||||||||||||||||||||||

GRACILIN F

Source: *Spongionella gracilis*
Mol. formula: $C_{19}H_{30}O_2$
Mol. wt.: 290
Solvent: $CDCl_3$

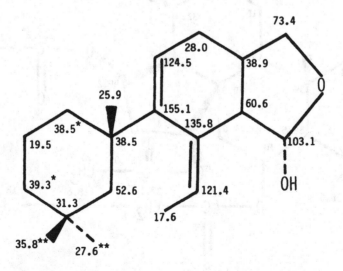

L. Mayol, V. Piccialli and D. Sica, <u>Tetrahedron</u>, **42** (19), 5369 (1986).

MEMBRANOLIDE

Source: *Dendrilla membranosa*
Mol. formula: $C_{21}H_{28}O_4$
Mol. wt.: 344
Solvent: $CDCl_3$

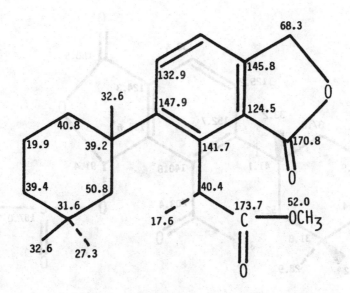

T.F. Molinski and D.J. Faulkner, J. Org. Chem., 52 (2), 296 (1987).

||||||||||||||||||||||||||||||||||||||||||||||||||||||||||||||||||||||||||||||||||||||||||||||||||||||||||||||||||||||||||||

SPONGIALACTONE A

Source: *Spongia arabica*
Mol. formula: $C_{20}H_{26}O_5$
Mol. wt.: 346
Solvent: $CDCl_3$

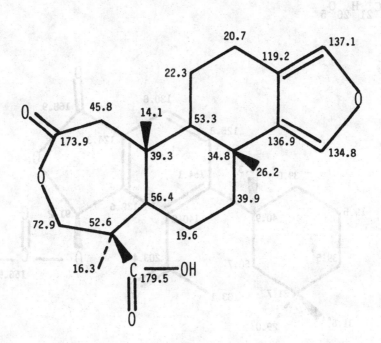

S. Hirsch and Y. Kashman, J. Nat. Prod., (Lloydia), 51 (6), 1243 (1988).

MACFARLANDIN A

Source: *Chromodoris macfarlandi*
Mol. formula: $C_{21}H_{26}O_5$
Mol. wt.: 358
Solvent: $CDCl_3$

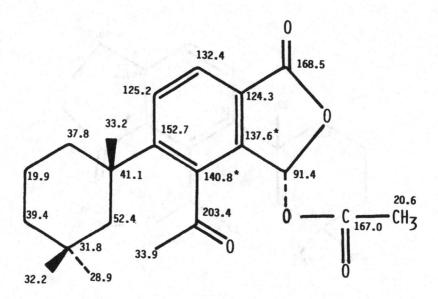

T.F. Molinski and D.J. Faulkner, <u>J. Org. Chem.</u>, 51(13), 2601 (1986).

MACFARLANDIN B

Source: *Chromodoris macfarlandi*
Mol. formula: $C_{21}H_{26}O_5$
Mol. wt.: 358
Solvent: $CDCl_3$

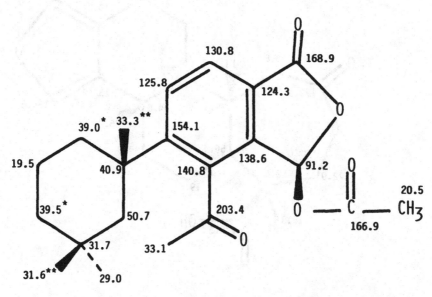

T.F. Molinski and D.J. Faulkner, <u>J. Org. Chem.</u>, 51(13), 2601 (1986).

APLYSULPHURIN

Source: *Aplysilla sulphurea*
Mol. formula: $C_{22}H_{28}O_5$
Mol. wt.: 372
Solvent: $CDCl_3$

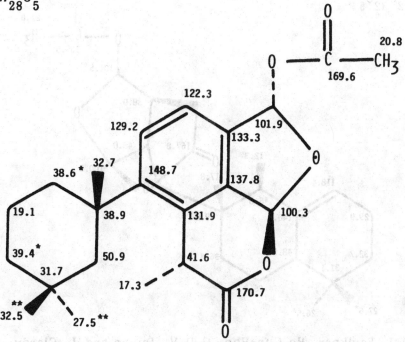

T.F. Molinski and D.J. Faulkner, J. Org.Chem., 51(13), 2601 (1986).

MACFARLANDIN C

Source: *Chromodoris macfarlandi*
Mol. formula: $C_{22}H_{32}O_5$
Mol. wt.: 376
Solvent: $CDCl_3$

T.F. Molinski, D.J. Faulkner, He Cun-heng, G.D.V. Duyne and J. Clardy, J. Org. Chem., **51** (24), 4564 (1986).

MACFARLANDIN D

Source: *Chromodoris macfarlandi*

Mol. formula: $C_{22}H_{32}O_5$

Mol. wt.: 376

Solvent: $CDCl_3$

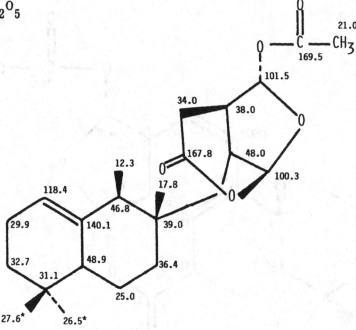

T.F. Molinski, D.J. Faulkner, He Cun-Hen, G.D.V. Duyne and J. Clardy, <u>J. Org. Chem.</u>, 51 (24), 4564 (1986).

POLYRHAPHIN C

Source: *Aplysilla polyrhaphis*

Mol. formula: $C_{22}H_{32}O_5$

Mol. wt.: 376

Solvent: $CDCl_3$

CH_3: 20.3, 21.0, 26.6 and 30.5

CH_2: 12.9, 17.6, 21.4, 33.5, 36.7, 38.8 and 42.5

CH: 25.7, 37.8, 46.9, 51.2, 100.8 and 101.4

C: 22.9, 31.7 and 34.6

$C=O$: 167.8 and 169.4

S.C. Bobzin and D.J. Faulkner, <u>J. Org. Chem.</u>, 54 (16), 3902 (1989).

434

TETRAHYDROAPLYSULPHURIN-I

Source: *Darwinella oxeata* Bergquist
Mol. formula: $C_{22}H_{32}O_5$
Mol. wt.: 376
Solvent: $CDCl_3$

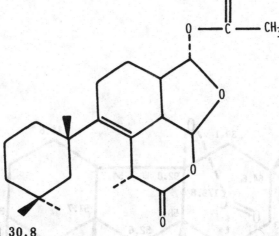

CH_3: 14.6, 20.9, 28.1 and 30.8
CH_2: 20.5, 23.8, 24.8, 38.8, 39.5 and 50.7
CH: 37.8, 40.5, 42.3, 100.2 and 102.5
C: 31.4, 39.3, 121.1 and 146.1
C=O: 169.6 and 170.7

P. Karuso, P.R. Bergquist, R.C. Cambie, J.S. Buckleton, G.R. Clark and C.E. F. Rickard, <u>Aust. J. Chem.</u>, 39(10), 1643 (1986).

||||||||||||||||||||||||||||||||||||||||||||||||||||||||||||||||||||||||||||||||||||||||||||||||||||||||||

GIBBANE

Basic skeleton:

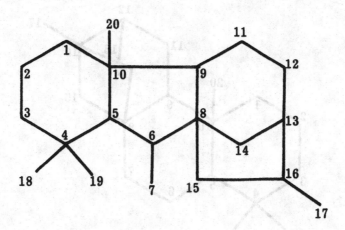

435

GIBBERELLIN A$_{40}$

Source: *Gibberella fujikuroi*

Mol. formula: $C_{19}H_{24}O_5$

Mol. wt.: 332

Solvent: C_5D_5N

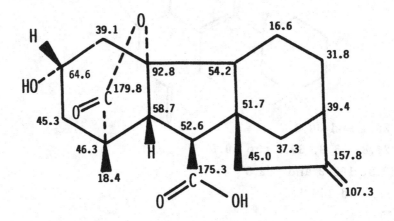

I. Yamaguchi, M. Miyamoto, H. Yamane, N. Murofushi, N. Takahashi and K. Fujita, J. Chem. Soc. Perkin Trans. I, 996 (1975).

TRACHYLOBANE

Basic skeleton:

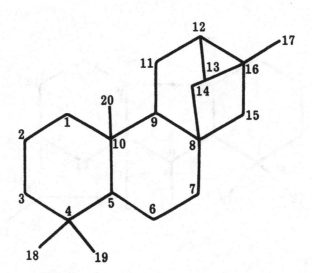

CILIARIC ACID

Source: *Helianthus niveus* (A.Gray) Heisev
Mol. formula: $C_{20}H_{30}O_3$
Mol. wt.: 318
Solvent: C_5D_5N

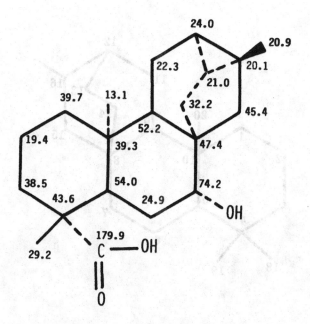

N. Ohno and T.J. Mabry, <u>Phytochemistry</u>, 19(4) 609 (1980).

<u>ENT</u>-TRACHYLOBAN-3β-ACETOXY

Source: *Xylopia aromatica* Lam. (Mart.)
Mol. formula: $C_{22}H_{34}O_2$
Mol. wt.: 330
Solvent: $CDCl_3$

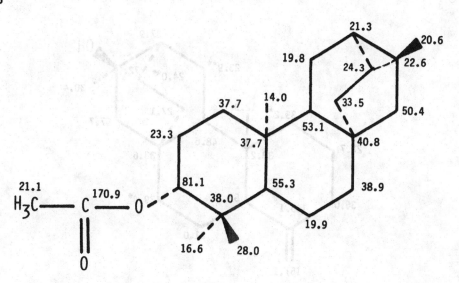

M.P.L. Moraes and N.F. Roque, <u>Phytochemistry</u>, 27 (10), 3205 (1988).

Basic skeleton:

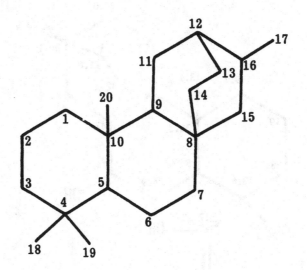

ENT-19-NOR-ATISAN-4(18)-EN-16α-OL

Source: *Xylopia aromatica* Lam. (Mart.)

Mol. formula: $C_{19}H_{30}O$

Mol. wt.: 274

Solvent: $CDCl_3$

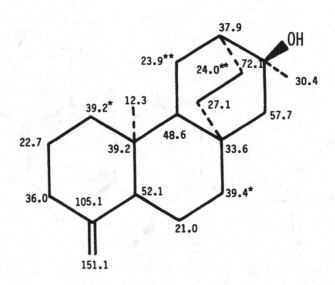

M.P.L. Moraes and N.F. Roque, <u>Phytochemistry</u>, 27(10), 3205 (1988).

DITERPENE HA-VII

Source: *Helianthus annus*
Mol. formula: $C_{20}H_{34}O$
Mol. wt.: 290
Solvent: $CDCl_3$

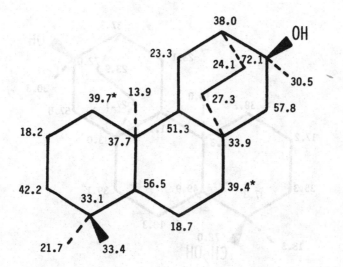

M.P.L. Moraes and N.F. Roque, <u>Phytochemistry</u>, 27(10), 3205 (1988).

||||||||||||||||||||||||||||||||||||||||||||||||||||||||||||||||||||||||||||||||||||||||||||||||||||||||||||||

<u>ENT-18-NOR-ATISAN-4β,16α-DIOL</u>

Source: *Xylopia aromatica* Lam. (Mart.)
Mol. formula: $C_{19}H_{32}O_2$
Mol. wt.: 292
Solvent: $CDCl_3$

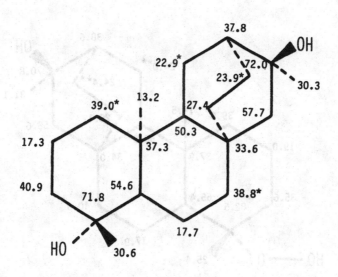

M.P.L. Moraes and N.F. Roque, <u>Phytochemistry</u>, 27(10), 3205 (1988).

ENT-ATISAN-16α,18-DIOL

Source: *Xylopia aromatica* Lam. (Mart.)
Mol. formula: $C_{20}H_{34}O_2$
Mol. wt.: 306
Solvent: $CDCl_3$

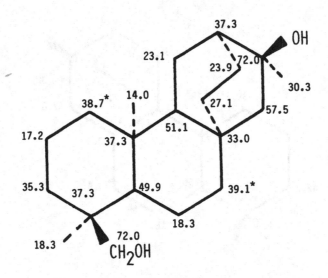

M.P.L. Moraes and N.F. Roque, <u>Phytochemistry</u>, 27(10), 3205 (1988).

||||||||||||||||||||||||||||||||||||||||||||||||||||||||||||||||||||||||||||||||||||||||||||||||||||||||||||||||||||||||||||||

ENT-18-NOR-ATISAN-4β-HYDROPEROXIDE-16α-OL

Source: *Xylopia aromatica* Lam. (Mart.)
Mol. formula: $C_{19}H_{32}O_3$
Mol. wt.: 308
Solvent: C_5D_5N

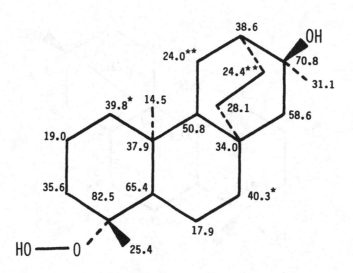

M.P.L. Moraes and N.F. Roque, <u>Phytochemistry</u>, 27(10), 3205 (1988).

3-OXOATISANE-16α,17-DIOL

Source: *Euphorbia acaulis* Roxh.

Mol. formula: $C_{20}H_{32}O_3$

Mol. wt.: 320

Solvent: $CDCl_3$

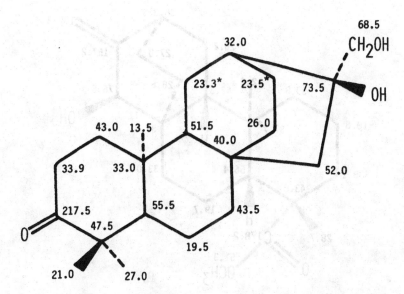

N.K. Satti, O.P. Suri, R.K. Thaper and P.L. Kachroo, Phytochemistry, 27 (5), 1530 (1988).

||||||||||||||||||||||||||||||||||||||||||||||||||||||||||||||||||||||||||||||||||||||||||||||||||||||||||||||||||||||||||||||||||||||||

ENT-ATISANE-3β,16α,17-TRIOL

Source: *Euphorbia acaulis* Roxh.

Mol. formula: $C_{20}H_{34}O_3$

Mol. wt.: 322

Solvent: C_5D_5N

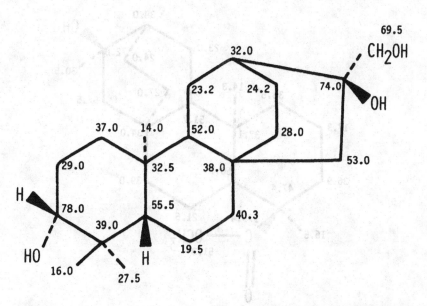

N.K. Satti, O.P. Suri, R.K. Thaper and P.L. Kachroo, Phytochemistry, 27 (5), 1530 (1988).

ENT-15α-HYDROXYATIS-16-EN-19-OIC ACID (METHYL ESTER)

Source: *Elaeoselinum foetidum* (L) Boiss.

Mol. formula: $C_{21}H_{32}O_3$

Mol. wt.: 332

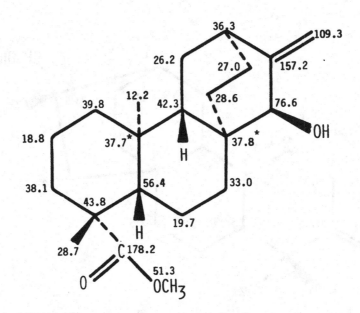

M. Pinar and M.P. Galan, <u>J. Nat. Prod.</u>, (Lloydia), **49**(2), 334 (1986).

||||||||||||||||||||||||||||||||||||||||||||||||||||||||||||||||||||||||||||||||||||||||||||||||||||||||||||||||||||||||||||

ENT-ATISAN-16α-OL-18-OIC METHYL ESTER

Source: *Xylopia aromatica* Lam. (Mart.)

Mol. formula: $C_{21}H_{34}O_3$

Mol. wt.: 334

Solvent: $CDCl_3$

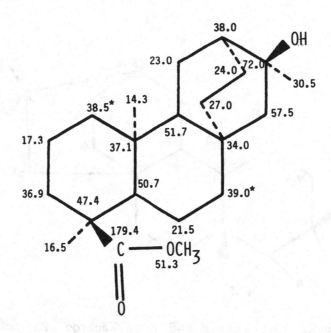

M.P.L. Moraes and N.F. Roque, <u>Phytochemistry</u>, **27**(10), 3205 (1988).

ENT –13(S)–ANGELOXYATIS–16–EN–19–OIC ACID (METHYL ESTER)

Source: *Helianthus decapetalus* Dart

Mol. formula: $C_{26}H_{38}O_4$

Mol. wt.: 414

Solvent: $CDCl_3$

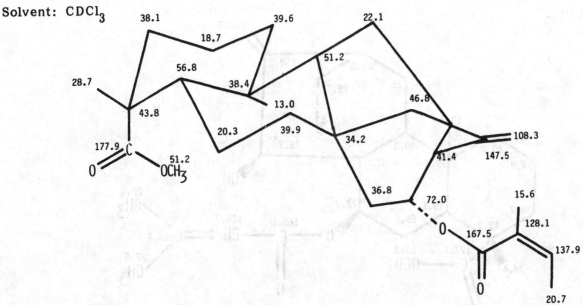

M.H. Beale, J.R. Bearder, J. MacMillan, A. Matsuo and B.O. Phinney, <u>Phytochemistry</u>, 22(4), 875 (1983).

ENT-15α-SENECIOXY-ATIS-16-EN-19-OIC ACID (METHYL ESTER)

Source: *Elaeoselinum foetidum* (L.) Boiss.

Mol. formula: $C_{26}H_{38}O_4$

Mol. wt.: 414

Solvent: $CDCl_3$

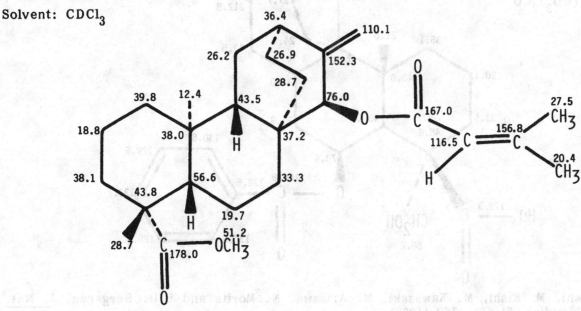

M. Pinar, M. Rico, B. Rodriguez and B. Fernandez, <u>Phytochemistry</u>, 23(1), 195 (1984).

ENT-7α-SENECIOXY-15 α-HYDROXY-ATIS-16-EN-19-OIC ACID

Source: *Elaeoselinum foetidum* (L.) Boiss.

Mol. formula: $C_{26}H_{38}O_5$

Mol. wt.: 430

Solvent: $CDCl_3$

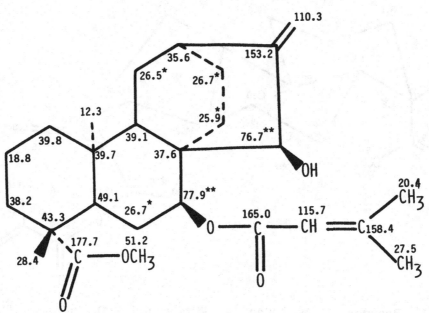

M. Pinar, <u>Phytochemistry</u>, 23 (9), 2075 (1984).

SCOPADULCIC ACID A

Source: *Scoparia dulcis* L.

Mol. formula: $C_{27}H_{34}O_6$

Mol. wt.: 454

Solvent: $(CD_3)_2CO$

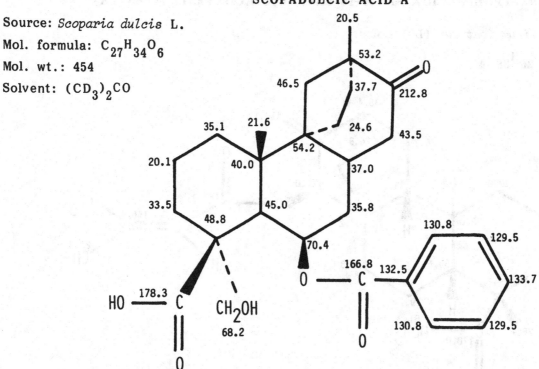

T. Hayashi, M. Kishi, M. Kawasaki, M. Arisawa, N. Morita and L.H. Berganza, <u>J. Nat. Prod.</u>, (Lloydia), 51 (2), 360 (1988).

17-ACETOXY-13α-HYDROXY-19-(3-METHYLVALERYLOXY)VILLANOVAN

Source: *Villanova titicaensis* Meyer et Walp.

Mol. formula: $C_{28}H_{46}O_5$

Mol. wt.: 462

Solvent: $CDCl_3$

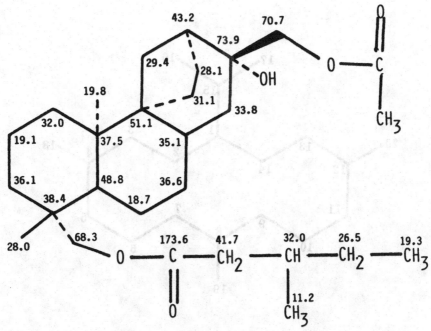

F. Bohlmann, C. Zdero, R.M. King and H. Robinson, <u>Leibigs Ann. Chem.</u>, 250 (1984).

||||||||||||||||||||||||||||||||||||||||||||||||||||||||||||||||||||||||||||||||||||||||||||||||||||||||||||||||||||||||||||||||||||

MARGOTIANIN

Source: *Margotia gummifera* (Desf.) Lange. (syn. *Elaeoselinum gummiferum* (Desf.) Tutin)

Mol. formula: $C_{28}H_{40}O_6$

Mol. wt.: 472

Solvent: $CDCl_3$

B. Rodriguez and M. Pinar, <u>Phytochemistry</u>, 18 (5), 891 (1979).

CEMBRANE

Basic skeleton:

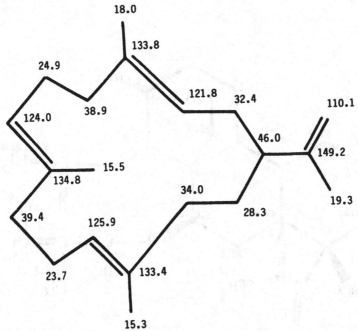

||||||||||||||||||||||||||||||||||||||||||||||||||||||||||||||||||||||||||||||||||||||||||||||||||||||||||||||||||||||||||||||||||||||||||||||||||

CEMBRENE A

Source: *Cubitermes umbratus* Williams
Mol. formula: $C_{20}H_{32}$
Mol. wt.: 272
Solvent: $CDCl_3$

D.F. Wiemer, J. Meinwald, G.D. Prestwich and I. Miura, J. Org. Chem., 44(22), 3950 (1979).

(3Z)-CEMBRENE A

Source: *Cubitermes umbratus* Williams
Mol. formula: $C_{20}H_{32}$
Mol.wt.: 272
Solvent: $CDCl_3$

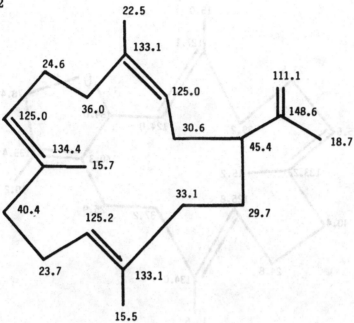

D.F. Wiemer, J. Meinwald, G.D. Prestwich and I. Miura, J. Org. Chem., **44**(22), 3950 (1979).

||||||||||||||||||||||||||||||||||||||||||||||||||||||||||||||||||||||||||||||||||||||||||||||||||||||||||||||

(3E,7E,11E)-13-OXO-3,7,11,15-CEMBRATETRAEN

Source: *Sarcophytum* sp.
Mol. formula: $C_{20}H_{30}O$
mol. wt.: 286
Solvent: $CDCl_3$

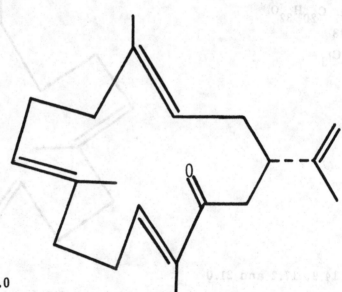

CH$_3$: 10.7, 15.7, 18.8 and 22.2
CH$_2$: 110.0
CH: 122.0, 126.5 and 141.2
C: 132.5, 134.5, 135.9 and 148.0
C=O: 191.1
26.0, 28.0, 31.7, 34.5, 37.5 and 45.6

B.N. Ravi and D.J. Faulkner, J. Org. Chem., **43** (11), 2127 (1978).

447

SARCOPHYTONIN A

Source: *Sarcophyton glaucum*

Mol. formula: $C_{20}H_{30}O$

Mol. wt.: 286

Solvent: $CDCl_3$

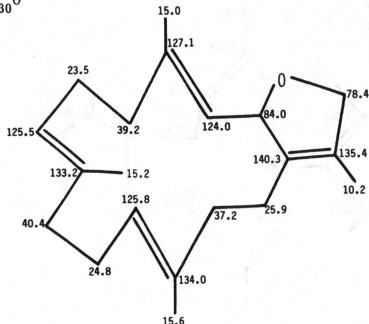

M. Kobayashi, T. Nakagawa and H. Mitsuhashi, Chem. Pharm. Bull. (Tokyo), 27 (10), 2382 (1979).

||||||||||||||||||||||||||||||||||||||||||||||||||||||||||||||||||||||||||||||||||||||||||||||||

(7E,11E)-3,4-EPOXY-7,11,15-CEMBRATRIENE

Source: *Sarcophytum* sp.

Mol. formula: $C_{20}H_{32}O$

Mol. wt.: 288

Solvent: $CDCl_3$

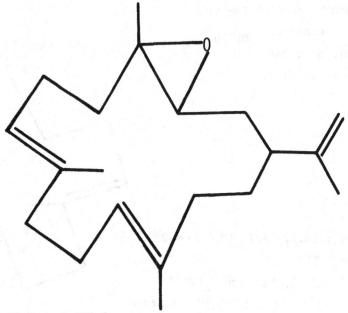

CH_3: 14.9, 14.9, 17.2 and 21.0

CH_2: 24.2, 24.4, 24.7, 29.8, 32.6, 36.3, 38.7 and 110.0

CH: 45.0, 61.0, 124.4 and 126.2

C: 61.0, 132.6, 135.0 and 147.5

B.N. Ravi and D.J. Faulkner, J. Org. Chem., 43 (11), 2127 (1978).

SARCOPHYTOL A

Source: *Sarcophyton glaucum*
Mol. formula: $C_{20}H_{32}O$
Mol. wt.: 288
Solvent: $CDCl_3$

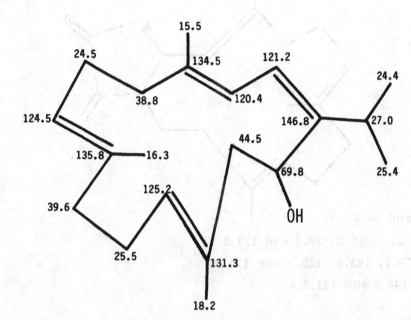

M. Kobayashi, T. Nakagawa and H. Mitsuhashi, <u>Chem. Pharm. Bull.</u> (Tokyo), 27 (10), 2382 (1979).

||||||||||||||||||||||||||||||||||||||||||||||||||||||||||||||||||||||||||||||||||||||||||||||||||||||||||||||||||||||||

DITERPENE SM-I

Source: *Sinularia mayi* Luttschw.
Mol. formula: $C_{20}H_{28}O_2$
Mol. wt.: 300
Solvent: $CDCl_3$

CH_3: 15.2, 15.2 and 15.8
CH_2: 23.5, 24.6, 27.3, 36.4, 39.8, 40.0 and 120.3
CH: 43.5, 78.2, 120.3, 124.0 and 125.6
C: 133.7, 133.7, 139.2, 142.4 and 170.7

Y. Uchio, S. Eguchi, M. Nakayama and T. Hase, <u>Chem. Lett.</u>, 277 (1982).

449

Source: *Sinularia mayi* Luttschw.

Mol. formula: $C_{20}H_{28}O_2$

Mol. wt.: 300

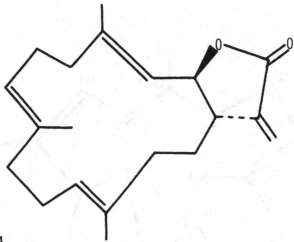

CH_3: 15.3, 15.6 and 16.4

CH_2: 24.0, 24.0, 32.1, 36.0, 38.4 and 121.3

CH: 38.9, 43.1, 79.1, 123.6, 125.3 and 125.9

C: 131.4, 133.6, 140.9 and 141.2

C=O: 170.4

Y. Uchio, S. Eguchi, M. Nakayama and T. Hase, <u>Chem. Lett.</u>, 277 (1982).

||||||||||||||||||||||||||||||||||||||||||||||||||||||||||||||||||||||||||||||||||||||||||||||||||||||||||||||||||||||||||||||||||||||||||

(1<u>S</u>*,3<u>S</u>*,4<u>S</u>*,7<u>E</u>,11<u>E</u>)-3,4-EPOXY-13-OXO-7,11,15-CEMBRATRIENE

Source: *Sarcophytum* sp.

Mol. formula: $C_{20}H_{30}O_2$

Mol. wt.: 302

Solvent: $CDCl_3$

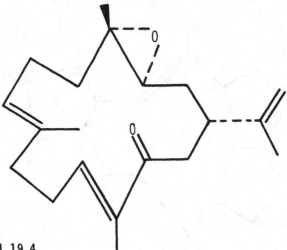

CH_3: 11.4, 15.7, 16.6 and 19.4

CH_2: 24.1, 25.5, 32.5, 38.2, 38.6, 40.9 and 110.8

CH: 43.7, 61.7, 125.5 and 143.4

C: 59.8, 134.2, 137.2 and 147.4

C=O: 189.0

B.N. Ravi and D.J. Faulkner, <u>J. Org. Chem.</u>, **43** (11), 2127 (1978).

(1S*,3S*,4S*,7E,11Z)-3,4-EPOXY-13-OXO-7,11,15-CEMBRATRIENE

Source: *Sarcophytum* sp.

Mol. formula: $C_{20}H_{30}O_2$

Mol. wt.: 302

Solvent: $CDCl_3$

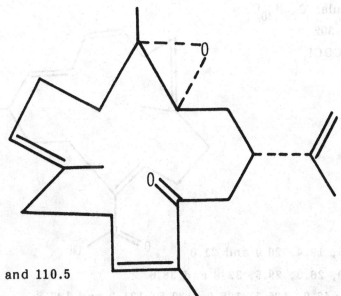

CH₃: 16.7, 16.7, 20.5 and 21.7

CH₂: 23.5, 29.2, 31.1, 39.0, 44.9 and 110.5

CH: 38.2, 60.4, 124.1 and 134.5

C: 59.5, 135.4, 137.3 and 147.2

C=O: 186.7

B.N. Ravi and D.J. Faulkner, J. Org. Chem., 43 (11), 2127 (1978).

|||||||||||||||||||||||||||||||||||||||||||||||||||||||||||||||||||||||||||||||||||||||||||||||||||||||||||||||||||||||||||||||||||||

(1S*,3S*,4S*,7E,11E)-3,4-EPOXY-14-OXO-7,11,15-CEMBRATRIENE

Source: *Sarcophytum* sp.

Mol. formula: $C_{20}H_{30}O_2$

Mol. wt.: 302

Solvent: $CDCl_3$

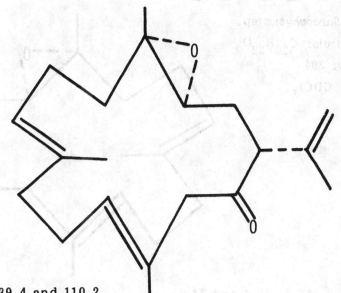

CH₃: 12.4, 15.0, 15.4 and 21.7

CH₂: 24.7, 25.4, 30.7 36.6, 37.0, 39.4 and 110.2

CH: 39.0, 58.5, 122.3 and 127.7

C: 63.6, 132.1, 136.2 and 148.4

C=O: 209.6

B.N. Ravi and D.J. Faulkner, J. Org. Chem., 43 (11), 2127 (1978).

Source: *Croton poilanei* Gagnep.

Mol. formula: $C_{20}H_{30}O_2$

Mol. wt.: 302

Solvent: $CDCl_3$

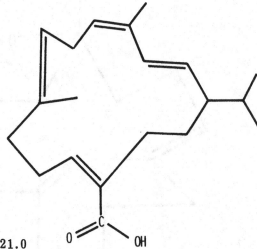

CH_3: 14.5, 19.4, 20.0 and 21.0

CH_2: 25.9, 26.3, 29.5, 32.8 and 38.6

CH: 32.1, 48.0, 125.7, 128.0, 130.5, 131.3 and 147.8

C: 128.9, 131.0 and 135.2

C=O: 173.7

A. Sato, M. Kurabayashi, A. Ogiso and H. Kuwano, <u>Phytochemistry</u>, **20** (8), 1915 (1981).

||||||||||||||||||||||||||||||||||||||||||||||||||||||||||||||||||||||||||||||||||||||||||||||||||||||||||||||||

(1R*,3S*,4S*,14R*,7E,11E)-3,4-EPOXY-14-HYDROXY-7,11,15-CEMBRATRIENE

Source: *Sarcophytum* sp.

Mol. formula: $C_{20}H_{32}O_2$

Mol. wt.: 304

Solvent: $CDCl_3$

CH_3: 15.8, 16.6, 19.4 and 24.8

CH_2: 24.8, 33.2, 33.8, 36.1, 36.5, 38.7 and 110.9

CH: 45.4, 59.5, 70.6, 123.6 and 126.3

C: 60.8, 133.0, 135.9 and 148.3

B.N. Ravi and D.J. Faulkner, <u>J. Org. Chem.</u>, **43** (11), 2127 (1978).

SARCOPHYTOL C

Source: *Sarcophyton glaucum* Q. et. G.

Mol. formula: $C_{20}H_{32}O_2$

Mol. wt.: 304

Solvent: C_5D_5N

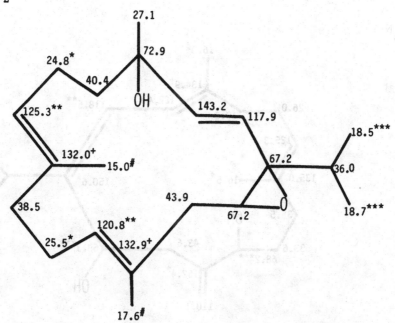

T. Nakagawa, M. Kobayashi, K. Hayashi and H. Mitsuhashi, <u>Chem. Pharm. Bull.</u> (Tokyo),
29 (1), 82 (1981).

SARCOPHYTOL D

Source: *Sarcophyton glaucum* Q. et G.

Mol. formula: $C_{20}H_{32}O_2$

Mol. wt.: 304

Solvent: C_5D_5N

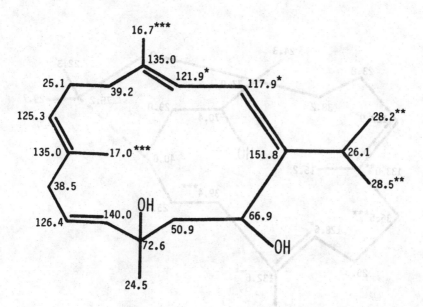

T. Nakagawa, M. Kobayashi, K. Hayashi and H. Mitsuhashi, <u>Chem. Pharm. Bull.</u> (Tokyo),
29 (1), 82 (1981).

SARCOPHYTOL E

Source: *Sarcophyton glaucum* Q. et G.

Mol. formula: $C_{20}H_{32}O_2$

Mol. wt.: 304

Solvent: C_5D_5N

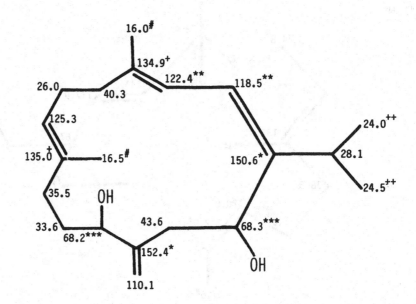

T. Nakagawa, M. Kobayashi, K. Hayashi and H. Mitsuhashi, <u>Chem. Pharm. Bull.</u> (Tokyo), **29** (1), 82 (1981).

||||||||||||||||||||||||||||||||||||||||||||||||||||||||||||||||||||||||||||||||||||||||||||||||||||||||||||||||||||||||||||||||||

DECARYIOL

Source: *Sarcophyton decaryi*

Mol. formula: $C_{20}H_{34}O_2$

Mol. wt.: 306

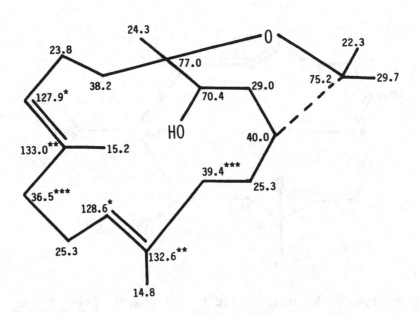

S. Carmely, A. Groweiss and Y. Kashman, <u>J. Org. Chem.</u>, **46** (21), 4279 (1981).

454

3,4-EPOXYNEPHTHENOL

Source: *Sarcophyton decaryi*

Mol. formula: $C_{20}H_{34}O_2$

Mol. wt.: 306

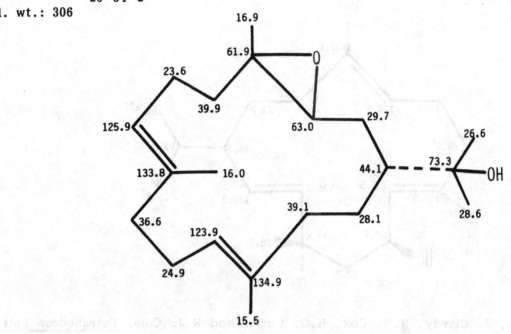

S. Carmely, A. Groweiss and Y. Kashman, J. Org. Chem., **46** (21), 4279 (1981).

||||||||||||||||||||||||||||||||||||||||||||||||||||||||||||||||||||||||||||||||||||||||||||||||||||||||||

(7E,11E,1R*,3aS*,4S*,12aR*)-11-ISOPROPYL-1,4,8-TRIMETHYL-1,2,3,3a,4,5,6,9,10,12a-DECAHYDROCYCLOPENTACYCLOUNDECENE-1,4-DIOL

Source: *Cespitularia* sp.

Mol. formula: $C_{20}H_{34}O_2$

Mol. wt.: 306

Solvent: $CDCl_3$

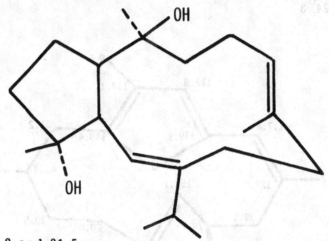

CH$_3$: 18.3, 22.2, 24.2, 24.2 and 31.5

CH$_2$: 22.9, 23.4, 29.6, 34.2, 36.0 and 39.3

CH: 33.2, 50.2, 56.7, 124.1 and 127.7

C: 74.8, 81.5, 134.7 and 149.0

B.F. Bowden, J.C. Coll, J.M. Gulbis, M.F. Mackay and R.H. Willis, Aust. J. Chem., **39** (5), 803 (1986).

4,8,13-DUVATRIENE-1,3-DIOL

Source: *Nicotiana tabacum*

Mol. formula: $C_{20}H_{34}O_2$

Mol. wt.: 306

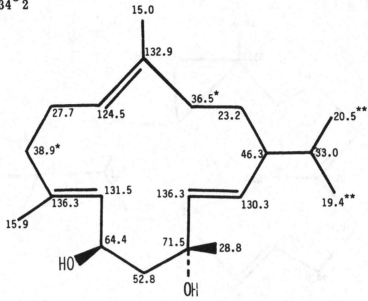

1. J.P. Springer, J. Clardy, R.H. Cox, H.G. Cutler and R.J. Cole, <u>Tetrahedron Lett.</u>, (32), 2737 (1975).

2. D.L. Roberts and R.L. Rowland, <u>J. Org. Chem.</u>, 27 (11), 3989 (1962).

||||||||||||||||||||||||||||||||||||||||||||||||||||||||||||||||||||||||||||||||||||||||||||||||||||||||||

RUBIFOLIDE

Source: *Gersemia rubiformis*

Mol. formula: $C_{20}H_{24}O_3$

Mol. wt.: 312

Solvent: $CDCl_3$

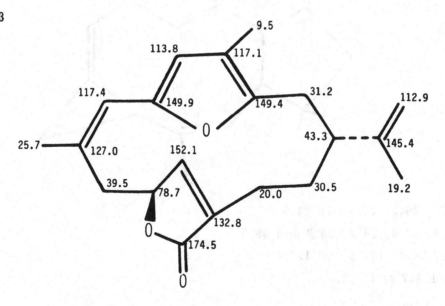

D. Williams, R.J. Andersen, G.D.V. Duyne and J. Clardy, <u>J. Org. Chem.</u>, 52 (3), 332 (1987).

(1Z,5S,9E,11E)-5,9-DIMETHYL-12-ISOPROPYL-6-OXOCYCLOTETRADECA-1,9,11-TRIENE-1,5-CARBOLACTONE

Source: *Sarcophyton glaucum*

Mol. formula: $C_{20}H_{28}O_3$

Mol. wt.: 316

Solvent: $CDCl_3$

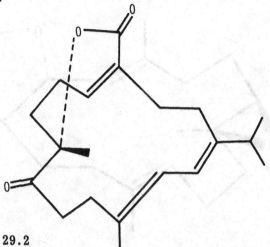

CH_3: 18.7, 21.9, 22.8 and 29.2

CH_2: 26.9, 26.9, 29.5, 31.7, 33.7 and 36.6

CH: 35.1, 119.0, 119.4 and 142.3

C: 86.1, 130.9, 133.5 and 144.7

C=O: 165.6 and 208.4

B.F. Bowden, J.C. Coll and R.H. Willis, <u>Aust. J. Chem.</u>, 35 (3), 621 (1982).

||||||||||||||||||||||||||||||||||||||||||||||||||||||||||||||||||||||||||||||||||||||||||||||||||||||||

(1R,2R,3E,7R,8R,11E)-7,8-EPOXYCEMBRA-3,11,15-TRIEN-16,2-OLIDE

Source: *Efflatounaria variabilis*

Mol. formula: $C_{20}H_{28}O_3$

Mol. wt.: 316

Solvent: $CDCl_3$

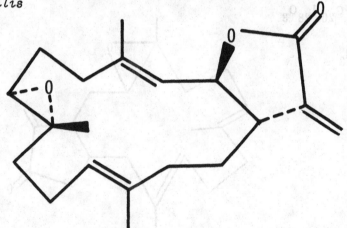

CH_3: 15.5, 17.5 and 17.6

CH_2: 22.4, 25.2, 31.3, 35.8, 35.8, 37.2 and 120.8

CH: 42.7, 62.6, 79.2, 123.1 and 126.6

C: 60.2, 132.6, 140.5 and 142.5

C=O: 170.1

B.F. Bowden, J.C. Coll, L.M. Engelhardt, G.V. Meehan, G.G. Pegg, D.M. Tapiolas, A.H. White and R.H. Willis, <u>Aust. J. Chem.</u>, 39(1), 123 (1986).

(1R,2S,3E,7S,8S,11E)-7,8-EPOXYCEMBRA-3,11,15-TRIEN-16,2-OLIDE

Source: *Efflatounaria variabilis*

Mol. formula: $C_{20}H_{28}O_3$

Mol. wt.: 316

Solvent: $CDCl_3$

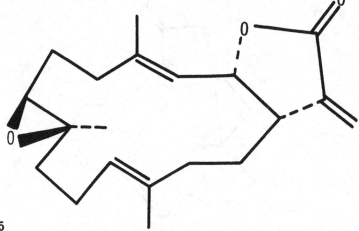

CH_3: 15.0, 16.3 and 16.5

CH_2: 23.7, 25.3, 28.0, 36.9, 37.5, 40.0 and 120.8

CH: 42.7, 61.9, 77.6, 120.0 and 123.9

C: 59.6, 135.0, 138.4 and 141.5

C=O: 169.5

B.F. Bowden, J.C. Coll, L.M. Engelhardt, G.V. Meehan, G.G. Pegg, D.M. Tapiolas, A.H. White and R.H. Willis, <u>Aust. J. Chem.</u>, **39** (1), 123 (1986).

||||||||||||||||||||||||||||||||||||||||||||||||||||||||||||||||||||||||||||||||||||||||||||||||||||||||||||||||||||||||

ISOLOBOPHYTOLIDE

Source: *Lobophytum crassum*

Mol. formula: $C_{20}H_{28}O_3$

Mol. wt.: 316

Solvent: $CDCl_3$

CH_3: 15.5, 17.0 and 17.0

CH_2: 24.2, 24.9, 32.3, 37.8, 39.0, 45.2 and 122.5

CH: 44.5, 63.0, 79.9, 124.5 and 130.3

C: 60.1, 129.3, 134.5 and 140.0

C=O: 169.6

B.F. Bowden, J.A. Brittle, J.C. Coll, N. Liyanage, S.J. Mitchell and G.J. Stokie, <u>Tetrahedron Lett.</u>, (41), 3661 (1977).

SARCOPHYTOLIDE

Source: *Sarcophyta elegans*

Mol. formula: $C_{20}H_{28}O_3$

Mol. wt.: 316

Solvent: $CDCl_3$

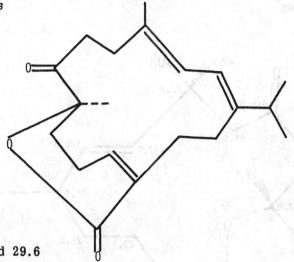

CH₃: 18.8, 22.0, 22.9 and 29.6 → CH_3: 18.8, 22.0, 22.9 and 29.6

CH_2: 27.2, 27.2, 29.8, 31.9, 33.8 and 36.8

CH: 35.3, 119.0, 119.7 and 143.1

C: 86.8, 131.1, 133.9 and 145.5

C=O: 166.5 and 209.3

Y. Uchio, M. Nitta, M. Nakayama, T. Iwagawa and T. Hase, <u>Chem. Lett.</u>, (4), 613 (1983).

||||||||||||||||||||||||||||||||||||||||||||||||||||||||||||||||||||||||||||||||||||||||||||||||||||||||||||||||||||||||||||||||||||

(1<u>Z</u>,5<u>S</u>,6<u>R</u>,9<u>E</u>,11<u>E</u>)-5,9-DIMETHYL-6-HYDROXY-12-ISOPROPYLCYCLOTETRADECA-1,9,11-TRIENE-1,5-CARBOLACTONE

Source: *Sarcophyton glaucum*

Mol. formula: $C_{20}H_{30}O_3$

Mol. wt.: 318

Solvent: $CDCl_3$

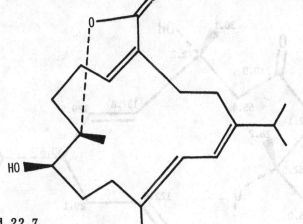

CH_3: 18.9, 21.9, 22.1 and 22.7

CH_2: 26.1, 26.7, 27.2, 31.3, 34.4 and 37.4

CH: 35.5, 65.8, 118.8, 120.7 and 140.5

C: 83.7, 132.3, 132.9 and 145.4

C=O: 166.8

B.F. Bowden, J.C. Coll and R.H. Willis, <u>Aust. J. Chem.</u>, 35(3), 621 (1982).

(1S,2E,4S,8R,11S)-8,11-EPOXY-4-HYDROXY-2,12(20)-CEMBRADIEN-6-ONE

Source: *Nicotianum tobacum*

Mol. formula: $C_{20}H_{32}O_3$

Mol. wt.: 320

Solvent: $CDCl_3$

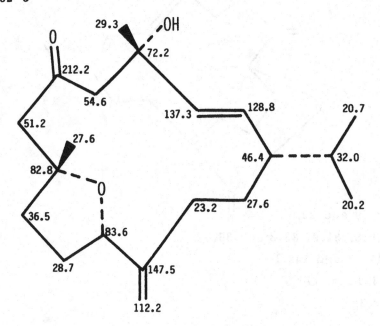

I. Wahlberg, I. Forsblom, C. Vogt, A.-M. Eklund, T. Nishida, C.R. Enzell and J.-E. Berg, J. Org. Chem., **50** (23), 4527 (1985).

(1S,2E,4S,8R,11S,12E)-8,11-EPOXY-4-HYDROXY-2,12-CEMBRADIEN-6-ONE

Source: *Nicotianum tobacum*

Mol. formula: $C_{20}H_{32}O_3$

Mol. wt.: 320

Solvent: $CDCl_3$

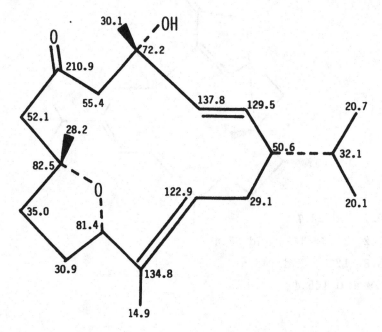

I. Wahlberg, I. Forsblom, C. Vogt, A.-M. Eklund, T. Nishida, C.R. Enzell and J.-E. Berg, J. Org. Chem., **50** (23), 4527 (1985).

(1S,2E,4S,7E,11S,12S)-11,12-EPOXY-4-HYDROXY-2,7-CEMBRADIENE-6-ONE

Source: *Nicotiana tabacum* Linn

Mol. formula: $C_{20}H_{32}O_3$

Mol. wt.: 320

Solvent: $CDCl_3$

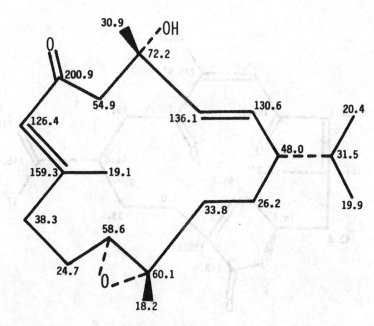

I. Wahlberg, I. Forsblom, C. Vogt, A.-M. Eklund, T. Nishida, C.R. Enzell and J.-E. Berg, J. Org. Chem., **50** (23), 4527 (1985).

||||||||||||||||||||||||||||||||||||||||||||||||||||||||||||||||||||||||||||||||||||||||||||||||||||||||||||||||||||||||||||||||||||

(-)-CORALLOIDOLIDE A

Source: *Alcyonium coralloides*

Mol. formula: $C_{20}H_{24}O_4$

Mol. wt.: 328

Solvent: $(CD_3)_2CO$

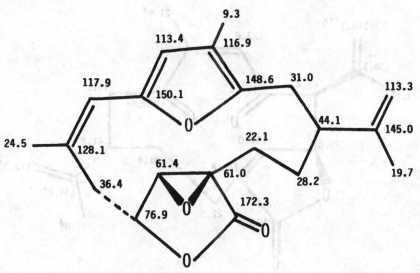

M.D. Ambrosio, D. Fabbri, A. Guerriero and F. Pietra, Helv. Chim. Acta, 70 (1), 63 (1987).

EPILOPHODIONE

Source: *Gersemia rubiformis*

Mol. formula: $C_{20}H_{24}O_4$

Mol. wt.: 328

Solvent: $CDCl_3$

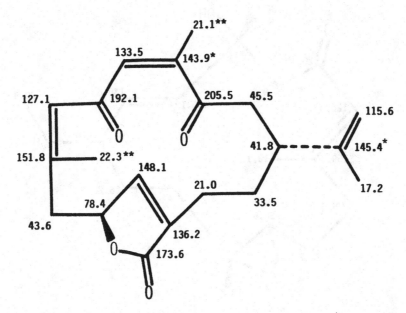

D. Williams, R.J. Andersen, G.D.V. Duyne and J. Clardy, J. Org. Chem., 52(3), 332 (1987).

IIIIIIIIIIIIIIIIIIIIIIIIIIIIIIIIIIIIIIIIIIIIIIIIIIIIIIIIIIIIIIIIIIIIIIIIIIIIIIIIIIIIIIIIIIIIIIIIIIIIIII

GERSEMOLIDE

Source: *Gersemia rubiformis*

Mol. formula: $C_{20}H_{24}O_4$

Mol. wt.: 328

Solvent: $CDCl_3$

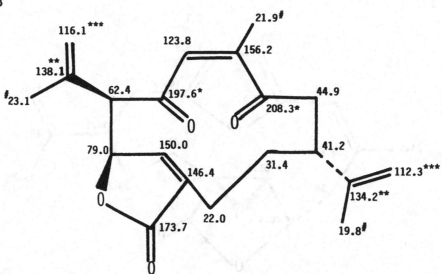

D. Williams, R.J. Andersen, G.D.V. Duyne and J. Clardy, J. Org. Chem., 52(3), 332 (1987).

ISOLOPHODIONE

Source: *Lophogorgia alba* (Duch. and Mich.)

Mol. formula: $C_{20}H_{24}O_4$

Mol. wt.: 328

Solvent: $CDCl_3$

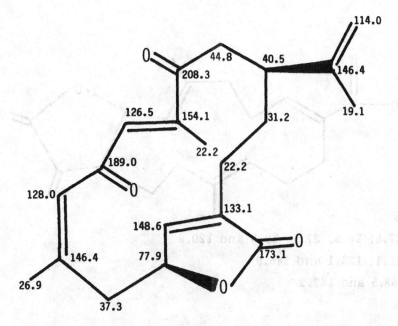

M.M. Bandurraga, B. McKittrick, W. Fenical, E. Arnold and J. Clardy, <u>Tetrahedron</u>, **38** (2), 305 (1982).

||||||||||||||||||||||||||||||||||||||||||||||||||||||||||||||||||||||||||||||||||||||||||||||||||||||

LOBOHEDLEOLIDE

Source: *Lobophytum hedleyi* Whitelegge

Mol. formula: $C_{20}H_{26}O_4$

Mol. wt.: 330

Solvent: $CDCl_3$

CH_3: 15.3 and 16.1

CH_2: 24.9, 26.7, 27.0, 35.1, 36.1, 39.8 and 120.6

CH: 43.0, 78.1, 120.6, 122.9 and 148.2

C: 129.1, 135.4, 138.8 and 142.0

C=O: 170.7 and 173.2

Y. Uchio, J. Toyota, H. Nozaki, M. Nakayama, Y. Nishizono and T. Hase, <u>Tetrahedron Lett.</u>, **22** (41), 4089 (1981).

(7Z)-LOBOHEDLEOLIDE

Source: *Lobophytum hedleyi* Whitelegge

Mol. formula: $C_{20}H_{26}O_4$

Mol. wt.: 330

Solvent: $CDCl_3$

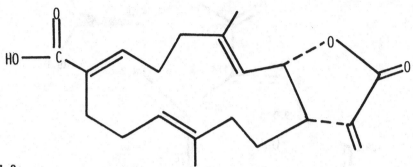

CH_3: 15.0 and 17.3

CH_2: 24.9, 26.3, 27.5, 28.6, 37.1, 39.5 and 120.9

CH: 43.3, 77.9, 121.1, 123.1 and 146.9

C: 130.4, 135.5, 138.5 and 142.2

C=O: 170.7 and 172.9

Y. Uchio, J. Toyota, H. Nozaki, M. Nakayama, Y. Nishizono and T. Hase, <u>Tetrahedron Lett.</u>, **22** (41), 4089 (1981).

||||||||||||||||||||||||||||||||||||||||||||||||||||||||||||||||||||||||||||||||||||||||||||||||||||||||

KERICEMBRENOLIDE E

Source: *Clavularia koellikeri*

Mol. formula: $C_{20}H_{28}O_4$

Mol. wt.: 332

Solvent: $CDCl_3$

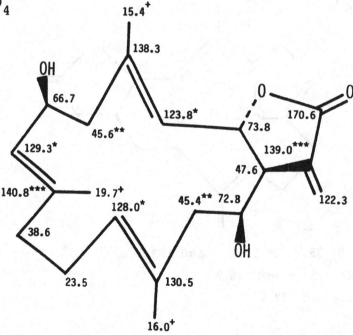

M. Kobayashi, B.W. Son, Y. Kyogoku and I. Kitagawa, <u>Chem. Pharm. Bull.</u> (Tokyo), **34**(5), 2306 (1986).

MAYOLIDE B

Source: *Sinularia mayi*

Mol. formula: $C_{20}H_{28}O_4$

Mol. wt.: 332

Solvent: $CDCl_3$

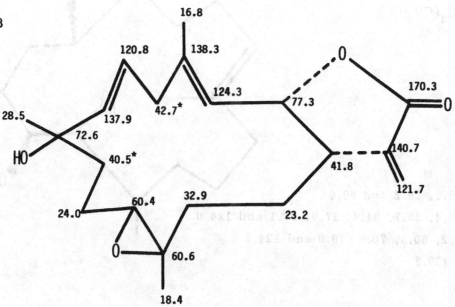

M. Kobayashi, <u>Chem. Pharm. Bull.</u> (Tokyo), 36 (2), 488 (1988).

||||||||||||||||||||||||||||||||||||||||||||||||||||||||||||||||||||||||||||||||||||||||||||||||||||||||||||||||||||||||

NORCEMBRENOLIDE-II

Source: *Sinularia numerosa* Tixier - Durivault

Mol. formula: $C_{19}H_{24}O_5$

Mol. wt.: 332

Solvent: $CDCl_3$

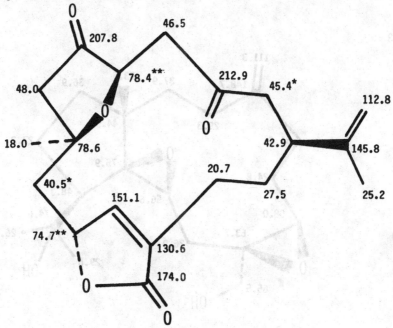

A. Sato, W. Fenical, Z.Qi-Tai and J. Clardy, <u>Tetrahedron</u>, 41(19), 4303 (1985).

12,13-BISEPIEUPALMERIN

Source: *Eunicea succinea*

Mol. formula: $C_{20}H_{30}O_4$

Mol. wt.: 334

Solvent: $CDCl_3/CD_3OD$

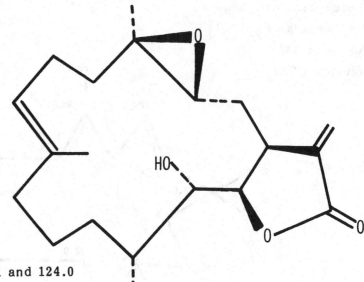

CH_3: 12.0, 15.5, 16.2 and 60.4

CH_2: 23.1, 23.4, 30.7, 31.4, 37.0, 38.1 and 124.0

CH: 31.5, 39.2, 60.5, 70.8, 79.0 and 124.2

C: 136.3 and 139.2

O=C: 171.0

Y. Gopichand, L.S. Ciereszko, F.J. Schmitz, D. Switzner, A. Rahman, M.B. Hossain and D. Van Der Helm, J. Nat. Prod., (Lloydia), 47 (4), 607 (1984).

||||||||||||||||||||||||||||||||||||||||||||||||||||||||||||||||||||||||||||||||||||||||||||||||||||||||||||||||

STOLONIDIOL

Source: *Clavularia* sp.

Mol. formula: $C_{20}H_{32}O_4$

Mol. wt.: 336

Solvent: $CDCl_3$

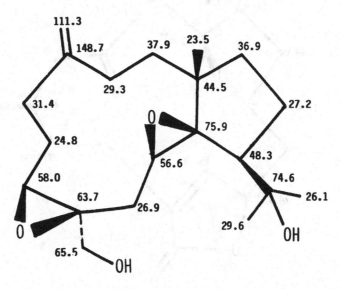

K. Mori, K. Iguchi, N. Yamada, Y. Yamada and Y. Inouye, Tetrahedron Lett., 28(46), 5673 (1987).

(1S,2E,4S,8R,11S,12R)-4,12-DIHYDROXY-8,11-EPOXY-2-CEMBREN-6-ONE

Source: *Nicotianum tobacum*

Mol. formula: $C_{20}H_{34}O_4$

Mol. wt.: 338

Solvent: $CDCl_3$

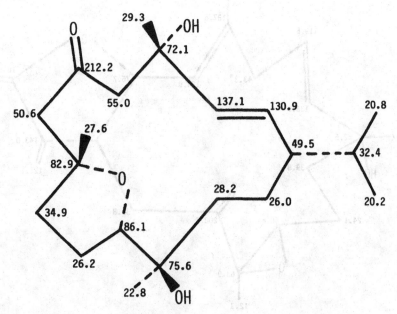

I. Wahlberg, I. Forsblom, C. Vogt, A.-M. Eklund, T. Nishida, C.R. Enzell and J.-E. Berg, J. Org. Chem., **50** (23), 4527 (1985).

||||||||||||||||||||||||||||||||||||||||||||||||||||||||||||||||||||||||||||||||||||||||||||||||||||||||||||||||||||||

EPOXYLOPHODIONE

Source: *Lophogorgia alba* (Duch. and Mich.)

Mol. formula: $C_{20}H_{24}O_5$

Mol. wt.: 344

Solvent: $CDCl_3$

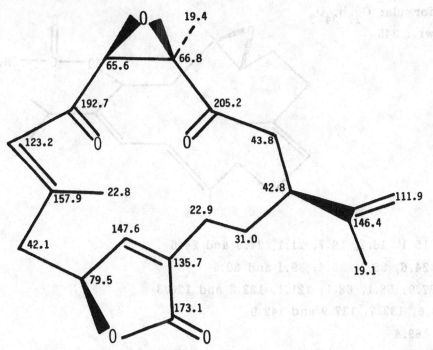

M.M. Bandurraga, B. McKittrick, W. Fenical, E. Arnold and J. Clardy, Tetrahedron, **38** (2), 305 (1982).

Source: *Sinularia mayi*

Mol. formula: $C_{20}H_{26}O_5$

Mol. wt.: 346

Solvent: $CDCl_3$

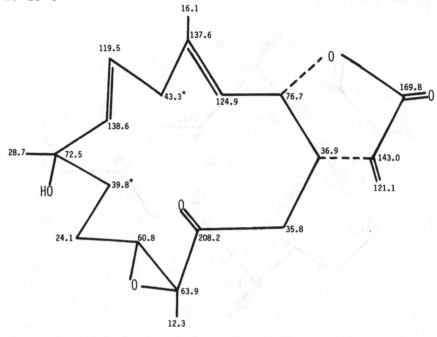

M. Kobayashi, <u>Chem. Pharm. Bull.</u> (Tokyo), 36 (2), 488 (1988).

||||||||||||||||||||||||||||||||||||||||||||||||||||||||||||||||||||||||||||||||||||||||||||||||||||||

(14<u>S</u>,1<u>E</u>,3<u>E</u>,7<u>E</u>)-14-ACETOXY-11,12-EPOXYCEMBRA-1,3,7-TRIENE

Source: *Lobophytum* sp.

Mol. formula: $C_{22}H_{34}O_3$

Mol. wt.: 346

CH_3: 15.1, 16.2, 19.7, 21.1, 24.6 and 24.6

CH_2: 24.6, 25.7, 36.4, 39.1 and 39.8

CH: 27.9, 58.1, 68.1, 121.1, 122.3 and 126.3

C: 58.6, 133.7, 137.9 and 142.9

C=O: 169.4

B.F. Bowden, J.C. Coll and D.M. Tapiolas, <u>Aust. J. Chem.</u>, 36(11), 2289 (1983).

NOR DITERPENE SL-I

Source: *Sinularia leptoclados* and *S. foeta*

Mol. formula: $C_{19}H_{24}O_6$

Mol. wt.: 348

Solvent: $CDCl_3 + CD_3OD$

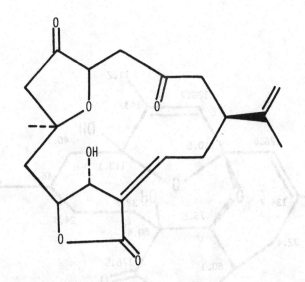

C: 80.3, 84.9, 111.6, 133.7, 146.0 and 148.7

C=O: 171.1, 207.9 and 217.4

22.2, 27.1, 29.1, 40.6, 43.1, 44.1, 47.0, 52.2, 75.5 and 76.2

V. Lakshmi and F.J. Schmitz, <u>J. Nat. Prod.</u>, (Lloydia), 49 (4), 728 (1986).

|||||||||||||||||||||||||||||||||||||||||||||||||||||||||||||||||||||||||||||||||||||||||||||||||||||||||||||

KERICEMBRENOLIDE A

Source: *Clavularia koellikeri*

Mol. formula: $C_{22}H_{30}O_4$

Mol. wt.: 358

Solvent: $CDCl_3$

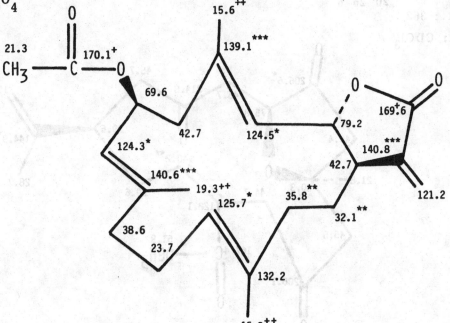

M. Kobayashi, B.W. Son, Y. Kyogoku and I. Kitagawa, <u>Chem. Pharm. Bull.</u> (Tokyo), 34(5), 2306 (1986).

(-)-CORALLOIDOLIDE B

Source: *Alcyonium coralloides*

Mol. formula: $C_{20}H_{26}O_6$

Mol. wt.: 362

Solvent: $(CD_3)_2CO$

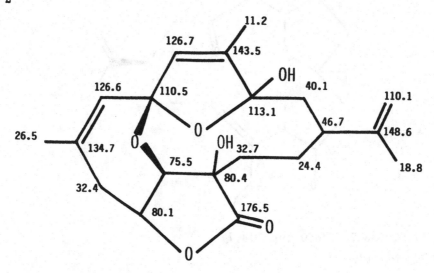

M.D. Ambrosio, D. Fabbri, A. Guerriero and F. Pietra, <u>Helv. Chim. Acta</u>, **70** (1), 63 (1987).

NORCEMBRENE

Source: *Sinularia querciformis* (Pratt)

Mol. formula: $C_{20}H_{26}O_6$

Mol. wt.: 362

Solvent: $CDCl_3$

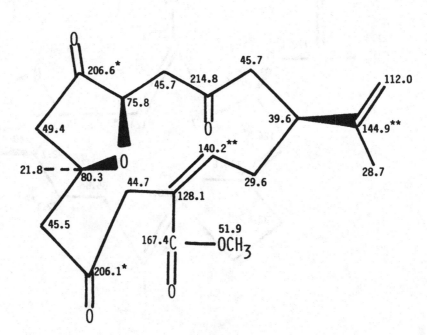

A. Sato, W. Fenical, Z. Qi-Tai and J. Clardy, <u>Tetrahedron</u>, **41** (19), 4303 (1985).

470

KERICEMBRENOLIDE B

Source: *Clavularia koellikeri*
Mol. formula: $C_{22}H_{30}O_4$
Mol. wt.: 358
Solvent: $CDCl_3$

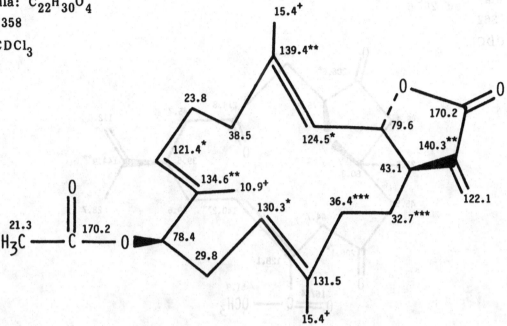

M. Kobayashi, B.W. Son, Y. Kyogoku and I. Kitagawa, <u>Chem. Pharm. Bull.</u> (Tokyo), **34** (5), 2306 (1986).

||||||||||||||||||||||||||||||||||||||||||||||||||||||||||||||||||||||||||||||||||||||||||||||||||||||||||||||||||||||||||||||||||||||||||||

KETOEMBLIDE

Source: *Sarcophyta elegans*
Mol. formula: $C_{21}H_{28}O_5$
Mol. wt.: 360
Solvent: $CDCl_3$

CH_3: 21.7, 22.8, 29.0 and 51.4
CH_2: 24.4, 27.0, 27.9, 33.2, 34.0 and 35.1
CH: 35.1, 120.7, 136.2 and 143.8
C: 86.8, 125.1, 130.7 and 155.5
C=O: 166.9, 168.0 and 208.3

Y. Uchio, M. Nitta, M. Nakayama, T. Iwagawa and T. Hase, <u>Chem. Lett.</u>, (4), 613 (1983).

NORDITERPENE SQ-V

Source: *Sinularia querciformis* (Pratt)

Mol. formula: $C_{20}H_{26}O_6$

Mol. wt.: 362

Solvent: $CDCl_3$

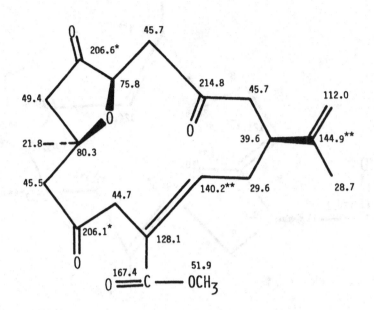

A. Sato, W. Fenical, Z.Qi-Tai and J. Clardy, <u>Tetrahedron</u>, 41(19), 4303 (1985).

||||||||||||||||||||||||||||||||||||||||||||||||||||||||||||||||||||||||||||||||||||||||||||||||||||||||||||

FLACCIDOXIDE

Source: *Alcyonium flaccidum*

Mol. formula: $C_{22}H_{34}O_4$

Mol. wt.: 362

Solvent: $CDCl_3$

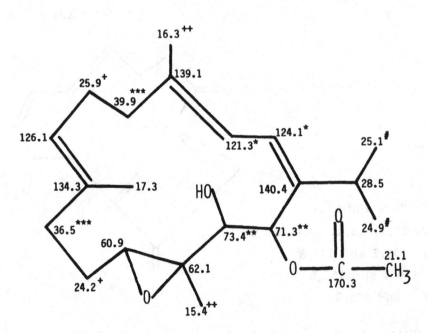

Y. Kashman, S. Carmely and A. Groweiss, <u>J. Org. Chem.</u>, **46** (18), 3592 (1981).

PUKALIDE

Source: *Sinularia abrupta*
Mol. formula: $C_{21}H_{24}O_6$
Mol. wt.: 372
Solvent: $CDCl_3$

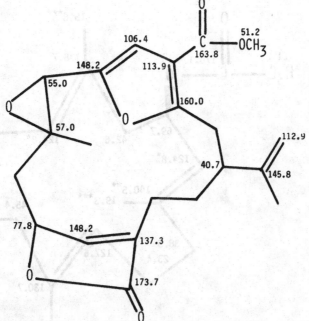

CH$_3$: 18.7 and 19.8
CH$_2$: 22.8, 32.5, 32.5 and 40.0

M.G. Missakian, B.J. Burreson and P.J. Scheuer, <u>Tetrahedron</u>, 31 (20), 2513 (1975).

||||||||||||||||||||||||||||||||||||||||||||||||||||||||||||||||||||||||||||||||||||||||||||||||||||||||||||||||

(7<u>E</u>,11<u>E</u>,1<u>R</u>,2<u>S</u>,3<u>R</u>,4<u>R</u>,14<u>S</u>)-14-ACETOXY-3,4-EPOXYCEMBRA-7,11,15-TRIEN-17,2-OLIDE

Source: *Lobophytum cristigalli*
Mol. formula: $C_{22}H_{30}O_5$
Mol. wt.: 374

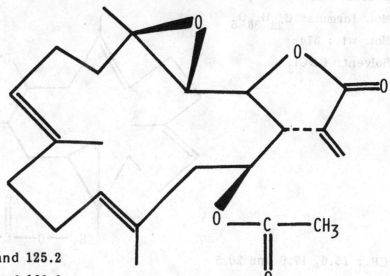

CH$_3$: 15.6, 16.6, 16.9 and 20.8
CH$_2$: 24.3, 24.7, 37.2, 39.6, 39.6 and 125.2
CH: 41.5, 65.4, 72.8, 80.5, 123.4 and 129.0
C: 59.8, 130.0, 133.4 and 136.1
C=O: 169.4 and 170.4

B.F. Bowden, J.C. Coll, M.S.L De Costa, M. F. Mackay, M. Mahendran, E.D. De Silva and R.H. Willis, <u>Aust. J.Chem</u>., 37(3), 545 (1984).

KERICEMBRENOLIDE D

Source: *Clavularia koellikeri*

Mol. formula: $C_{22}H_{30}O_5$

Mol. wt.: 374

Solvent: $CDCl_3$

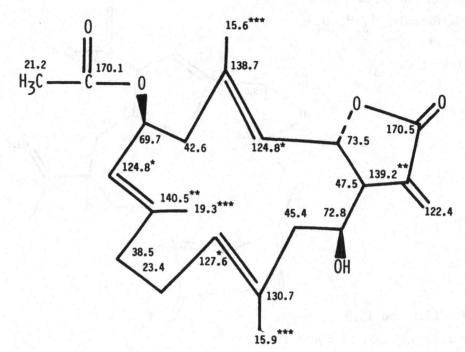

M. Kobayashi, B.W. Son, Y. Kyogoku and I. Kitagawa, <u>Chem. Pharm. Bull.</u> (Tokyo), 34 (5), 2306 (1986).

|||||||||||||||||||||||||||||||||||||||||||||||||||||||||||||||||||||||||||||||||||||||||||||||||||||||||||||

LOBOLIDE

Source: *Lobophytum* sp.

Mol. formula: $C_{22}H_{30}O_5$

Mol. wt.: 374

Solvent: $CDCl_3$

CH_3: 15.6, 17.0 and 20.5

CH_2: 23.2, 24.8, 31.8, 32.4, 38.7, 45.3, 64.3 and 122.7

CH: 45.0, 61.9, 79.6, 124.5 and 130.2

C: 60.4, 129.5, 135.0 and 139.7

C=O: 169.4 and 170.4

Y. Kashman and A. Groweiss, <u>Tetrahedron Lett.</u>, 19 (13), 1159 (1977).

CRASSIN ACETATE

Source: *Pseudoplexaura porosa* (syn. *Plexaura crassa* Ellis)

Mol. formula: $C_{22}H_{32}O_5$

Mol. wt.: 376

Solvent: $CDCl_3$

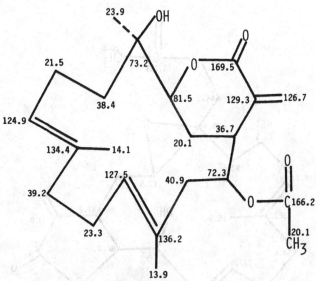

1. G.E. Martin, J.A. Matson, J.C. Turley and A.J. Weinheimer, J. Am. Chem. Soc., **101** (7), 1888 (1979).

2. A.J. Weinheimer and J.A. Matson, J. Nat. Prod., (Lloydia), **38**, 378 (1975).

||||||||||||||||||||||||||||||||||||||||||||||||||||||||||||||||||||||||||||||||||||||||||||||||||||||||||||||||||||||||||||||||||||||

11β,12β-EPOXYPUKALIDE

Source: *Leptogorgia setacea* Pallas

Mol. formula: $C_{21}H_{24}O_7$

Mol. wt.: 388

Solvent: $CDCl_3$

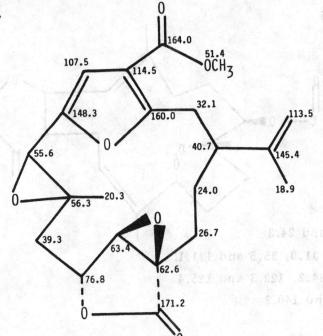

M.B. Ksebati, L.S. Ciereszko and F.J. Schmitz, J. Nat. Prod., (Lloydia), **47** (6), 1009 (1984).

(7E,11E)-18-ACETOXY-3,4-EPOXY-13-HYDROXY-7,11,15(17)-CEMBRATRIEN-16,14-OLIDE

Source: *Lobophytum crassum*

Mol. formula: $C_{22}H_{30}O_6$

Mol. wt.: 390

Solvent: $CDCl_3$

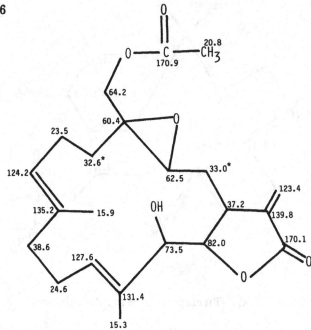

Y. Kashman, S. Carmely and A. Groweiss, <u>J. Org. Chem.</u>, **46** (18), 3592 (1981).

|||||||||||||||||||||||||||||||||||||||||||||||||||||||||||||||||||||||||||||||||||||||||||||||||||||

DENTICULATOLIDE

Source: *Lobophytum denticulatum* (Tixie-Durivault)

Mol. formula: $C_{22}H_{30}O_6$

Mol. wt.: 390

Solvent: $CDCl_3$

CH_3: 16.4, 19.5, 21.0 and 24.3

CH_2: 24.7, 26.0, 28.0, 31.0, 35.5 and 121.1

CH: 44.5, 70.1, 79.3, 84.2, 123.3 and 125.4

C: 80.0, 135.5, 138.4 and 140.2

C=O: 170.3 and 170.5

Y. Uchio, S. Eguchi, J. Kuramoto, M. Nakayama and T. Hase, <u>Tetrahedron Lett.</u>, **26** (37), 4487 (1985).

MAYOLIDE C

Source: *Sinularia mayi*
Mol. formula: $C_{22}H_{30}O_6$
Mol. wt.: 390
Solvent: $CDCl_3$

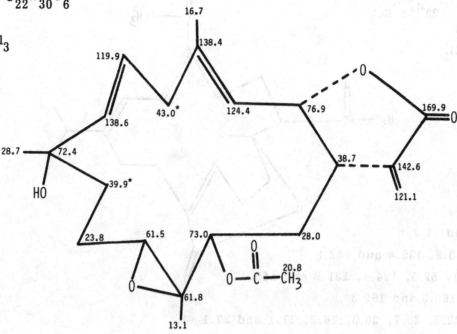

M. Kobayashi, <u>Chem. Pharm. Bull.</u> (Tokyo), 36 (2), 488 (1988).

||||||||||||||||||||||||||||||||||||||||||||||||||||||||||||||||||||||||||||||||||||||||||||||||||||||||||||||||||||||||||||||||||

DITERPENE SF-1

Source: *Sinularia flexibilis*
Mol. formula: $C_{22}H_{32}O_6$
Mol. wt.: 392
Solvent: $CDCl_3$

CH_3: 16.4, 17.1, 21.1 and 29.3
CH_2: 26.7, 27.3, 29.8, 33.2, 33.6, 34.3, 34.9 and 123.6
CH: 35.7, 60.8, 77.9 and 83.4
C: 59.9, 85.4, 88.2 and 145.4
C=O: 169.5 and 171.2

K. Mori, S. Suzuki, K. Iguchi and Y. Yamada, <u>Chem. Lett.</u>, (10), 1515 (1983).

EMBLIDE

Source: *Sarcophyton glaucum*

Mol. formula: $C_{23}H_{32}O_6$

Mol. wt.: 404

Solvent: $CDCl_3$

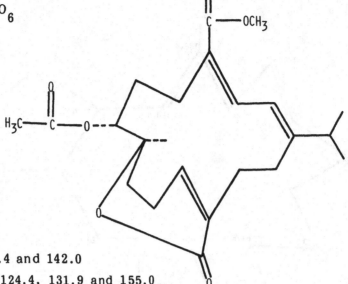

CH_3: 35.9 and 51.2

CH: 68.2, 120.8, 135.4 and 142.0

C: 34.3, 37.1, 82.3, 124.4, 131.9 and 155.0

C=O: 166.1, 168.0 and 169.3

20.9, 21.9, 22.7, 23.7, 25.3, 26.3, 27.1 and 27.1

J.A. Toth, B.J. Burreson, P.J. Scheuer, J. Finer-Moore and J. Clardy, <u>Tetrahedron</u>, 36 (10), 1307 (1980).

KERICEMBRENOLIDE C

Source: *Clavularia koellikeri*

Mol. formula: $C_{24}H_{32}O_6$

Mol. wt.: 416

Solvent: $CDCl_3$

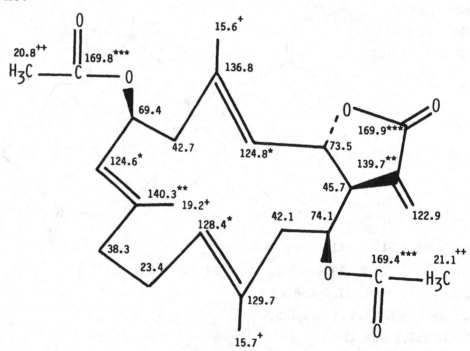

M. Kobayashi, B.W. Son, Y. Kyogoku and I. Kitagawa, <u>Chem. Pharm. Bull.</u> (Tokyo), 34(5), 2306 (1986).

Basic skeleton

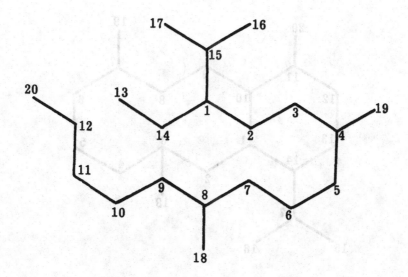

MAYOLIDE A

Source: *Sinularia mayi*
Mol. formula: $C_{20}H_{30}O_4$
Mol. wt.: 334
Solvent: $CDCl_3$

M. Kobayashi, <u>Chem. Pharm. Bull.</u> (Tokyo), **36** (2), 488 (1988).

Basic skeleton

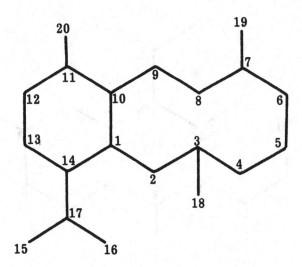

SCLEROPHYTIN F

Source: *Sclerophytum capitalis*

Mol. formula: $C_{20}H_{34}O_4$

Mol. wt.: 338

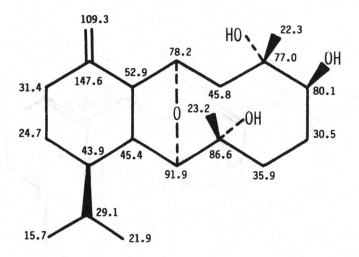

M. Alam, P. Sharma, A.S. Zektzer, G.E. Martin, X. Ji and D.V. der Helm, J. Org. Chem., 54 (8), 1896 (1989).

SCLEROPHYTIN B

Source: *Sclerophytum capitalis*

Mol. formula: $C_{22}H_{34}O_4$

Mol. wt.: 362

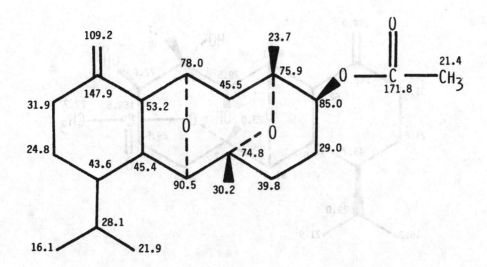

P. Sharma and M. Alam, <u>J. Chem. Soc. Perkin Trans. 1</u>, (9), 2537 (1988).

||||||||||||||||||||||||||||||||||||||||||||||||||||||||||||||||||||||||||||||||||||||||||||||||||||||||||||||||||||||||||||||||||||||

SCLEROPHYTIN E

Source: *Sclerophytum capitalis*

Mol. formula: $C_{22}H_{36}O_5$

Mol. wt.: 380

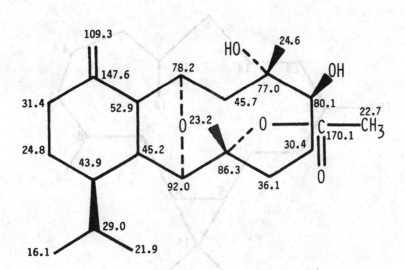

M. Alam, P. Sharma, A.S. Zektzer, G.E. Martin, X. Ji and D.V. der Helm, <u>J. Org. Chem.</u>, 54 (8), 1896 (1989).

Source: *Sclerophytum capitalis*

Mol. formula: $C_{22}H_{36}O_6$

Mol. wt.: 396

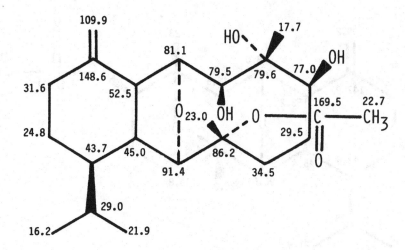

M. Alam, P. Sharma, A.S. Zektzer, G.E. Martin, X. Ji and D.V. der Helm, <u>J. Org. Chem.</u>, 54 (8), 1896 (1989).

CASBANE

Basic skeleton

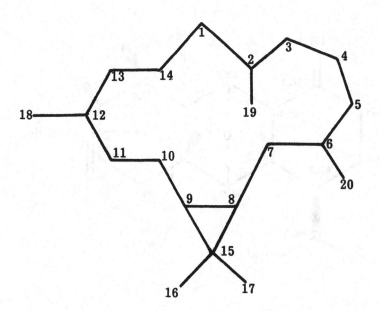

14-DEHYDROAGROSTISTACHIN

Source: *Agrostistachys hookeri* Benth. and Hook. f.

Mol. formula: $C_{20}H_{28}O_3$

Mol. wt.: 316

Solvent: $CDCl_3$

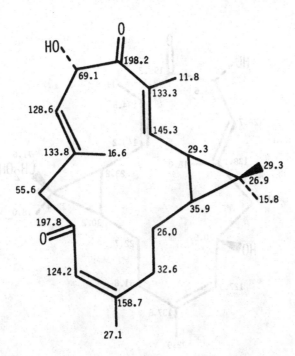

Y.-H. Choi, J.M. Pezzuto, A.D. Kinghorn and N.R. Farnsworth, J. Nat. Prod., (Lloydia), 51 (1), 110 (1988).

AGROSKERIN

Source: *Agrostistachys hookeri* Benth. and Hook. f.

Mol. formula: $C_{20}H_{30}O_3$

Mol. wt.: 318

Solvent: $CDCl_3$

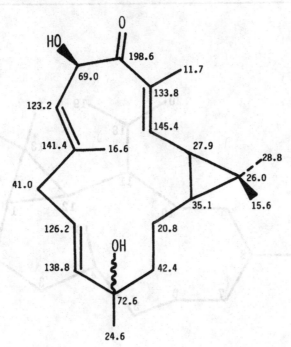

Y.-H. Choi, J.M. Pezzuto, A.D. Kinghorn and N.R. Farnsworth, J. Nat. Prod., (Lloydia), 51 (1), 110 (1988).

17-HYDROXYAGROSTISTACHIN

Source: *Agrostistachys hookeri* Benth. and Hook. f.

Mol. formula: $C_{20}H_{30}O_4$

Mol. wt.: 334

Solvent: $CDCl_3$

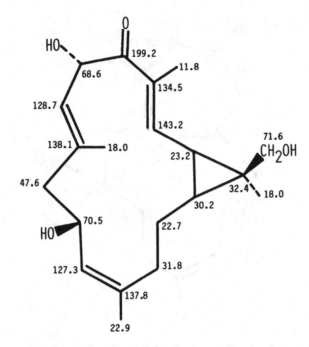

Y.-H. Choi, J.M. Pezzuto, A.D. Kinghorn and N.R. Farnsworth, J. Nat. Prod., (Lloydia),51 (1), 110 (1988).

BERDIMERANE

Basic skeleton

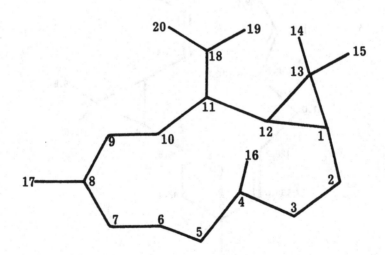

DITERPENE BD-II (METHYL ESTER)

Source: *Bertya dimerostigma* F. Muell.

Mol. formula: $C_{21}H_{32}O_4$

Mol. wt.: 348

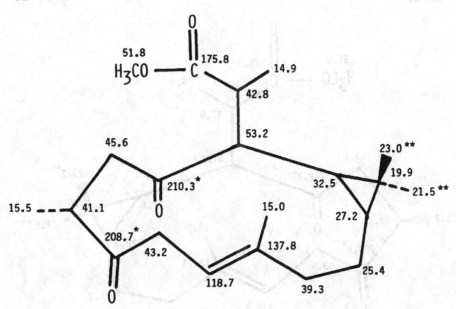

E.L. Ghisalberti, P.R. Jefferies, T.A. Mori, B.W. Skelton and A.H. White, <u>Tetrahadron</u>, **41** (12), 2517 (1985).

||||||||||||||||||||||||||||||||||||||||||||||||||||||||||||||||||||||||||||||||||||||||||||||||||||||||||||||||||||||||||

DITERPENE BD-III (METHYL ESTER)

Source: *Bertya dimerostigma* F.Muell.

Mol. formula: $C_{21}H_{32}O_4$

Mol. wt.: 348

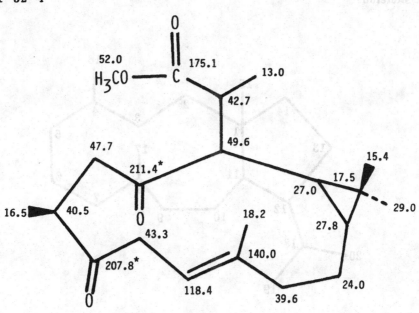

E.L. Ghisalberti, P.R. Jefferies, T.A. Mori, B.W. Skelton and A.H. White, <u>Tetrahedron</u>, **41**(12), 2517 (1985).

DITERPENE BD-IV (METHYL ESTER)

Source: *Bertya dimerostigma* F. Muell.

Mol. formula: $C_{21}H_{32}O_4$

Mol. wt.: 348

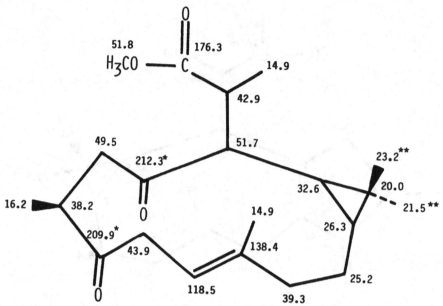

E.L. Ghisalberti, P.R. Jefferies, T.A. Mori, B.W. Skelton and A.H. White, <u>Tetrahedron</u>, 41(12), 2517 (1985).

DOLLABELLANE

Basic skeleton

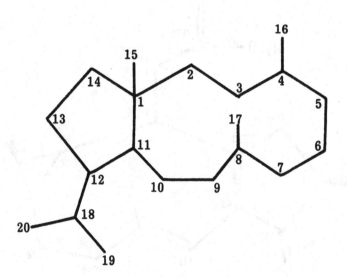

CLAENONE

Source: *Clavularia viridis*

Mol. formula: $C_{20}H_{30}O_2$

Mol. wt.: 302

Solvent: $CDCl_3$

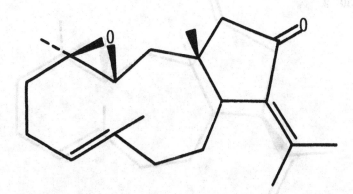

CH_3: 15.5, 16.6, 21.3, 23.3 and 24.5

CH_2: 24.4, 27.4, 38.5, 40.9 and 55.5

CH: 42.3, 63.8 and 128.3

C: 37.2, 37.5, 61.3, 133.0, 137.2 and 148.6

C=O: 206.0

K. Mori, K. Iguchi, N. Yamada, Y. Yamada and Y. Inouye, <u>Chem. Pharm. Bull.</u>, (Tokyo), **36**(8), 2840 (1988).

||||||||||||||||||||||||||||||||||||||||||||||||||||||||||||||||||||||||||||||||||||||||||||||||||||||||||||||||||||||||||||||||||

DITERPENE EC-1

Source: *Eunicea calyculata* (Ellis and Solander)

Mol. formula: $C_{20}H_{30}O_2$

Mol. wt.: 302

Solvent: $CDCl_3$

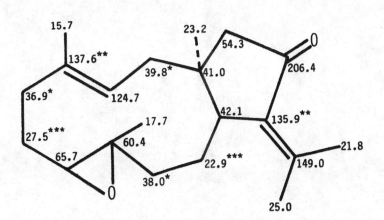

S.A. Look and W. Fenical, <u>J. Org. Chem.</u>, **47** (21), 4129 (1982).

3,4-EPOXY-14-OXO-7,18-DOLABELLADIENE

Source: *Dictyota dichotoma*

Mol. formula: $C_{20}H_{30}O_2$

Mol. wt.: 302

Solvent: $CDCl_3$

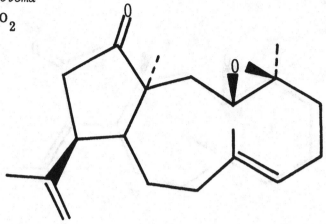

CH_3: 15.8, 15.9, 19.0 and 23.0

CH_2: 21.9, 24.2, 36.2, 37.0, 38.9, 40.2 and 112.1

CH: 40.0, 44.3, 63.1 and 127.2

C: 52.1, 61.3, 133.3 and 143.9

C=O: 221.0

V. Amico, G. Oriente, M. Piattelli, C. Tringali, E. Fattorusso, S. Magno and L. Mayol, Tetrahedron, **36** (10), 1409 (1980).

(1S*,3E,6Z,11S*,12S*)-12-HYDROXYDOLABELLA-3,6-DIEN-9-ONE

Source: *Dictyota dichotoma* (Huds.) Lamouroux

Mol. formula: $C_{20}H_{32}O_2$

Mol. wt.: 304

Solvent: $CDCl_3$

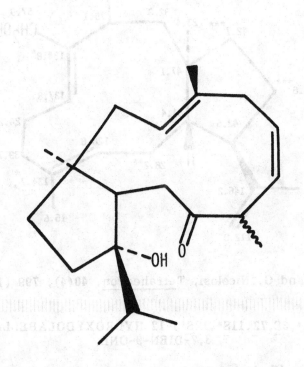

1)

CH_3: 17.2, 17.6, 17.9, 18.0 and 22.6

CH_2: 29.5, 32.5, 36.1, 40.2 and 42.4

CH: 35.6, 45.4, 46.5, 122.8, 129.8 and 132.5

C: 45.3, 85.8 and 135.0

C=O: 213.4

2)

CH_3: 15.8, 17.0, 17.6, 17.9 and 20.9

CH_2: 33.9, 36.8, 38.4, 40.8 and 42.2

CH: 36.2, 47.3, 48.8, 122.7, 127.2 and 134.8

C: 45.9, 86.0 and 136.7

C=O: 213.7

Note: The data are given for 1:1 mixture of 8R and 8S compounds.

C.B. Rao, K.C. Pullaiah, R.K. Surapaneni, B.W. Sullivan, K.F. Albizati, D.J. Faulkner, He. Cun-Heng and J. Clardy, J. Org. Chem., 51 (14), 2736 (1986).

3(R*),16-DIHYDROXY-1(S*),11(R*),12(S*)-DOLABELL-4(E),8(E),18-TRIENE

Source: *Dictyota* sp.

Mol. formula: $C_{20}H_{32}O_2$

Mol. wt.: 304

Solvent: $CDCl_3$

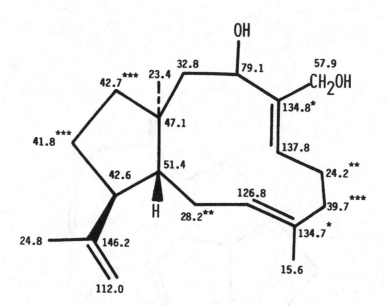

C. Tringali, M. Piattelli and G. Nicolosi, <u>Tetrahedron</u>, 40(4), 799 (1984).

(1S*,3E,7Z,11S*,12S*)-12-HYDROXYDOLABELLA-3,7-DIEN-9-ONE

Source: *Dictyota dichotoma* (Huds.) Lamouroux

Mol. formula: $C_{20}H_{32}O_2$

Mol. wt.: 304

Solvent: $CDCl_3$

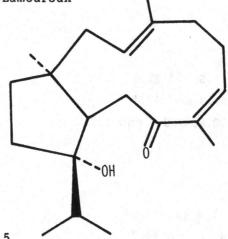

CH_3: 17.4, 17.6, 18.1, 21.2 and 24.5

CH_2: 24.9, 31.0, 37.1, 37.6, 40.6 and 43.0

CH: 34.2, 47.9, 122.2 and 135.6

C: 44.4, 87.6, 134.9 and 137.8

C=O: 207.4

C.B. Rao, K.C. Pullaiah, R.K. Surapaneni, B.W. Sullivan, K.F. Albizati, D.J. Faulkner, He. Cun-Heng and J. Clardy, <u>J. Org. Chem.</u>, 51 (14), 2736 (1986).

18-HYDROXYDOLABELL-7E-ENONE

Source: *Barbilophozia floerkei* (Web. et Mohr) Loeske

Mol. formula: $C_{20}H_{34}O_2$

Mol. wt.: 306

Solvent: $CDCl_3$

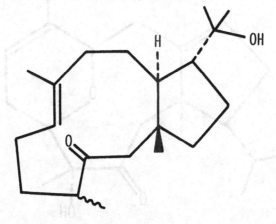

CH_3: 16.6, 17.8, 24.2, 26.6, 29.3 and 34.1

CH_2: 25.6, 25.6, 29.3, 37.7, 39.5, 43.3 and 49.0

CH: 43.6, 58.4 and 126.1

C: 44.5, 74.0 and 136.7

C=O: 214.6

S. Huneck, G.A. Baxter, A.F. Cameron, J.D. Connolly, L.J. Harrison, W.R. Phillips, D.S. Rycroft and G.A. Sim, J. Chem. Soc. Perkin Trans. I, (5), 809 (1986).

||||||||||||||||||||||||||||||||||||||||||||||||||||||||||||||||||||||||||||||||||||||||||||||||||||||||||||||||||

(1S*,3S*,4R*,7E,11S*,12S*)-3,4-EPOXY-12-HYDROXYDO-LABELLA-7-EN-9-ONE

Source: *Dictyota dichotoma* (Huds.) Lamouroux

Mol. formula: $C_{20}H_{32}O_3$

Mol. wt.: 320

Solvent: $CDCl_3$

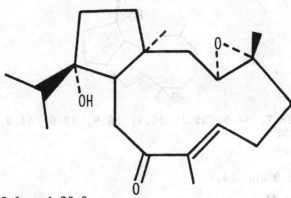

CH_3: 11.9, 15.4, 17.7, 18.1 and 23.8

CH_2: 24.5, 29.6, 36.3, 37.9, 41.0 and 43.5

CH: 34.2, 48.6, 63.0 and 142.3

C: 42.9, 61.8, 87.1 and 135.5

C=O: 207.1

C.B. Rao, K.C. Pullaiah, R.K. Surapaneni, B.W. Sullivan, K.F. Albizati, D.J. Faulkner, H. Cun-Heng and J. Clardy, J. Org. Chem., 51(14), 2736 (1986).

Source: *Cespitularia* sp.

Mol. formula: $C_{20}H_{30}O_3$

Mol. wt.: 318

Solvent: C_5D_5N

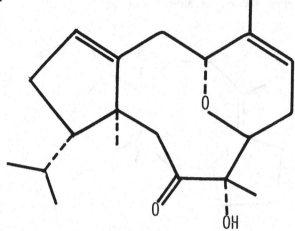

CH$_3$: 16.8, 19.2, 22.3, 22.7 and 22.8

CH$_2$: 25.7, 28.7, 34.5 and 40.6

CH: 29.6, 49.4, 72.3, 81.1, 121.4 and 124.8

C: 53.8, 78.7, 136.7 and 149.2

C=O: 209.3

B.F. Bowden. J.C. Coll, J.M. Gulbis, M.F. Mackay and R.H. Willis, Aust. J. Chem., 39 (5), 803 (1986).

||||||||||||||||||||||||||||||||||||||||||||||||||||||||||||||||||||||||||||||||||||||||||||||||||||||||||||||||||||||||||||||||||||

(1S*,2E,4R*,6Z,11S*,12S*)-4,12-DIHYDROXYDOLABELLA-2,6-DIEN-9-ONE

Source: *Dictyota dichotoma* (Huds.) Lamouroux

Mol. formula: $C_{20}H_{32}O_3$

Mol. wt.: 320

Solvent: CDCl$_3$

1)

17.3, 17.9, 18.2, 21.7, 28.7, 34.5, 35.1, 36.7, 38.9, 42.6, 44.8, 49.6, 53.4, 77.3, 83.9, 127.8, 131.4, 135.2 and 137.6

2)

CH$_3$: 17.0, 17.4, 17.7, 18.0 and 24.3

CH$_2$: 35.4, 36.8, 38.6 and 41.1

CH: 47.1, 53.6, 124.9, 132.1, 135.1 and 137.3

C: 47.3, 73.5 and 84.2

C=O: 214.0

Note: The data are given for 1:1 mixture of 8R and 8S compounds.

C.B. Rao, K.C. Pullaiah, R.K. Surapaneni, B.W. Sullivan, K.F. Albizati, D.J. Faulkner, He. Cun-Heng and J. Clardy, J. Org. Chem., 51 (14), 2736 (1986).

(1S*,2E, 4R*, 7E, 11S*, 12S*)-4,12-DIHYDROXYDOLABELLA-2,7-DIEN-9-ONE

Source: *Dictyota dichotoma* (Huds.) Lamouroux

Mol. formula: $C_{20}H_{32}O_3$

Mol. wt.: 320

Solvent: CD_3OD

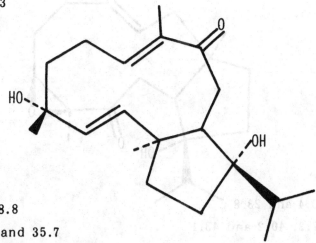

CH$_3$: 12.2, 17.9 and 18.8

CH$_2$: 26.0, 32.2, 34.0 and 35.7

CH: 24.7, 50.5, 135.6, 141.6 and 147.8

C: 39.6, 42.5, 73.6, 86.9 and 133.4

C=O: 208.5

C.B. Rao, K.C. Pullaiah, R.K. Surapaneni, B.W. Sullivan, K.F. Albizati, D.J. Faulkner, H. Cun-Heng and J. Clardy, J. Org. Chem., 51 (14), 2736 (1986).

||||||||||||||||||||||||||||||||||||||||||||||||||||||||||||||||||||||||||||||||||||||||||||||||||||

(1S*,2E,4R*,7Z,11S*,12S*)-4,12-DIHYDROXYDOLABELLA-2,7-DIEN-9-ONE

Source: *Dictyota dichotoma* (Huds.) Lamouroux

Mol. formula: $C_{20}H_{32}O_3$

Mol. wt.: 320

Solvent: $CDCl_3$

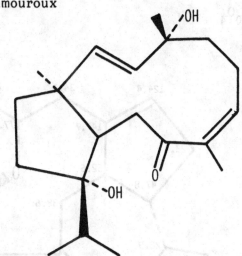

CH$_3$: 17.1, 18.0, 18.0, 20.1 and 25.5

CH$_2$: 23.1, 35.4, 35.7, 39.3 and 40.3

CH: 36.9, 52.9, 132.3, 138.6 and 139.4

C: 46.5, 73.7, 84.5 and 136.0

C=O: 206.4

C.B. Rao, K.C. Pullaiah, R.K. Surapaneni, B.W. Sullivan, K.F. Albizati, D.J. Faulkner, H. Cun-Heng and J. Clardy, J. Org. Chem., 51 (14), 2736 (1986).

(1S*,3E,7R*,8R*,11S*,12S*)-7,8-EPOXY-12-HYDROXYDO-LABELLA-3-EN-9-ONE

Source: *Dictyota dichotoma* (Huds.) Lamouroux

Mol. formula: $C_{20}H_{32}O_3$

Mol. wt.: 320

Solvent: $CDCl_3$

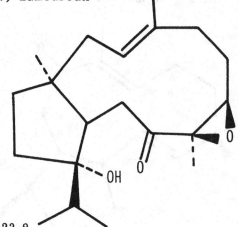

CH_3: 16.5, 17.2, 18.0, 20.4 and 23.8

CH_2: 23.5, 32.7, 34.1, 37.2, 40.2 and 43.1

CH: 34.9, 46.2, 66.5 and 121.6

C: 44.3, 62.7, 86.3 and 135.2

C=O: 205.8

C.B. Rao, K.C. Pullaiah, R.K. Surapaneni, B.W. Sullivan, K.F. Albizati, D.J. Faulkner, H. Cun-Heng and J. Clardy, J. Org. Chem., 51 (14), 2736 (1986).

||||||||||||||||||||||||||||||||||||||||||||||||||||||||||||||||||||||||||||||||||||||||||||||||||||||

NEODOLABELLINE

Source: *Clavularia koellikeri*

Mol. formula: $C_{20}H_{30}O_4$

Mol. wt.: 334

Solvent: $DMSO-d_6$

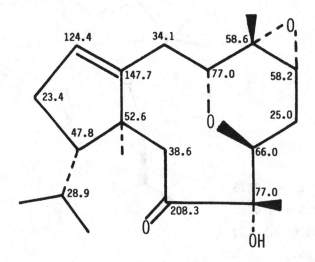

CH_3: 15.9, 20.7, 21.6, 22.3 and 22.8

M. Kobayashi, B.W. Son, T. Fujiwara, Y. Kyogoku and I. Kitagawa, Tetrahedron Lett., 25 (48), 5543 (1984).

(1R*,3E,7E,9S*,11S*)-9-ACETOXYDOLABELLA-3,7, 12-TRIEN-16-AL

Source: *Dictyota dichotoma* (Huds.) Lamouroux

Mol. formula: $C_{22}H_{32}O_3$

Mol. wt.: 344

Solvent: $CDCl_3$

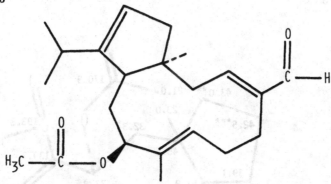

CH$_3$: 11.1, 21.5, 21.5, 22.2 and 25.9

CH$_2$: 23.1, 27.6, 30.0, 40.8 and 46.3

CH: 27.0, 46.3, 79.8, 119.3, 133.2 and 154.7

C: 43.7, 131.5, 142.5 and 151.5

C=O: 170.3 and 195.2

C.B. Rao, K.C. Pullaiah, R.K. Surapaneni, B.W. Sullivan, K.F. Albizati, D.J. Faulkner, H. Cun-Heng and J. Clardy, J. Org. Chem., 51 (14), 2736 (1986).

||||||||||||||||||||||||||||||||||||||||||||||||||||||||||||||||||||||||||||||||||||||||||||||||||||||||||||||||||||||||||||

3(R*)-ACETOXY-1(S*),11(R*),12(S*)-DOLABELL-4(Z), 8(E),18-TRIEN-16-AL

Source: *Dictyota* sp.

Mol. formula: $C_{22}H_{32}O_3$

Mol. wt.: 344

Solvent: $CDCl_3$

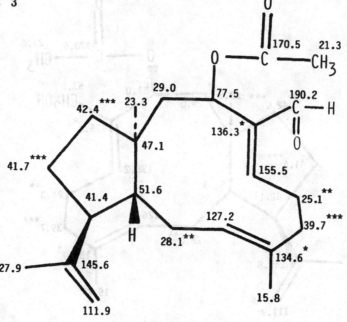

C. Tringali, M. Piattelli and G. Nicolosi, Tetrahedron, 40(4), 799 (1984).

3(<u>S</u>)-ACETOXY-1(<u>R</u>),11(<u>S</u>),12(<u>R</u>)-DOLABELL-4(<u>E</u>),8(<u>E</u>),18-TRIEN-16-AL

Source: *Dictyota* sp.

Mol. formula: $C_{22}H_{32}O_3$

Mol. wt.: 344

Solvent: $CDCl_3$

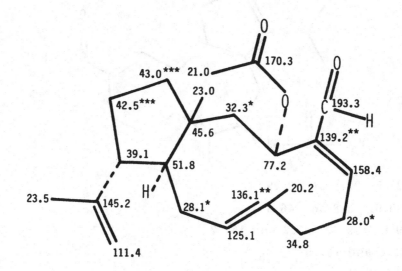

C. Tringali, G. Oriente, M. Piattelli and G. Nicolosi, <u>J. Jat. Prod.</u>, (Lloydia), 47 (4), 615 (1984).

||||||||||||||||||||||||||||||||||||||||||||||||||||||||||||||||||||||||||||||||||||||||||||||||||||

3(<u>R</u>*)-ACETOXY-16-HYDROXY-1(<u>S</u>*),11(<u>R</u>*),12(<u>S</u>*)-DOLABELL-4(<u>E</u>),8(<u>E</u>),18-TRIENE

Source: *Dictyota* sp.

Mol. formula: $C_{22}H_{34}O_3$

Mol. wt.: 346

Solvent: $CDCl_3$

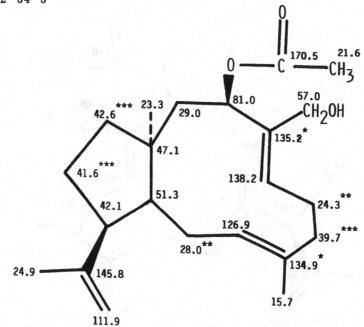

C. Tringali, M. Piattelli and G. Nicolosi, <u>Tetrahedron</u>, 40(4), 799 (1984).

3(S)-HYDROXY-16-ACETOXY-1(R),11(S),12(R)-DOLABELL-4(E),8(E),18-TRIENE

Source: *Dictyota* sp.

Mol. formula: $C_{22}H_{34}O_3$

Mol. wt.: 346

Solvent: $CDCl_3$

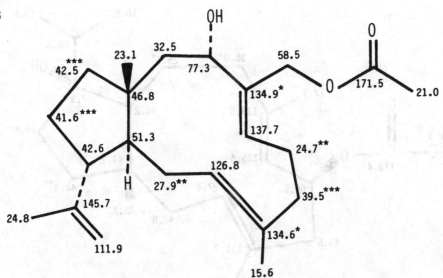

C. Tringali, G. Oriente, M. Piattelli and G. Nicolosi, J. Nat. Prod., (Lloydia), 47 (4), 615 (1984).

(1S*,3E,7E,11S*,12S*)-9-ACETOXYDOLABELLA-3, 7-DIEN-12-OL

Source: *Dictyota dichotoma* (Huds.) Lamouroux

Mol. formula: $C_{22}H_{36}O_3$

Mol. wt.: 348

Solvent: $CDCl_3$

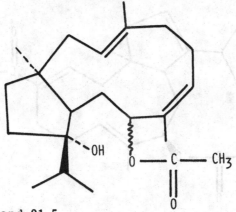

CH_3: 12.9, 15.6, 18.0, 18.7 and 21.5

CH_2: 23.9, 24.2, 29.7, 30.4, 38.6, 38.9, 40.4 and 44.3

CH: 35.9, 79.9, 124.0 and 126.8

C: 44.5, 86.1, 134.1 and 134.2

C=O: 171.4

C.B. Rao, K.C. Pullaiah, R.K. Surapaneni, B.W. Sullivan, K.F. Albizati, D.J. Faulkner, H. Cun-Heng and J. Clardy, J. Org. Chem., 51 (14), 2736 (1986).

(+)-ACETOXYODONTOSCHISMENOL

Source: *Odontoschisma denudatum* (Nees) Dum.

Mol. formula: $C_{22}H_{36}O_3$

Mol. wt.: 348

Solvent: $CDCl_3$

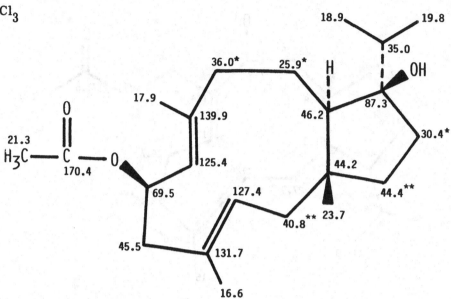

A. Matsuo, KI.-I. Kamio, K. Uohama, K.-I. Yoshida, J.D. Connolly and G.A. Sim, Phytochemistry, 27(4), 1153 (1988).

‖‖‖

(1R*,3E,7E,9S*,11S*)-9-ACETOXYDOLABELLA-3,7, 12-TRIEN-16-OIC ACID

Source: *Dictyota dichotoma* (Huds.) Lamouroux

Mol. formula: $C_{22}H_{32}O_4$

Mol. wt.: 360

Solvent: C_6D_6

CH_3: 11.8, 21.4, 22.6, 23.0 and 27.9

CH_2: 23.4, 32.7, 35.8, 42.7 and 49.4

CH: 26.5, 48.9, 80.3, 119.8, 133.0 and 148.0

C: 45.5, 130.6, 134.1 and 153.3

C=O: 170.1 and 174.2

C.B. Rao, K.C. Pullaiah, R.K. Surapaneni, B.W. Sullivan, K.F. Albizati, D.J. Faulkner, H. Cun-Heng and J. Clardy, J. Org. Chem., 51 (14), 2736 (1986).

(1R*,2R*,3E,7E,11S*,12S*)-2-ACETOXY-12-HYDROXYDOLA-BELLA-3,7-DIEN-9-ONE

Source: *Dictyota dichotoma* (Huds.) Lamouroux

Mol. formula: $C_{22}H_{34}O_4$

Mol. wt.: 362

Solvent: $CDCl_3$

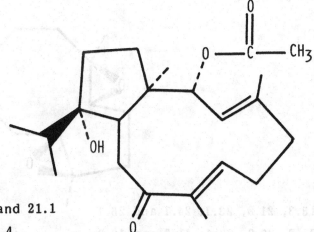

CH$_3$: 11.8, 15.6, 17.5, 18.0, 18.0 and 21.1

CH$_2$: 23.7, 29.6, 35.4, 36.4 and 39.4

CH: 33.9, 46.3, 75.6, 122.8 and 143.6

C: 48.4, 86.2, 133.7 and 140.6

C=O: 170.4 and 208.0

C.B. Rao, K.C. Pullaiah, R.K. Surapaneni, B.W. Sullivan, K.F. Albizati, D.J. Faulkner, H. Cun-Heng and J. Clardy, J. Org. Chem., 51(14), 2736 (1986).

(1S*,3E,11S*,12S*)-9-ACETOXY-7,8-EPOXYDOLABELLA-3-EN-12-OL

Source: *Dictyota dichotoma* (Huds.) Lamouroux

Mol. formula: $C_{22}H_{36}O_4$

Mol. wt.: 364

Solvent: $CDCl_3$

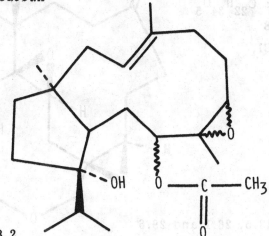

CH$_3$: 11.4, 15.6, 17.9, 18.7, 21.4 and 23.2

CH$_2$: 24.4, 27.3, 30.9, 36.9, 37.6 and 37.6

CH: 36.9, 44.5, 61.0, 81.9 and 123.3

C: 44.0, 61.9, 84.6 and 133.6

C=O: 172.1

C.B. Rao, K.C. Pullaiah, R.K. Surapaneni, B.W. Sullivan, K.F. Albizati, D.J. Faulkner, H.Cun-Heng and J. Clardy, J. Org. Chem., 51(14), 2736 (1986).

10-DEACETOXYBARBILYCOPODIN

Source: *Barbilophozia floerkei* (Web. et Mohr) Loeske

Mol. formula: $C_{22}H_{36}O_4$

Mol. wt.: 364

Solvent: $CDCl_3$

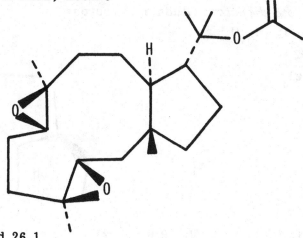

CH_3: 17.1, 18.3, 21.9, 23.1, 24.1 and 26.1

CH_2: 22.8, 25.8, 30.9, 37.1, 41.5 and 43.2

CH: 32.2, 59.7, 61.6 and 64.9

C: 44.1, 61.1, 61.5 and 84.9

C=O: 170.3

S. Huneck, G.A. Baxter, A.F. Cameron, J.D. Connolly, L.J. Harrison, W.R. Phillips, D.S. Raycroft and G.A. Sim, J. Chem. Soc. Perkin Trans. I, (5), 809 (1986).

||||||||||||||||||||||||||||||||||||||||||||||||||||||||||||||||||||||||||||||||||||||||||||||||||||||||||||

STOLONIDIOL MONOACETATE

Source: *Clavularia viridis*

Mol. formula: $C_{22}H_{34}O_5$

Mol. wt.: 378

Solvent: $CDCl_3$

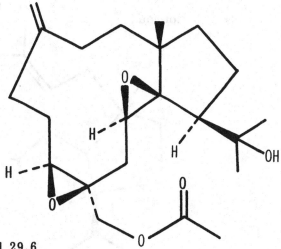

CH_3: 21.8, 23.5, 26.1 and 29.6

CH_2: 24.6, 26.6, 27.4, 29.4, 31.1, 37.1, 38.9, 67.9 and 110.2

CH: 48.5, 56.3 and 58.6

C: 44.5, 60.9, 74.5, 76.2 and 148.2

C=O: 170.5

K. Mori, K. Iguchi, N. Yamada, Y. Yamada and Y. Inouye, Chem. Pharm. Bull., (Tokyo), 36(8), 2840 (1988).

10R,18-DIACETOXY-3S,4S-EPOXYDOLABELL-7E-ENE

Source: *Barbilophozia floerkei* (Web. et Mohr) Loeske

Mol. formula: $C_{24}H_{38}O_5$

Mol. wt.: 406

Solvent: $CDCl_3$

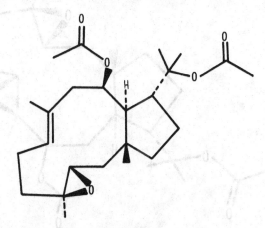

CH$_3$: 16.2, 16.8, 21.5, 22.8, 23.0, 23.9 and 25.8

CH$_2$: 24.3, 25.7, 38.6, 42.0, 42.2 and 45.7

CH: 44.4, 54.8, 63.8, 76.6 and 128.6

C: 44.4, 62.3, 84.6 and 130.4

C=O: 170.4 and 170.8

S. Huneck, G.A. Baxter, A.F. Cameron, J.D. Connolly, L.J. Harrison, W.R. Phillips, D.S. Rycroft and G.A. Sim, J. Chem. Soc. Perkin Trans. I, (5), 809 (1986).

|||||||||||||||||||||||||||||||||||||||||||||||||||||||||||||||||||||||||||||||||||||||||||||||||||||||||||||

DITERPENE AD-1

Source: *Aplysia dactylomela*

Mol. formula: $C_{24}H_{38}O_5$

Mol. wt.: 406

Solvent: $CDCl_3$

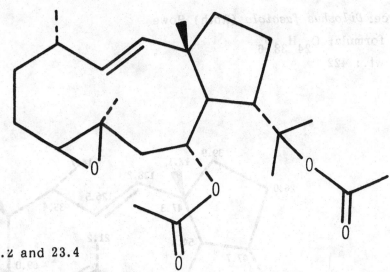

CH$_3$: 15.9, 16.8, 20.7, 21.1, 23.2 and 23.4

CH$_2$: 22.8, 32.3, 39.9 and 44.5

CH: 33.5, 44.3, 50.6, 63.4, 69.9, 134.2 and 136.0

C: 48.9, 60.4 and 84.8

C=O: 170.3

A.G. Gonzalez, F. Cataldo, J. Fernandez and M. Norte, J. Nat. Prod.. (Lloydia), 50 (6), 1158 (1987).

BARBILYCOPODIN

Source: *Barbilophozia floerkei* (Web. et Mohr) Loeske

Mol. formula: $C_{24}H_{38}O_6$

Mol. wt.: 422

Solvent: $CDCl_3$

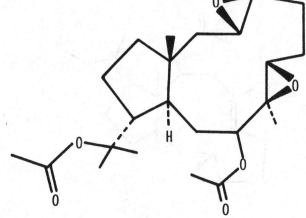

CH_3: 17.0, 17.5, 21.3, 22.7, 23.6, 23.6 and 26.0

CH_2: 23.2, 25.5, 37.2, 40.4, 41.6 and 45.2

CH: 45.2, 55.5, 62.4, 65.6 and 75.4

C: 44.3, 58.8, 61.3 and 84.4

C=O: 170.2 and 170.3

S. Huneck, G.A. Baxter, A.F. Cameron, J.D. Connolly, L.J. Harrison, W.R. Phillips, D.S. Rycroft and G.A. Sim, J. Chem. Soc. Perkin Trans. I, (5), 809 (1986).

||||||||||||||||||||||||||||||||||||||||||||||||||||||||||||||||||||||||||||||||||||||||||||||||||||||||

(1R,2E,4S,5R,6S,7E,10S,11S,12R)-5,6-DIACETOXY-10,18-DIHYDROXY-2,7-DOLABELLADIENE

Source: *Dilophus fasciola* (Roth) Howe

Mol. formula: $C_{24}H_{38}O_6$

Mol. wt.: 422

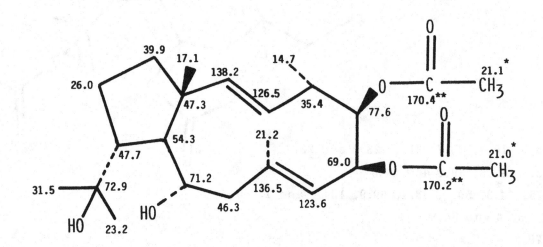

S.De Rosa, S.De Stefano, S. Macura, E. Trivellone and N. Zavodnik, Tetrahedron, 40(23), 4991 (1984).

(1R,2E,4S,5R,6S,7E,10S,11S,12R)-5,6,10,18-TETRAACETOXY-2,7-DOLABELLADIENE

Source: *Dilophus fasciola* (Roth) Howe

Mol. formula: $C_{28}H_{42}O_8$

Mol. wt.: 506

Solvent: $CDCl_3$

CH_3: 15.2, 16.8, 21.2, 23.0, 23.4 and 26.4

CH_2: 26.7, 39.9 and 41.6

CH: 35.4, 44.5, 52.4, 69.1, 72.9, 77.7, 125.3, 128.2 and 137.4

C: 48.2, 84.7 and 134.6

C=O: 170.1, 170.2, 170.3 and 170.6

S.De Rosa, S.De Stefano, S. Macura, E. Trivellone and N. Zavodnik, Tetrahedron, 40(23), 4991 (1984).

||||||||||||||||||||||||||||||||||||||||||||||||||||||||||||||||||||||||||||||||||||||||||||||||||||||||||||||||||||||

(1R,2E,4S,5R,6S,7E,10S,11S,12R)-5,6,10-TRIACETOXY-18-HYDROXY-2,7-DOLABELLADIENE

Source: *Dilophus fasciola* (Roth) Howe

Mol. formula: $C_{26}H_{40}O_7$

Mol. wt.: 464

Solvent: $CDCl_3$

CH_3: 15.1, 16.8, 20.9, 21.2, 22.8 and 31.6

CH_2: 26.8, 40.0 and 41.5

CH: 35.3, 48.8, 52.4, 69.1, 73.5, 77.6, 125.1, 127.9 and 137.5

C: 47.8, 72.7 and 134.6

C=O: 169.1, 169.9 and 170.3

S.De Rosa, S.De Stefano, S. Macura, E. Trivellone and N. Zavodnik, Tetrahedron, 40(23), 4991 (1984).

Basic skeleton

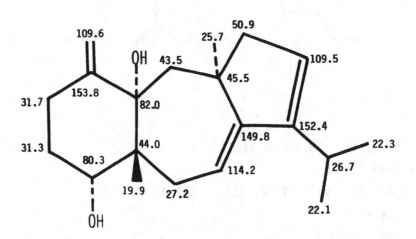

||||||||||||||||||||||||||||||||||||||||||||||||||||||||||||||||||||||||||||||||||||||||||||||||||||||||

(4R,14S)-DIHYDROXYDOLAST-1(15),7,9-TRIENE

Source: *Dictyota* sp.

Mol. formula: $C_{20}H_{30}O_2$

Mol. wt.: 302

Solvent: $CDCl_3$

A.G. Gonzalez, J.D. Martin, M. Norte, P. Rivera, A. Perales and J. Fayos, <u>Tetrahedron</u>, 39(20), 3355 (1983).

(5S*,8S*,9S*,12S*,14R*)-9-HYDROXYDOLASTA-1, 3-DIEN-6-ONE

Source: *Dictyota dichotoma* (Huds.) Lamouroux

Mol. formula: $C_{20}H_{30}O_2$

Mol. wt.: 302

Solvent: $CDCl_3$

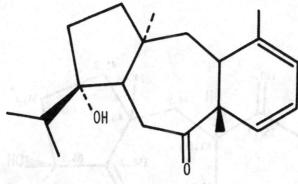

CH_3: 17.6, 18.4, 22.1, 23.3 and 25.1

CH_2: 35.8, 38.2, 40.1 and 41.4

CH: 37.5, 44.4, 45.7, 117.8, 121.6 and 126.7

C: 43.3, 53.2, 83.3 and 136.4

C=O: 213.1

C.B. Rao, K.C. Pullaiah, R.K. Surapaneni, B.W. Sullivan, K.F. Albizati, D.J. Faulkner, H. Cun-Heng and J. Clardy, J. Org. Chem., 51(14), 2736 (1986).

||||||||||||||||||||||||||||||||||||||||||||||||||||||||||||||||||||||||||||||||||||||||||||||||

(4R,9R,14S)-4,9,14-TRIHYDROXYDOLAST-1(15),7-DIENE

Source: *Dictyota* sp.

Mol. formula: $C_{20}H_{32}O_3$

Mol. wt.: 320

Solvent: $CDCl_3$

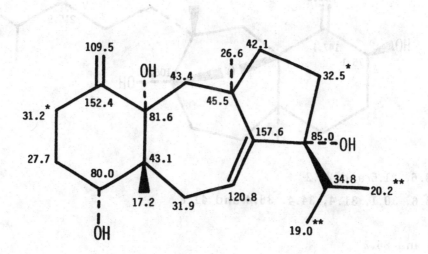

A.G. Gonzalez, J.D. Martin, M. Norte, P. Rivera, A. Perales and J. Fayos, Tetrahedron, 39(20), 3355 (1983).

(4R,9S,14S)-4,9,14-TRIHYDROXYDOLAST-1(15),7-DIENE

Source: *Dictyota* sp.

Mol. formula: $C_{20}H_{32}O_3$

Mol. wt.: 320

Solvent: $CDCl_3$

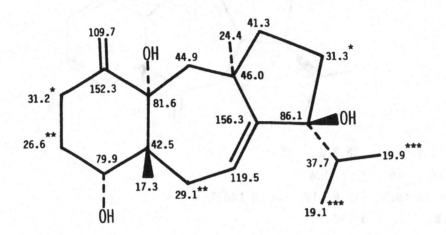

A.G. Gonzalez, J.D. Martin, M. Norte, P. Rivera, A. Perales and J. Fayos, Tetrahedron, 39(20), 3355 (1983).

||||||||||||||||||||||||||||||||||||||||||||||||||||||||||||||||||||||||||||||||||||||

LINEAROL

Source: *Dictyota linearis*

Mol. formula: $C_{20}H_{32}O_4$

Mol. wt.: 336

Solvent: $CDCl_3$

CH_3: 18.5, 18.6, 21.6 and 23.2

CH_2: 27.9, 28.6, 30.1, 31.4, 34.4, 36.4 and 41.8

CH: 41.3

C: 37.5, 43.9 and 86.2

M. Ochi, I. Miura and T. Tokoroyama, J. Chem. Soc. Chem. Commun., (3), 100 (1981).

DITERPENE DC-II

Source: *Dictyota cervicornis*
Mol. formula: $C_{20}H_{34}O_4$
Mol. wt.: 338
Solvent: $CDCl_3$

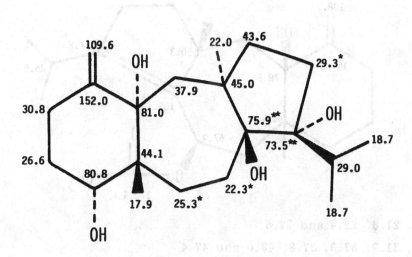

A. Kelecom and V. Teixeira, <u>Phytochemistry</u>, 27(9), 2907 (1988).

||||||||||||||||||||||||||||||||||||||||||||||||||||||||||||||||||||||||||||||||||||||||||||||||||||||||||||||||||||||||||||||||||||||||||

4(S*)-ACETOXY-14(S*)-HYDROXYDOLAST-1(15),7,9-TRIENE

Source: *Dictyota divaricata* Lamouroux
Mol. formula: $C_{22}H_{32}O_3$
Mol. wt.: 344
Solvent: $CDCl_3$

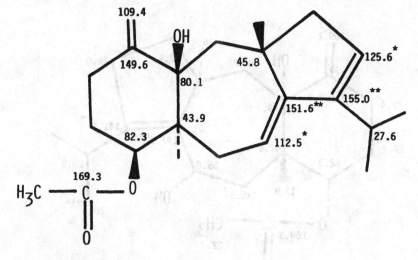

CH_3: 20.2, 21.3, 22.1, 22.1 and 25.6
CH_2: 26.9, 28.4, 30.8, 41.9 and 51.1

H.H. Sun, O.J. McConnell, W. Fenical, K. Hirotsu and J. Clardy, <u>Tetrahedron</u>, 37(6), 1237 (1981).

7(S*)-ACETOXY-14(S*)-HYDROXYDOLAST-1(15),8-DIENE

Source: *Dictyota divaricata* Lamouroux

Mol. formula: $C_{22}H_{34}O_3$

Mol. wt.: 346

Solvent: $CDCl_3$

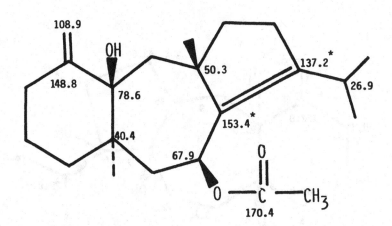

CH$_3$: 17.8, 20.3, 21.6, 22.4 and 27.6

CH$_2$: 27.4, 29.7, 31.9, 37.3, 37.8, 43.0 and 47.4

H.H. Sun, O.J. McConnell, W. Fenical, K. Hirotsu and J. Clardy, <u>Tetrahedron</u>, 37(6), 1237 (1981).

(4R,7R,14S)-4-ACETOXY-7,14-DIHYDROXYDOLAST-1(15),8-DIENE

Source: *Dictyota* sp.

Mol. formula: $C_{22}H_{34}O_4$

Mol. wt.: 362

Solvent: $CDCl_3$

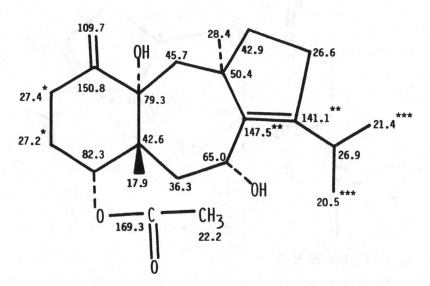

A.G. Gonzalez, J.D. Martin, M. Norte, P. Rivera, A. Perales and J. Fayos, <u>Tetrahedron</u>, 39(20), 3355 (1983).

(4R,9S,14S)-4-ACETOXY-9,14-DIHYDROXYDOLAST-1(15),7-DIENE

Source: *Dictyota* sp.

Mol. formula: $C_{22}H_{34}O_4$

Mol. wt.: 362

Solvent: $CDCl_3$

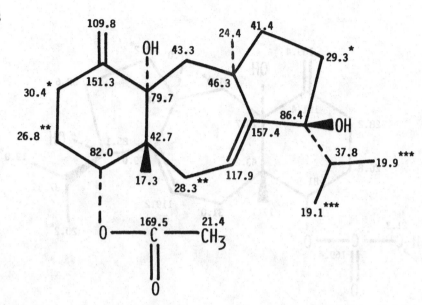

A.G. Gonzalez, J.D. Martin, M. Norte, P. Rivera, A. Perales and J. Fayos, Tetrahedron, 39(20), 3355 (1983).

||||||||||||||||||||||||||||||||||||||||||||||||||||||||||||||||||||||||||||||||||||||||||||||||||||||||||||||||||

7(S*)-ACETOXY-4(S*),14(S*)-DIHYDROXYDOLAST-1(15),8-DIENE

Source: *Dictyota divaricata* Lamouroux

Mol. formula: $C_{22}H_{34}O_4$

Mol. wt.: 362

Solvent: $CDCl_3$

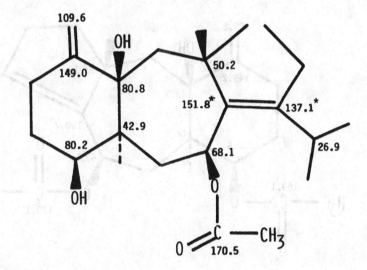

CH_3: 17.8, 20.3, 21.6, 21.6 and 27.4

CH_2: 26.5, 27.4, 30.2, 34.3, 42.9 and 47.1

H.H. Sun, O.J. McConnell, W. Fenical, K. Hirotsu and J. Clardy, Tetrahedron, 37(6), 1237 (1981).

Source: *Dictyota cervicornis*

Mol. formula: $C_{22}H_{34}O_4$

Mol. wt.: 362

Solvent: $CDCl_3$

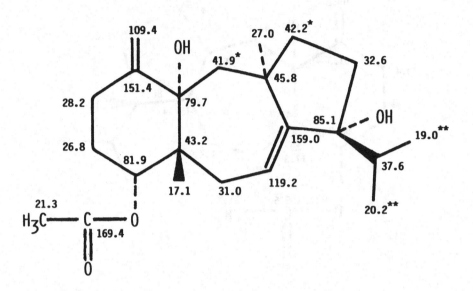

A. Kelecom and V.L. Teixeira, <u>Phytochemistry</u>, **27**(9), 2907 (1988).

4(S*),7(S*)-DIACETOXY-14(S*)-HYDROXYDOLAST-1(15),8-DIENE

Source: *Dictyota divaricata* Lamouroux

Mol. formula: $C_{24}H_{36}O_5$

Mol. wt.: 404

Solvent: $CDCl_3$

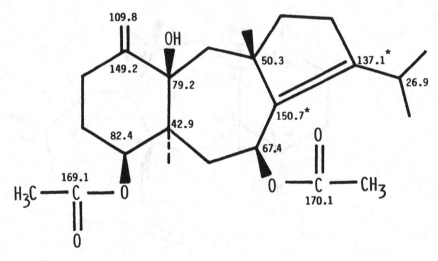

CH_3: 17.9, 20.2, 21.4, 21.5, 21.5 and 27.6

CH_2: 26.5, 27.1, 27.4, 33.3, 42.9 and 45.7

H.H. Sun, O.J. McConnell, W. Fenical, K. Hirotsu and J. Clardy, <u>Tetrahedron</u>, **37**(6), 1237 (1981).

Basic skeleton:

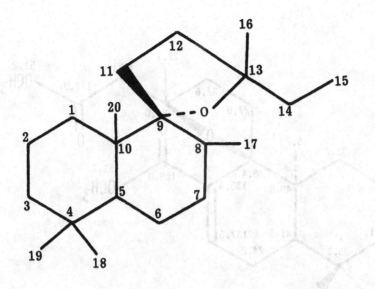

||||||||||||||||||||||||||||||||||||||||||||||||||||||||||||||||||||||||||||||||||||||||||||||||||||||

18-HYDROXYGRINDELIC ACID

Source: *Chrysothamnus nauseosus* (Pall.) Britt
Mol. formula: $C_{20}H_{32}O_4$
Mol. wt.: 336
Solvent: $CDCl_3$

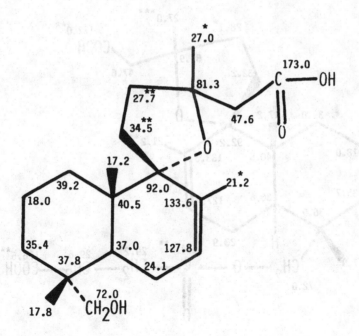

A.F. Rose, <u>Phytochemistry</u>, 19(12), 2689 (1980).

DIMETHYL-17-CARBOXYGRINDELATE

Source: *Grindelia camporum* Greene

Mol. formula: $C_{22}H_{34}O_5$

Mol. wt.: 378

Solvent: $CDCl_3$

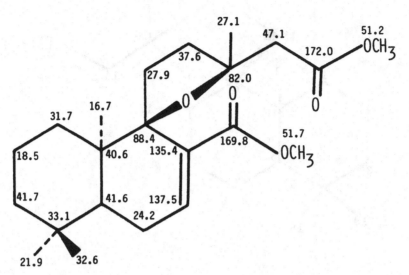

J.J. Hoffmann, S.D. Jolad, R.N. Timmermann, R.B. Bates and F.A. Camou, Phytochemistry, 27 (2), 493 (1988).

18-SUCCINYLOXYGRINDELIC ACID

Source: *Chrysothamnus nauseosus* (Pall.) Britt

Mol. formula: $C_{24}H_{36}O_7$

Mol. wt.: 436

Solvent: $CDCl_3$

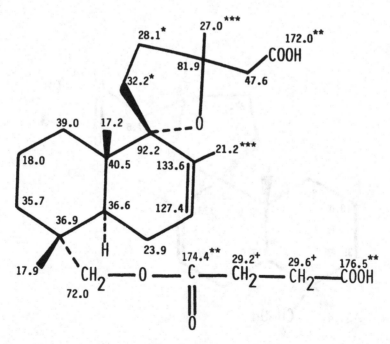

A.F. Rose, Phytochemistry, 19 (12), 2689 (1980).

Basic Skeleton

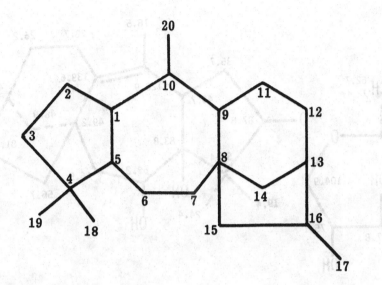

GRAYANOSIDE D

Source: *Leucothoe grayana*
Mol. formula: $C_{26}H_{42}O_9$
Mol. wt.: 498
Solvent: C_5D_5N

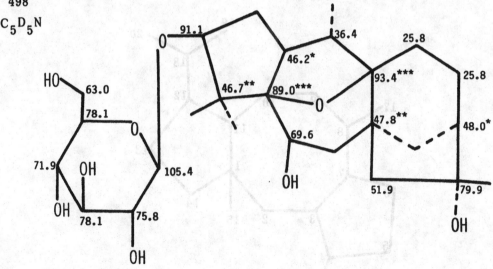

CH_3: 14.3, 19.8, 24.3 and 24.8
CH_2: 31.7, 40.3 and 40.6

J. Sakakibara and N. Shirai, Phytochemistry, 19 (10), 2159 (1980).

Source: *Pieris japonica* D.Don
Mol. formula: $C_{26}H_{42}O_9$
Mol. wt.: 498
Solvent: C_5D_5N

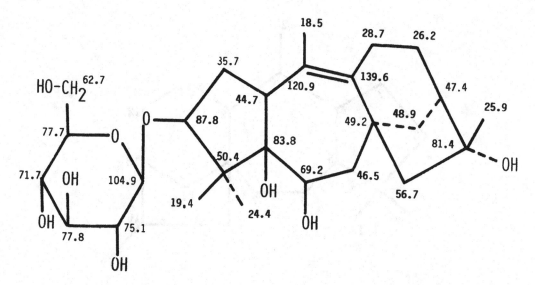

T. Kaiya and J. Sakakibara, <u>Chem. Pharm. Bull.</u>, (Tokyo), 33(10), 4637 (1985).

FUSICOCCIN

Basic skeleton

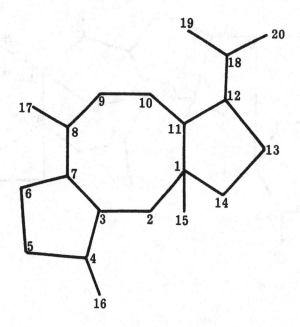

DITERPENE CT-I

Source: *Cercospora traversiana* Saccardo

Mol. formula: $C_{20}H_{28}O_3$

Mol. wt.: 316

CH_3: 17.0, 18.5 and 22.6

CH_2: 25.2, 28.5, 42.8, 43.9, 44.0 and 113.8

CH: 30.9, 49.8, 51.2, 54.9 and 167.3

C: 44.4, 79.1, 146.2 and 148.3

C=O: 193.6 and 220.6

A. Stoessl, J.B. Stothers and R.C. Zimmer, J. Chem. Soc. Chem. Commun., (7), 508 (1987).

|||||||||||||||||||||||||||||||||||||||||||||||||||||||||||||||||||||||||||||||||||||||||||||||||||||||||||||||||||||||||||||

ROSEADIONE

Source: *Hypoestes rosea* R. Br

Mol. formula: $C_{20}H_{30}O_3$

Mol. wt.: 318

CH_2: 25.5, 26.3, 37.0, 38.0 and 43.5

A.A. Adesomoju and J.I. Okogun, J. Nat. Prod., (Lloydia), 47 (2), 308 (1984).

515

ROSEANOLONE

Source: *Hypoestes rosea* R. Br

Mol. formula: $C_{20}H_{30}O_3$

Mol. wt.: 318

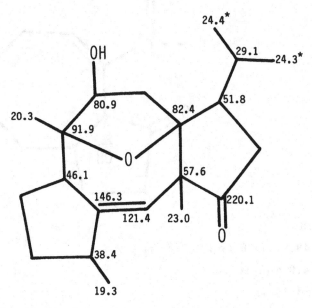

CH$_2$: 26.5, 31.4, 37.0 and 39.0

A.A. Adesomoju and J.I. Okogun, <u>J. Nat. Prod.</u>, (Lloydia), **49** (2), 308 (1984).

Seco-FUSICOCCIN

Basic skeleton

DICTYMAL

Source: *Dictyota dichotoma*
Mol. formula: $C_{20}H_{32}O$
Mol. wt.: 288
Solvent: C_6D_6

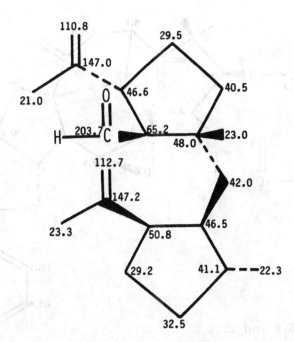

M. Segawa, N. Enoki, M. Ikura, K. Hikichi, R. Ishida, H. Shirahama and T. Matsumoto, Tetrahedron Lett., 28(32), 3703 (1987).

||||||||||||||||||||||||||||||||||||||||||||||||||||||||||||||||||||||||||||||||||||||||||||||||||

JATROPHANE

Basic skeleton

EUPHOHELIN D

Source: *Euphorbia helioscopia* L.
Mol. formula: $C_{31}H_{38}O_{10}$
Mol. wt.: 570
Solvent: $CDCl_3$

CH$_3$: 13.4, 15.2, 15.8, 16.9 and 22.4

S. Kosemura, Y. Shizuri and S. Yamamura, Bull. Chem. Soc. Jpn., **58** (11), 3112 (1985).

ENUKOKURIN

Source: *Euphorbia lateriflora* Schum and Thonner
Mol. formula: $C_{34}H_{44}O_{10}$
Mol. wt.: 612
Solvent: $CDCl_3$

C.O. Fakunle, J.D. Connolly and D.S. Rycroft, J. Nat. Prod., (Lloydia), 52 (2), 279 (1989).

EUPHOHELIN A

Source: *Euphorbia helioscopia* L.

Mol. formula: $C_{33}H_{44}O_{11}$

Mol. wt.: 616

Solvent: $CDCl_3$

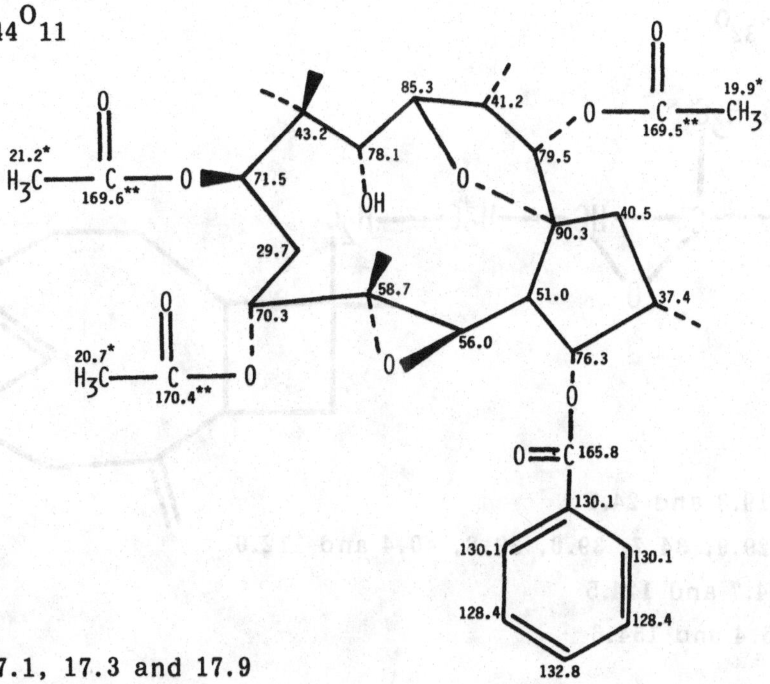

CH_3: 13.8, 14.1, 17.1, 17.3 and 17.9

S. Kosemura, Y. Shizuri and S. Yamamura, <u>Bull. Chem. Soc. Jpn.</u>, **58** (11), 3112 (1985).

‖‖

XENIAPHYLLANE

Basic skeleton:

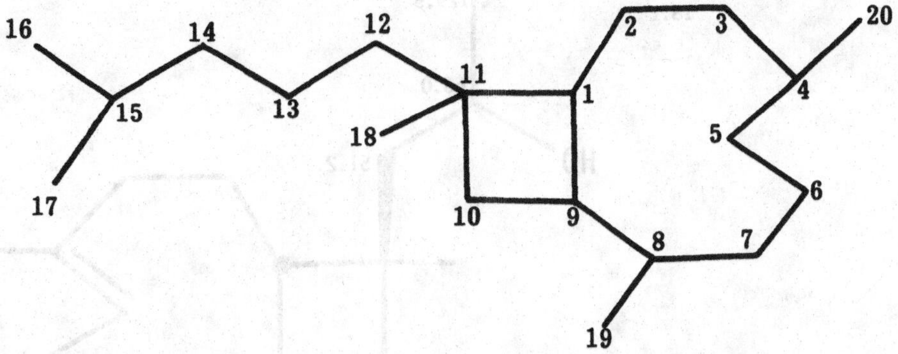

Source: *Nephthea chabrolii*

Mol. formula: $C_{20}H_{32}O$

Mol. wt.: 288

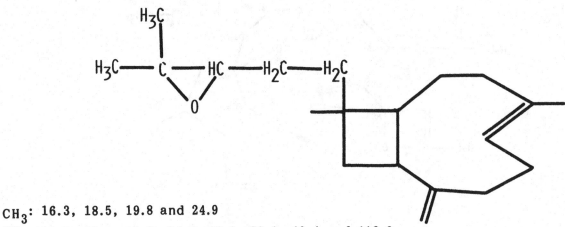

CH_3: 16.3, 18.5, 19.8 and 24.9

CH_2: 23.8, 28.4, 29.9, 34.7, 39.0, 39.8, 40.4 and 112.0

CH: 48.4, 52.3, 64.7 and 124.5

C: 35.8, 58.3, 135.4 and 154.6

A. Ahond, B.F. Bowden, J.C. Coll, J.-D. Fourneron and S.J. Mitchell, <u>Aust. J. Chem.</u>, **34**(12), 2657 (1981).

||||||||||||||||||||||||||||||||||||||||||||||||||||||||||||||||||||||||||||||||||||||||||||||||||||||||||||||

XENIAPHYLLENOL A

Source: *Xenia macrospiculata*

Mol. formula: $C_{20}H_{32}O$

Mol. wt.: 288

Solvent: $CDCl_3$ or C_6D_6

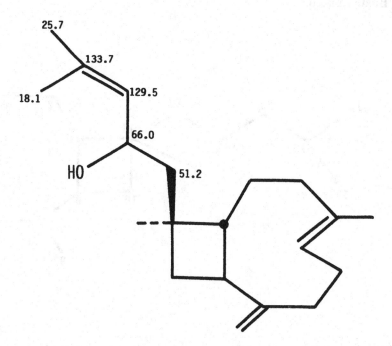

A. Groweiss and Y. Kashman, <u>Tetrahedron</u>, 39(20), 3385 (1983).

4,5,14,15-DIEPOXYXENIAPHYLL-8(19)-ENE (XENIAPHYLLENE-DIOXIDE)

Source: *Xenia macrospiculata*

Mol. formula: $C_{20}H_{32}O_2$

Mol. wt.: 304

Solvent: $CDCl_3$ or C_6D_6 [a]

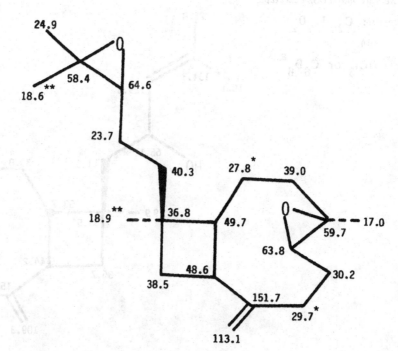

a: Not specified.

A. Groweiss and Y. Kashman, <u>Tetrahedron</u>, **39**(20), 3385 (1983).

XENIAPHYLLENOL B

Source: *Xenia macrospiculata*

Mol. formula: $C_{20}H_{32}O_2$

Mol. wt.: 304

Solvent: $CDCl_3$ or C_6D_6 [a]

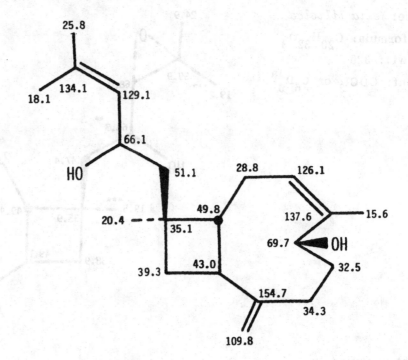

a: Not specified.

A. Groweiss and Y. Kashman, <u>Tetrahedron</u>, **39**(20), 3385 (1983).

XENIAPHYLLENOL C

Source: *Xenia macrospiculata*

Mol. formula: $C_{20}H_{32}O_2$

Mol. wt.: 304

Solvent: $CDCl_3$ or C_6D_6 [a]

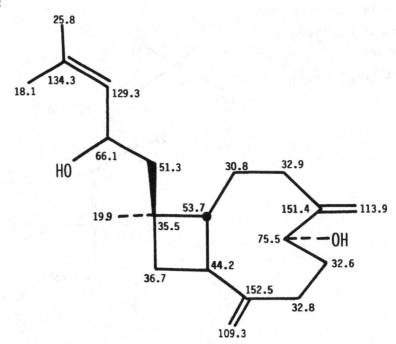

a: Not specified.

A. Groweiss and Y. Kashman, <u>Tetrahedron</u>, **39**(20), 3385 (1983).

||||||||||||||||||||||||||||||||||||||||||||||||||||||||||||||||||||||||||||||||||||||||||||||||||||||||

4,5,14,15-DIEPOXYXENIAPHYLL-8(19)-EN-13-OL
(DIEPOXY-XENIAPHYLLENOL-A)

Source: *Xenia lilielae*

Mol. formula: $C_{20}H_{32}O_3$

Mol. wt.: 320

Solvent: $CDCl_3$ or C_6D_6 [a]

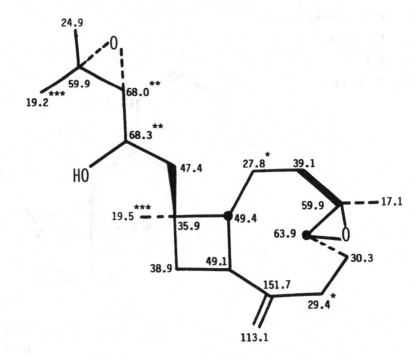

a: Not specified.

A. Groweiss and Y. Kashman, <u>Tetrahedron</u>, **39**(20), 3385 (1983).

EPOXYXENIAPHYLLANDIOL

Source: *Xenia macrospiculata*
Mol. formula: $C_{20}H_{34}O_3$
Mol. wt.: 322
Solvent: $CDCl_3$ or C_6D_6[a]

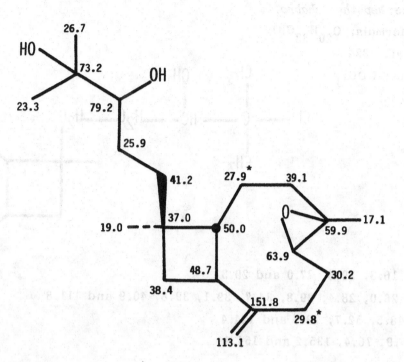

a: Not specified.

A. Groweiss and Y. Kashman, <u>Tetrahedron</u>, **39**(20), 3385 (1983).

|||||||||||||||||||||||||||||||||||||||||||||||||||||||||||||||||||||||||||||||||||||||||||||||||||||||||||||||||||||||||||||||||||||

XENIAPHYLLANTRIOL

Source: *Xenia obscuronata*
Mol. formula: $C_{20}H_{34}O_3$
Mol. wt.: 322
Solvent: $CDCl_3$ or C_6D_6

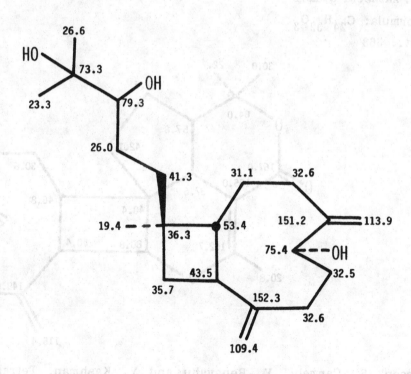

A. Groweiss and Y. Kashman, <u>Tetrahedron</u>, **39**(20), 3385 (1983).

Source: *Nephthea chabrolii*

Mol. formula: $C_{20}H_{33}ClO$

Mol. wt.: 324

Solvent: $CDCl_3$

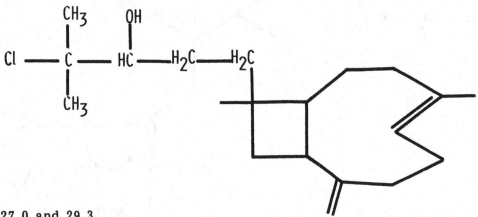

CH_3: 16.3, 19.8, 27.0 and 29.3

CH_2: 26.0, 28.4, 29.8, 34.7, 39.1, 39.8, 40.9 and 111.8

CH: 48.5, 52.7, 79.7 and 124.4

C: 35.9, 76.4, 135.2 and 154.4

A. Ahond, B.F. Bowden, J.C. Coll, J.-D. Fourneron and S.J. Mitchell, Aust. J. Chem., 34(12), 2657 (1981).

||||||||||||||||||||||||||||||||||||||||||||||||||||||||||||||||||||||||||||||||||||||||||||||||||||||||||

ANTHELIOLIDE A

Source: *Anthelia glauca*

Mol. formula: $C_{24}H_{32}O_3$

Mol. wt.: 368

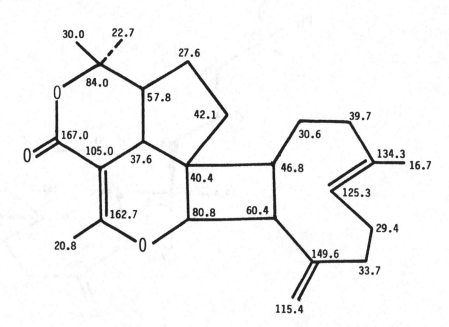

D. Green, S. Carmely, Y. Benayahu and Y. Kashman, Tetrahedron Lett., 29(13), 1605 (1988).

XENICIN

Basic skeleton:

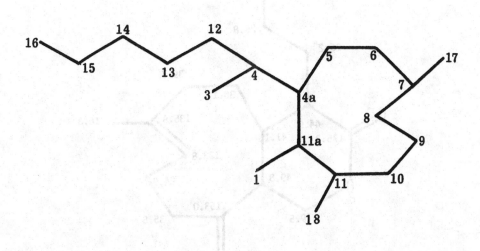

||||||||||||||||||||||||||||||||||||||||||||||||||||||||||||||||||||||||||||||||||||||||||||||||||||||||||||||||||||||||||||||||||

ACALYCIXENIOLIDE C

Source: *Acalycigorgia* sp.
Mol. formula: $C_{19}H_{24}O_2$
Mol. wt.: 284
Solvent: C_6D_6

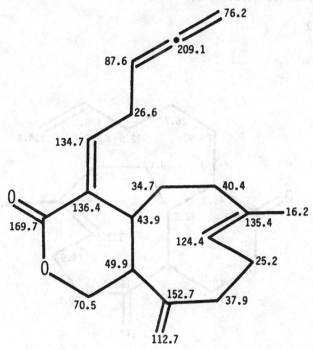

N. Fusetani, M. Asano, S. Matsunaga and K. Hashimoto, <u>Tetrahedron</u>, 45 (6), 1647 (1989).

ACALYCIXENIOLIDE B

Source: *Acalycigorgia inermis* (Hedlund)

Mol. formula: $C_{19}H_{26}O_2$

Mol. wt.: 286

Solvent: $CDCl_3$

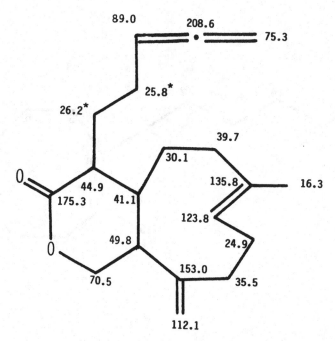

N. Fusetani, M. Asano, S. Matsunaga and K. Hashimoto, <u>Tetrahedron Lett.</u>, 28 (47), 5837 (1987).

ACALYCIXENIOLIDE A

Source: *Acalycigorgia inermis* (Hedlund)

Mol. formula: $C_{19}H_{28}O_2$

Mol. wt.: 288

Solvent: $CDCl_3$

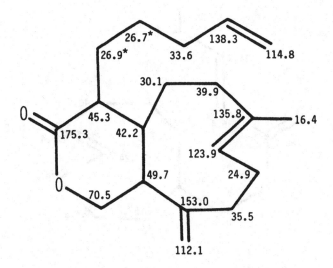

N. Fusetani, M. Asano, S. Matsunaga and K. Hashimoto, <u>Tetrahedron Lett.</u>, 28 (47), 5837 (1987).

CORAXENIOLIDE B

Source: *Corallium* sp.

Mol. formula: $C_{20}H_{28}O_2$

Mol. wt.: 300

Solvent: $CDCl_3$

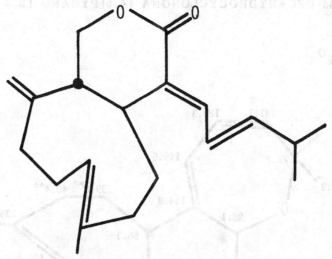

CH_3: 16.7, 22.3 and 25.3

CH_2: 34.7, 38.3, 40.7 and 113.5

CH: 32.1, 43.6, 49.8, 71.1, 121.1, 124.7, 138.3 and 151.4

C: 131.3, 137.0 and 151.7

C=O: 171.4

R.E. Schwartz, P.J. Scheuer, V. Zabel and W.H. Watson, Tetrahedron, 37(16), 2725 (1981).

||||||||||||||||||||||||||||||||||||||||||||||||||||||||||||||||||||||||||||||||||||||||||||||||||||||||||

CORAXENIOLIDE A

Source: *Corallium* sp.

Mol. formula: $C_{20}H_{30}O_2$

Mol. wt.: 302

Solvent: $CDCl_3$

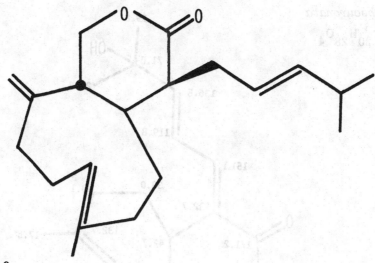

CH_3: 16.7, 22.9 and 22.9

CH_2: 25.3, 30.0, 35.8, 39.9, 70.8 and 112.2

CH: 31.3, 43.0, 44.4, 49.9, 123.5, 124.0 and 140.7

C: 140.0 and 153.1

C=O: 175.2

R.E. Schwartz, P.J. Scheuer, V. Zabel and W.H. Watson, Tetrahedron, 37(16), 2725 (1981).

(5aR*,7R*,7aS*,11E,14aR*)-4,4,12-TRIMETHYL-8-METHYLENE-4,5a,7,7a,8,9,
10,13,14,14a-DECAHYDROCYCLONONA [4,5]PYRANO [2,3-b] OXEPIN-7-OL

Source: *Xenia viridis*

Mol. formula: $C_{20}H_{28}O_3$

Mol. wt.: 316

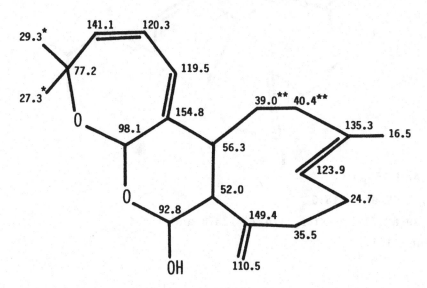

B.F. Bowden, J.C. Coll, E. Ditzel, S.J. Mitchell and W.T. Robinson, <u>Aust. J. Chem.</u>, 35(5), 997 (1982).

XENIOLIDE A

Source: *Xenia obscuronata*

Mol. formula: $C_{20}H_{28}O_4$

Mol. wt.: 332

Solvent: $CDCl_3$

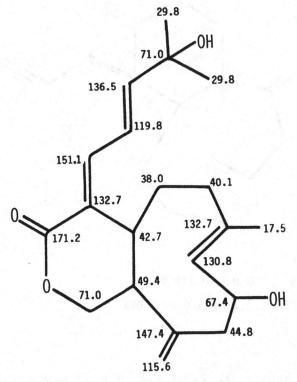

Y. Kashman and A. Groweiss, <u>J. Org. Chem.</u>, 45(19), 3814 (1980).

XENIALACTOL D

Source: *Xenia macrospiculata*, *X. obscuronata* and *X. lilielae*

Mol. formula: $C_{20}H_{30}O_4$

Mol. wt.: 334

Solvent: $CDCl_3$ or C_6D_6

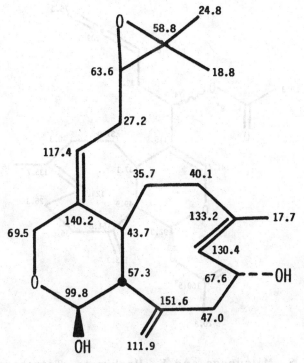

A. Groweiss and Y. Kashman, <u>Tetrahedron</u>, 39(20), 3385 (1983).

XENIALACTOL C

Source: *Xenia* sp.

Mol. formula: $C_{20}H_{30}O_4$

Mol. wt.: 334

Solvent: $CDCl_3$ or C_6D_6

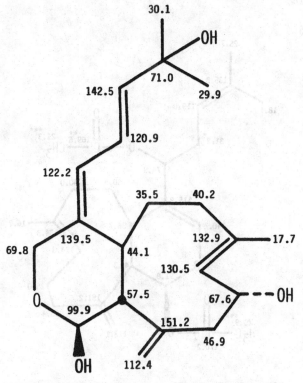

A. Groweiss and Y. Kashman, <u>Tetrahedron</u>, 39(20), 3385 (1983).

GINAMALLENE

Source: *Acalycigorgia* sp.
Mol. formula: $C_{23}H_{30}O_5$
Mol. wt.: 386
Solvent: C_6D_6

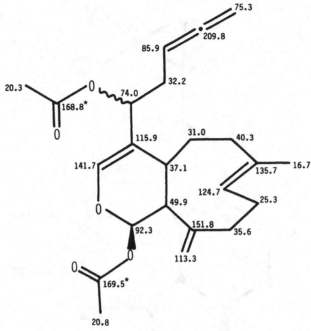

N. Fusetani, M. Asano, S. Matsunaga and K. Hashimoto, <u>Tetrahedron</u>, 45 (6), 1647 (1989).

||||||||||||||||||||||||||||||||||||||||||||||||||||||||||||||||||||||||||||||||||||||||||||||||||||||||||||||||

9-DEACETOXY-14,15-DEEPOXYXENICULIN

Source: *Xenia obscuronata*
Mol. formula: $C_{24}H_{34}O_5$
Mol. wt.: 402
Solvent: $CDCl_3$

Y. Kashman and A. Groweiss, <u>J. Org. Chem.</u>, 45(19), 3814 (1980).

9-DEACETOXYXENICIN

Source: *Xenia crassa*
Mol. formula: $C_{26}H_{36}O_7$
Mol. wt.: 460

B.F. Bowden, J.C. Coll, E. Ditzel, S.J. Mitchell and W.T. Robinson, <u>Aust. J. Chem.</u>, **35**(5), 997 (1982).

WAIXENICIN A

Source: *Anthelia edmondsoni*
Mol. formula: $C_{26}H_{36}O_7$
Mol. wt.: 460
Solvent: CD_2Cl_2

CH_3: 21.0,21.0 and 21.5
C=O: 169.5,170.1 and 170.6

S.J. Coval, P.J. Scheuer, G.K. Matsumoto and J. Clardy, <u>Tetrahedron</u>, **40**(19), 3823 (1984).

WAIXENICIN B

Source: *Anthelia edmondsoni*

Mol. formula: $C_{26}H_{36}O_8$

Mol. wt.: 476

Solvent: CD_2Cl_2

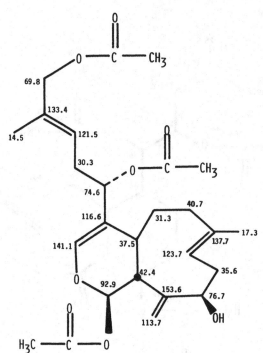

CH₃: 21.1, 21.2 and 21.5

C=O: 170.0, 170.5 and 170.6

S.J. Coval, P.J. Scheuer, G.K. Matsumoto and J. Clardy, <u>Tetrahedron</u>, **40**(19), 3823 (1984).

XENICULIN

Source: *Xenia macrospiculata*

Mol. formula: $C_{26}H_{36}O_8$

Mol. wt.: 476

Solvent: $CDCl_3$

CH₃: 21.0, 21.3 and 21.3

C=O: 169.3, 170.3 and 170.3

Y. Kashman and A. Groweiss, <u>J. Org. Chem.</u>, **45**(19), 3814 (1980).

11(19)-DESACETYL-13-HAVANNACHLORHYDRINE

Source: *Xenia membranacea* Schenk

Mol. formula: $C_{24}H_{33}ClO_9$

Mol. wt.: 500

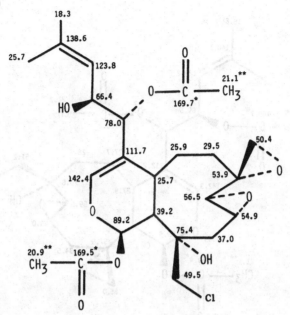

A. Almourabit, A. Ahond, A. Chiaroni, C. Poupat, C. Riche, P. Potier, P. Laboute and J.-L. Menou, J. Nat. Prod., (Lloydia), 51 (2), 282 (1988).

XENICIN

Source: *Xenia elongata*

Mol. formula: $C_{28}H_{38}O_9$

Mol. wt.: 518

Solvent: $CDCl_3$

Y. Kashman and A. Groweiss, J. Org. Chem., 45(19), 3814 (1980).

7(18)-HAVANNACHLORHYDRINE

Source: *Xenia membranacea* Schenk

Mol. formula: $C_{26}H_{35}ClO_{10}$

Mol. wt.: 542

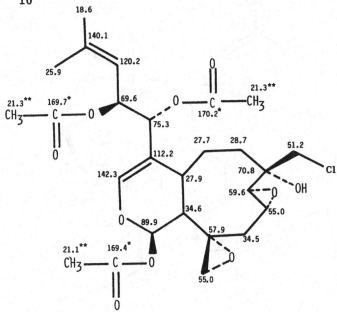

A. Almourabit, A. Ahond, A. Chiaroni, C. Poupat, C. Riche, P. Potier, P. Laboute and J.-L. Menou, J. Nat. Prod., (Lloydia), 51 (2), 282 (1988).

11(19)-HAVANNACHLORHYDRINE

Source: *Xenia membranacea* Schenk

Mol. formula: $C_{26}H_{35}ClO_{10}$

Mol. wt.: 542

A. Almourabit, A. Ahond, A. Chiaroni, C. Poupat, C. Riche, P. Potier, P. Laboute and J.-L. Menou, J. Nat. Prod., (Lloydia), 51 (2) 282 (1988).

7(18),11(19)-HAVANNADICHLORHYDRINE

Source: *Xenia membranacea* Schenk

Mol. formula: $C_{26}H_{36}Cl_2O_{10}$

Mol. wt.: 578

A. Almourabit, A. Ahond, A. Chiaroni, C. Poupat, C. Riche, P. Potier, P. Laboute and J.-L. Menou, J. Nat. Prod., (Lloydia), 51 (2), 282 (1988).

||||||||||||||||||||||||||||||||||||||||||||||||||||||||||||||||||||||||||||||||||||||||||||||||||||||||||||||||||||||||||||||||||||

GUAIANE ISOPRENOLOGUE

Basic skeleton:

DICTYOXIDE A

Source: *Dictyota binghamiae*
Mol. formula: $C_{20}H_{30}O$
Mol. wt.: 286
Solvent: $CDCl_3$

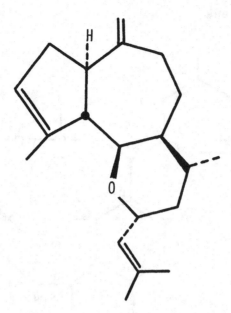

17.3, 18.3, 20.2, 25.1, 25.8, 29.4, 34.2, 34.3, 39.7, 45.5, 46.6, 53.6, 68.5, 79.5, 106.8, 124.9, 126.8, 133.5, 143.0 and 151.9

C. Pathirana and R.J. Andersen, Can. J. Chem., 62(9), 1666 (1984).

PACHYDICTYOL A

Source: *Pachydicton coriaceum* and *Aplysia depilans*
Mol. formula: $C_{20}H_{32}O$
Mol. wt.: 288
Solvent: $CDCl_3$

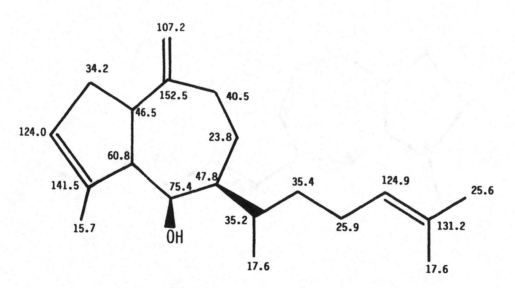

L. Minale and R. Riccio, Tetrahedron Lett., (31), 2711 (1976).

536

TORMESOL

Source: *Halimium viscosum*

Mol. formula: $C_{20}H_{34}O$

Mol. wt.: 290

Solvent: $CDCl_3$

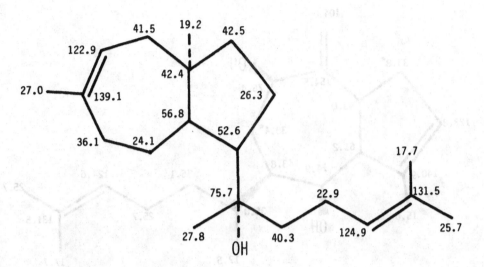

J.G. Urones, I.S. Marcos, N.M. Garrido, J. De P. Teresa and A.S.F. Martin, <u>Phyto-chemistry</u>, **28** (1), 183 (1989).

||||||||||||||||||||||||||||||||||||||||||||||||||||||||||||||||||||||||||||||||||||||||||||||||||||

DICTYOL A

Source: *Aplysia depilans*

Mol. formula: $C_{20}H_{30}O_2$

Mol. wt.: 302

Solvent: $CDCl_3$

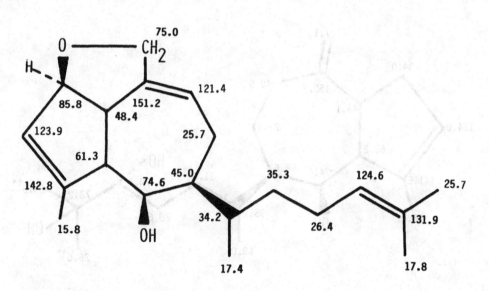

L. Minale and R. Riccio, <u>Tetrahedron Lett.</u>, (31), 2711 (1976).

DICTYOL B

Source: *Aplysia depilans*

Mol. formula: $C_{20}H_{32}O_2$

Mol. wt.: 304

Solvent: $CDCl_3$

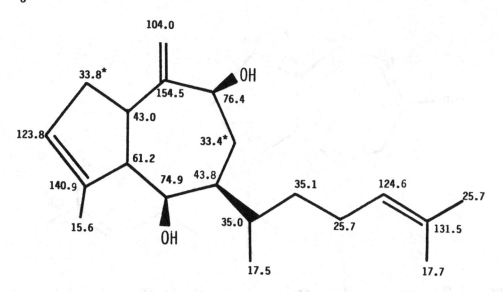

L. Minale and R. Riccio, Tetrahedron Lett., (31), 2711 (1976).

||||||||||||||||||||||||||||||||||||||||||||||||||||||||||||||||||||||||||||||||||||||||||||||||||||||||||||||||||||||||||||||||

ISODICTYTRIOL

Source: *Dictyota dichotoma*

Mol. formula: $C_{20}H_{34}O_3$

Mol. wt.: 322

Solvent: $CDCl_3$

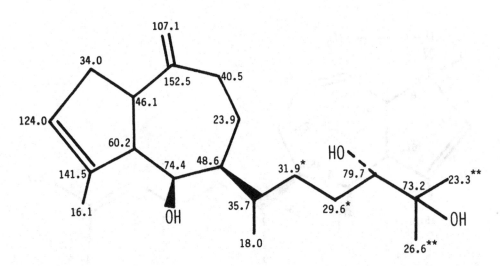

T. Kusumi, D.M. Nkongolo, M. Ishitsuka, Y. Inouye and H. Kakisawa, Chem. Lett., (7), 1241 (1986).

DICTYOL B ACETATE

Source: *Dictyota dichotoma* (Desf.) J.Ag.

Mol. formula: $C_{22}H_{34}O_3$

Mol. wt.: 346

Solvent: $CDCl_3$

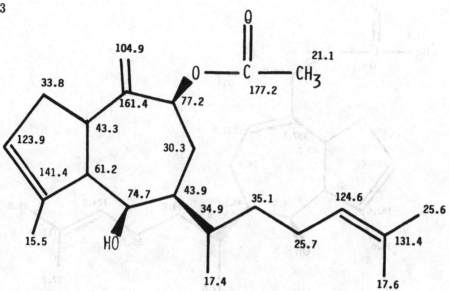

S.De Rosa, S.De Stefano and N. Zavodnik, <u>Phytochemistry</u>, 25(9), 2179 (1986).

||||||||||||||||||||||||||||||||||||||||||||||||||||||||||||||||||||||||||||||||||||||||||||||||||||

DICTYOL G ACETATE

Source: *Dictyota binghamiae*

Mol. formula: $C_{22}H_{34}O_3$

Mol. wt.: 346

Solvent: $CDCl_3$

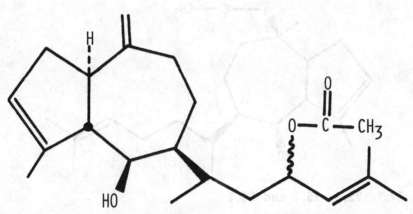

15.8, 18.1, 18.3, 21.2, 23.8, 25.6, 32.1, 34.0, 40.4, 40.4, 46.2, 48.0, 60.5, 70.6, 74.9, 107.2, 124.1, 124.5, 136.3, 141.4, 152.4 and 170.5

C. Pathirana and R.J. Andersen, <u>Can. J. Chem.</u>, 62(9), 1666 (1984).

DICTYOL-I-ACETATE

Source: *Dictyota dichotoma* (Desf.) J.Ag.

Mol. formula: $C_{22}H_{34}O_3$

Mol. wt.: 346

Solvent: $CDCl_3$

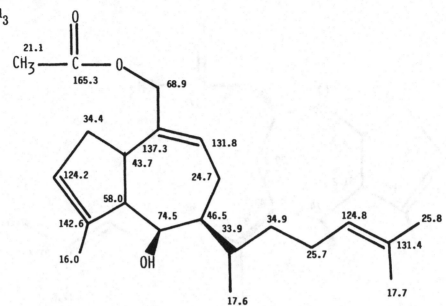

S. De Rosa, S. De Stefano and N. Zavodnik, <u>Phytochemistry</u>, 25(9), 2179 (1986).

‖‖

ACETYLDICTYOL C

Source: *Pachydictyon coriaceum* (Sanadagusa)

Mol. formula: $C_{22}H_{36}O_3$

Mol. wt.: 348

Solvent: $CDCl_3$

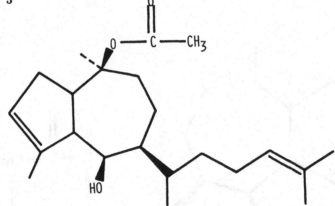

CH_3: 16.3, 17.5, 17.7, 22.5, 25.7 and 26.2

CH_2: 19.7, 25.6, 33.0, 34.9 and 40.5

CH: 34.5, 49.7, 51.7, 52.2, 74.6, 123.1 and 124.7

C: 84.4, 131.5 and 142.4

C=O: 170.4

M. Ishitsuka, T. Kusumi, J. Tanaka and H. Kakisawa, <u>Chem. Lett.</u>, 1517 (1982).

DICTYOL H

Source: *Dictyota dentata* Lamouroux
Mol. formula: $C_{22}H_{34}O_4$
Mol. wt.: 362
Solvent: $CDCl_3$

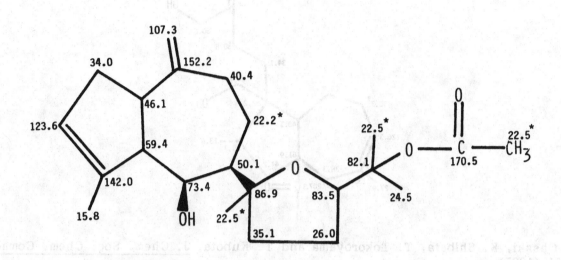

A.B. Alvarado and W.H. Gerwick, <u>J. Nat. Prod.</u>, (Lloydia), **48** (1), 132 (1985).

||||||||||||||||||||||||||||||||||||||||||||||||||||||||||||||||||||||||||||||||||||||||||||||||||||||||||||||||||||||||||||||||||

PORTULAL

Basic skeleton:

PORTULAL

Source: *Portulaca grandiflora* Hook.

Mol. formula: $C_{20}H_{32}O_4$

Mol. wt.: 336

Solvent: CD_3OD

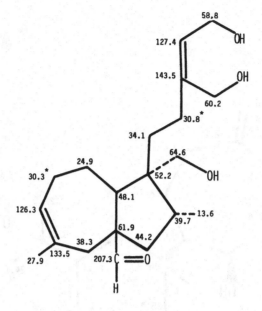

1. A. Ohsaki, K. Shibata, T. Tokoroyama and T. Kubota, J. Chem. Soc. Chem. Commun., (3), 151 (1987).

2. S. Yamazaki, S. Tamura, F. Marumo and Y. Saito, Tetrahedron Lett., (5), 359 (1969).

PORTULIC LACTONE

Source: *Portulaca grandiflora* Hook

Mol. formula: $C_{20}H_{32}O_5$

Mol. wt.: 352

Solvent: CD_3OD

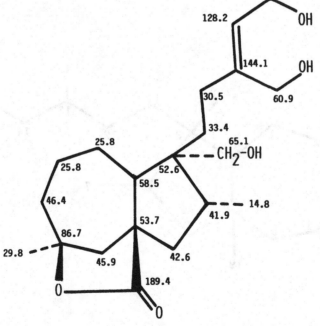

A. Ohsaki, K. Matsumoto, K. Shibata, T. Kubota and T. Tokoroyama, Chem. Pharm. Bull. (Tokyo), 33(5), 2171 (1985).

5-HYDROXYPORTULAL

Source: *Portulaca grandiflora* Hook
Mol. formula: $C_{20}H_{32}O_5$
Mol. wt.: 352
Solvent: CD_3OD

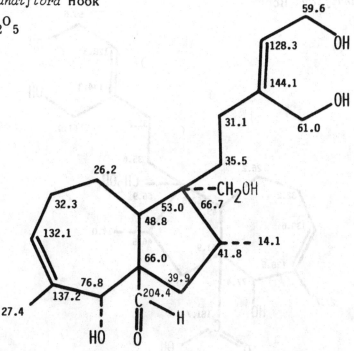

A. Ohsaki, K. Matsumoto, K. Shibata, T. Kubota and T. Tokoroyama, Chem. Pharm. Bull. (Tokyo), 33 (5), 2171 (1985).

3-HYDROXYPORTULOL ETHER

Source: *Portulaca grandiflora* Hook
Mol. formula: $C_{20}H_{34}O_5$
Mol. wt.: 354
Solvent: CD_3OD

A.Ohsaki, K.Matsumoto, K.Shibata, T.Kubota and T.Tokoroyama, Chem.Pharm.Bull. (Tokyo), 33(5), 2171 (1985).

Source: *Portulaca grandiflora* Hook
Mol. formula: $C_{20}H_{32}O_6$
Mol. wt.: 368
Solvent: CD_3OD

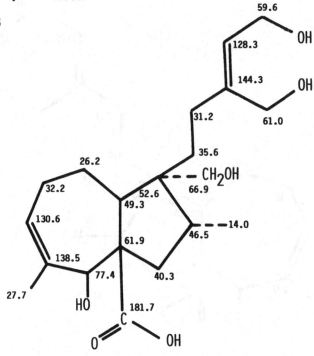

A.Ohsaki, K.Matsumoto, K.Shibata, T.Kubota and T.Tokoroyama, Chem.Pharm.Bull.
(Tokyo), 33(5), 2171 (1985).

UNDECANE

Basic skeleton:

PORTULENE

Source: *Portulaca grandiflora* Hook
Mol. formula: $C_{20}H_{32}O_4$
Mol. wt.: 336
Solvent: CD_3OD

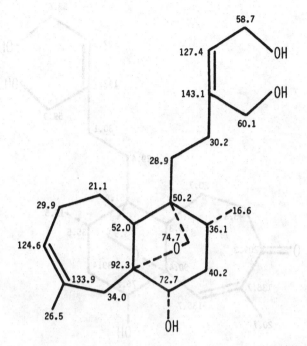

A. Ohsaki, K. Shibata, T. Tokoroyama, T. Kubota and H. Naoki, <u>Chem. Lett.</u>, (9), 1585 (1986).

JEWENOL B

Source: *Portulaca* cv Jewel
Mol. formula: $C_{20}H_{34}O_4$
Mol. wt.: 338
Solvent: CD_3OD

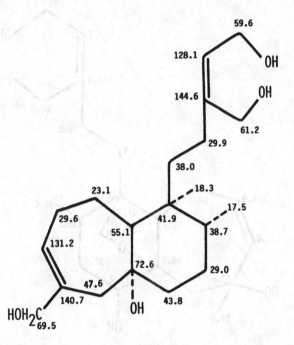

A. Ohsaki, N.Ohno, K. Shibata, T. Tokoroyama, T. Kubota, K. Hirotsu and T. Higuchi, <u>Phytochemistry</u>, **27**(7), 2171 (1988).

PORTULENONE

Source: *Portulaca grandiflora* Hook

Mol. formula: $C_{20}H_{30}O_5$

Mol. wt.: 350

Solvent: CD_3OD

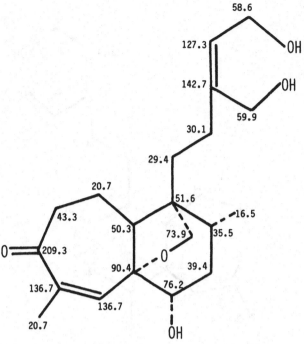

A. Ohsaki, K. Shibata, T. Tokoroyama, T. Kubota and H. Naoki, <u>Chem. Lett.</u>, (9), 1585 (1986).

PORTULENOL

Source: *Portulaca grandiflora* Hook

Mol. formula: $C_{20}H_{32}O_5$

Mol. wt.: 352

Solvent: CD_3OD

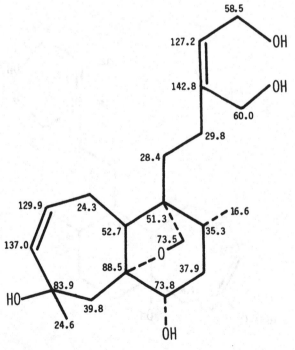

A. Ohsaki, K. Shibata, T. Tokoroyama, T. Kubota and H. Naoki, <u>Chem. Lett.</u>, (9), 1585 (1986).

DAPHNANE

Basic skeleton:

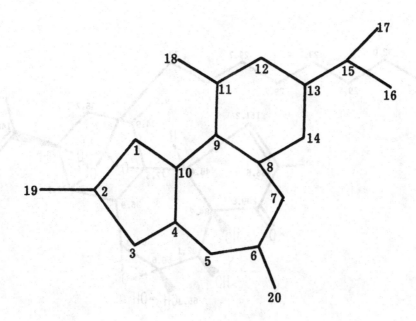

|||||||||||||||||||||||||||||||||||||||||||||||||||||||||||||||||||||||||||||||||||||||||||||||||||||||||||||||||||

EXCOECARIATOXIN

Source: *Diarthron vesiculosum* (Fisch. and C.A. Mey. ex Kar and Kir) Mey.
Mol. formula: $C_{30}H_{40}O_8$
Mol. wt.: 528
Solvent: $CDCl_3$

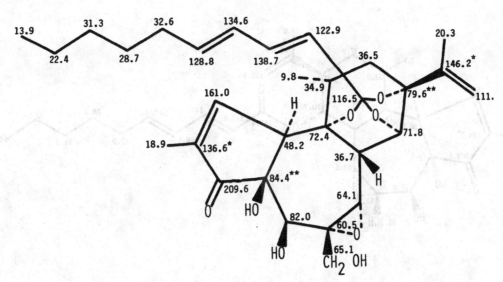

R.G. Powell, D. Weisleder and C.R. Smith, Jr., <u>J. Nat. Prod.</u>, (Lloydia), 48 (1), 102 (1985).

SIMPLEXIN

Source: *Diarthron vesiculosum* (Fisch and C.A. Mey. ex Kar and Kir.) Mey.

Mol. formula: $C_{30}H_{44}O_8$

Mol. wt.: 532

Solvent: $CDCl_3$

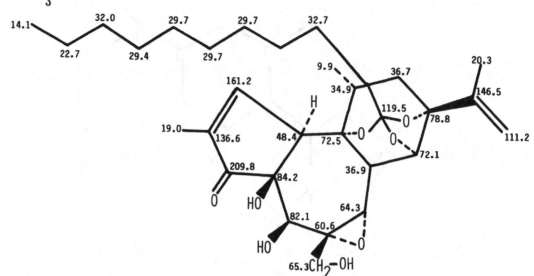

R.G. Powell, D. Weisleder and C.R. Smith, Jr., J. Nat. Prod., (Lloydia), 48 (1), 102 (1985).

VESICULOSIN

Source: *Diarthron vesiculosum* (Fisch and C.A. Mey. ex Kar and Kir) Mey.

Mol. formula: $C_{30}H_{42}O_9$

Mol. wt.: 546

Solvent: $CDCl_3$

R.G. Powell, D. Weisleder and C.R. Smith, Jr., J. Nat. Prod., (Lloydia), 48 (1), 102 (1985).

GENKWADAPHNIN

Source: *Daphne genkwa* Sieb et Zucc.

Mol. formula: $C_{34}H_{34}O_{10}$

Mol. wt.: 602

Solvent: $CDCl_3$

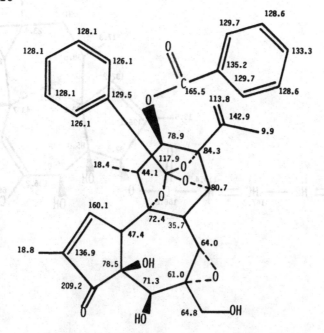

R. Kasai, K.-H. Lee and H.-C. Huang, <u>Phytochemistry</u>, 20(11), 2592 (1981).

TIGLIANE

Basic skeleton:

20-ACETYL-INGENOL-3-DECADIENOATE

Source: *Euphorbia broteri* Daveau

Mol. wt.: $C_{32}H_{44}O_7$

Mol. wt.: 540

Solvent: $CDCl_3$

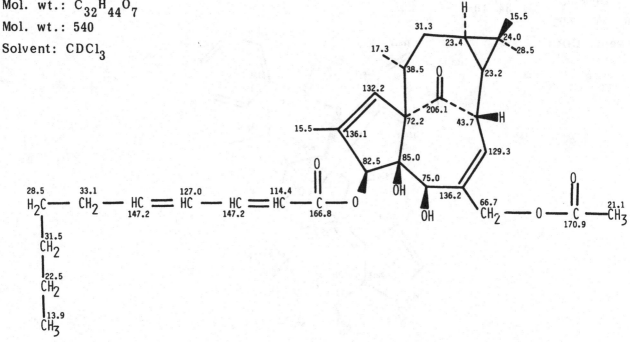

J.G. Urones, P.B. Barcala, M.J.S. Cuadrado and I.S. Marcos, <u>Phytochemistry</u>, 27 (1), 207 (1988).

12-O-(2Z,4E-OCTADIENOYL)-4-DEOXYPHORBOL-13,20-DIACETATE

Source: *Euphorbia broteri* Daveau

Mol. formula: $C_{32}H_{42}O_8$

Mol. wt.: 554

Solvent: $CDCl_3$

J.G. Urones, P.B. Barcala, M.J.S. Cuadrado and I.S. Marcos, <u>Phytochemistry</u>, 27 (1), 207 (1988).

12-O-(2Z,4E-OCTADIENOYL)-PHORBOL-13,20-DIACETATE

Source: *Euphorbia broteri* Daveau

Mol. formula: $C_{32}H_{42}O_9$

Mol. wt.: 570

Solvent: $CDCl_3$

J.G. Urones, P.B. Barcala, M.J.S. Cuadrado and I.S. Marcos, <u>Phytochemistry</u>, 27 (1), 207 (1988).

3-TETRADECANOATE-INGENOL(5,20-DIACETATE)

Source: *Euphorbia broteri* Daveau

Mol. formula: $C_{38}H_{58}O_8$

Mol. wt.: 642

Solvent: CDCl

J.G. Urones, P.B. Barcala, M.J.S. Cuadrado and I.S. Marcos, <u>Phytochemistry</u>, 27 (1), 207 (1988).

Basic skeleton:

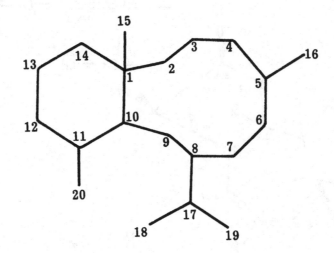

||||||||||||||||||||||||||||||||||||||||||||||||||||||||||||||||||||||||||||||||||||||||||||||||||||||||||||||||||||||||||

(-)-VERECYNARMIN B

Source: *Veretillum cynomorium* (Pallas).
Mol. formula: $C_{20}H_{26}O_2$
Mol. wt.: 298
Solvent: C_6D_6

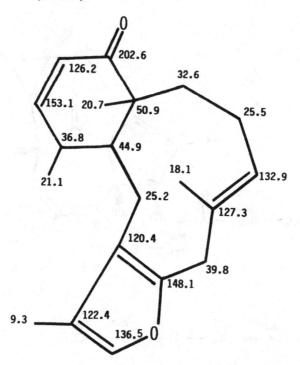

A. Guerriero, M.D. Ambrosio and F. Pietra, <u>Helv. Chim. Acta</u>, 71(2), 472 (1988).

(-)-VERECYNARMIN C

Source: *Veretillum cynomorium* (Pallas).

Mol. formula: $C_{20}H_{26}O_3$

Mol. wt.: 314

Solvent: C_6D_6

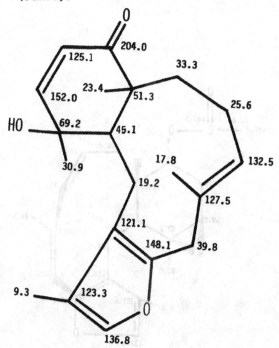

A. Guerriero, M.D. Ambrosio and F. Pietra, <u>Helv. Chim. Acta</u>, 71(2), 472 (1988).

(-)-VERECYNARMIN D

Source: *Veretillum cynomorium* (Pallas).

Mol. formula: $C_{20}H_{25}ClO_3$

Mol. wt.: 348

Solvent: $(CD_3)_2CO$

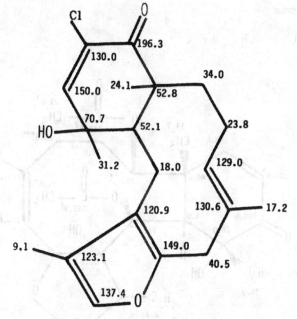

For chemical shifts in C_6D_6 see:

A. Guerriero, M.D. Ambrosio and F. Pietra, <u>Helv. Chim. Acta</u>, 71 (2), 472 (1988).

VERECYNARMIN A

Source: *Armina maculata* (Rafinesque), and *Veretillum cynomorium* (Pallas)

Mol. formula: $C_{22}H_{28}O_4$

Mol. wt.: 356

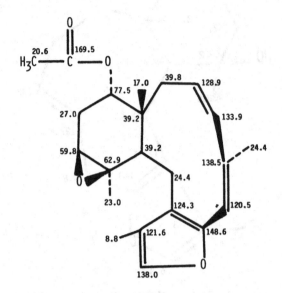

A. Guerriero, M.D. Ambrosio and F. Pietra, <u>Helv. Chim. Acta.</u>, **70**(4), 984 (1987).

||||||||||||||||||||||||||||||||||||||||||||||||||||||||||||||||||||||||||||||||||||||||||||||||||||||||||||||

RENILLAFOULIN A

Source: *Renilla reniformis*

Mol. formula: $C_{24}H_{32}O_9$

Mol. wt.: 464

Solvent: $CDCl_3$

P.A. Keifer, K.L. Rinehart, Jr. and I.R. Hooper, <u>J. Org. Chem.</u>, **51**(23), 4450 (1986).

RENILLAFOULIN B

Source: *Renilla reniformis*
Mol. formula: $C_{25}H_{34}O_9$
Mol. wt.: 478
Solvent: $CDCl_3$

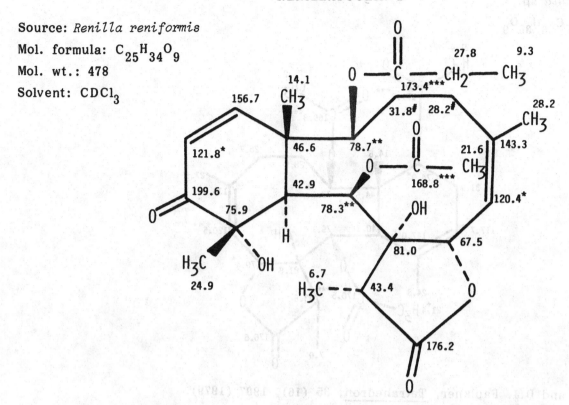

P.A. Keifer, K.L. Rinehart,Jr. and I.R. Hooper, <u>J. Org. Chem</u>., 51(23), 4450 (1986).

PTILOSARCENONE

Source: *Ptilosarcus gurneyi*
Mol. formula: $C_{24}H_{29}ClO_8$
Mol. wt.: 480
Solvent: $CDCl_3$

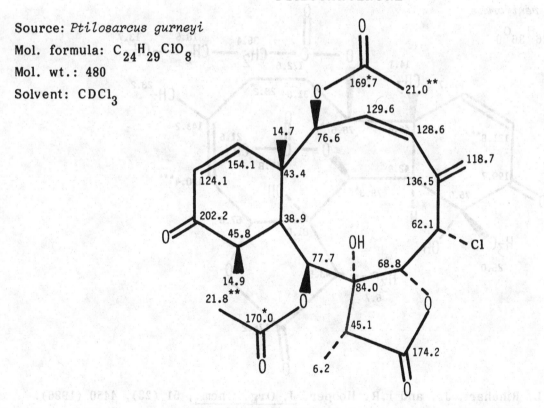

R.L. Hendrickson and J.H. Cardellina, <u>Tetrahedron</u>, 42 (23), 6565 (1986).

Source: *Stylatula* sp.

Mol. formula: $C_{26}H_{36}O_9$

Mol. wt.: 492

Solvent: $CDCl_3$

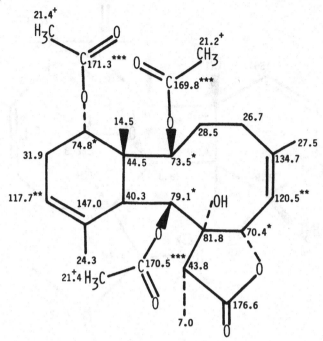

S.J. Wratten and D.J. Faulkner, <u>Tetrahedron</u>, **35** (16), 1907 (1979).

IIIIIIIIIIIIIIIIIIIIIIIIIIIIIIIIIIIIIIIIIIIIIIIIIIIIIIIIIIIIIIIIIIIIIIIIIIIIIIIIIIIIIIIIIIIIIIIIIIIIIII

RENILLAFOULIN C

Source: *Renilla reniformis*

Mol. formula: $C_{26}H_{36}O_9$

Mol. wt.: 492

Solvent: $CDCl_3$

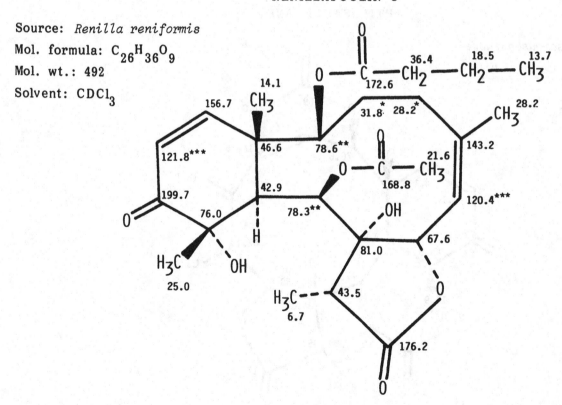

P.A. Keifer, K.L. Rinehart, Jr. and I.R. Hooper, <u>J. Org. Chem.</u>, **51** (23), 4450 (1986).

11-HYDROXYPTILOSARCENONE

Source: *Ptilosarcus gurneyi*

Mol. formula: $C_{24}H_{29}ClO_9$

Mol. wt.: 496

Solvent: $CDCl_3$

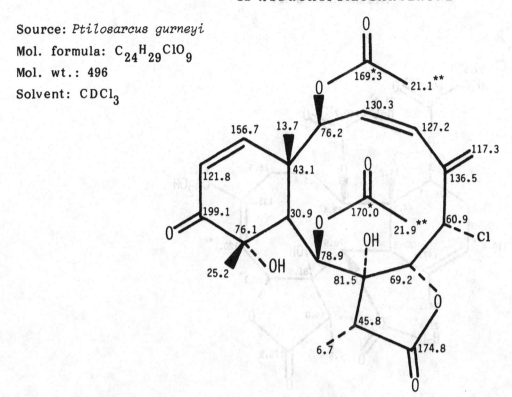

R.L. Hendrickson and J.H. Cardellina, <u>Tetrahedron</u>, 42 (23), 6565 (1986).

BRIANTHEIN X

Source: *Briareum polyanthes*

Mol. formula: $C_{24}H_{31}ClO_9$

Mol. wt.: 498

Solvent: $CDCl_3$

S.H. Grode, T.R. James, Jr., J.H. Cardellina and K.D. Onan, <u>J. Org. Chem.</u>, 48 (26), 5203 (1983).

DITERPENE S-XV

Source: *Stylatula* sp.
Mol. formula: $C_{26}H_{36}O_{10}$
Mol. wt.: 508
Solvent: $CDCl_3$

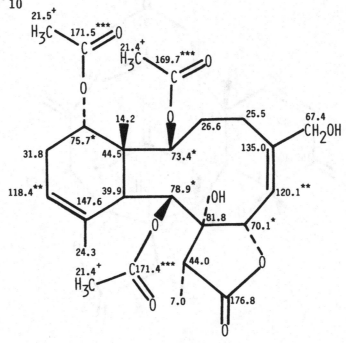

S.J. Wratten and D.J. Faulkner, <u>Tetrahedron</u>, **35** (16), 1907 (1979).

||||||||||||||||||||||||||||||||||||||||||||||||||||||||||||||||||||||||||||||||||||||||||||||||||||||||||||||||||||||||||||||||||||

JUNCEELLOLIDE B

Source: *Junceella fragilis* Ridley
Mol. formula: $C_{26}H_{33}ClO_9$
Mol. wt.: 524
Solvent: $CDCl_3$

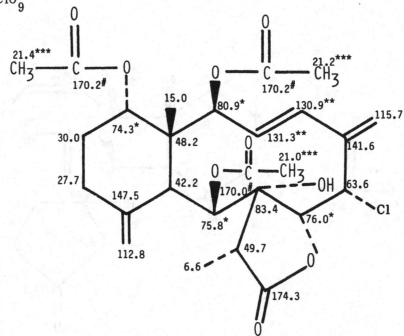

J. Shin, M. Park annd W. Fenical, <u>Tetrahedron</u>, **45** (6), 1633 (1989).

ERYTHROLIDE B

Source: *Erythropodium caribaeorum* (Duchassaing and Michelotti)

Mol. formula: $C_{26}H_{31}ClO_{10}$

Mol. wt.: 538

Solvent: $CDCl_3$

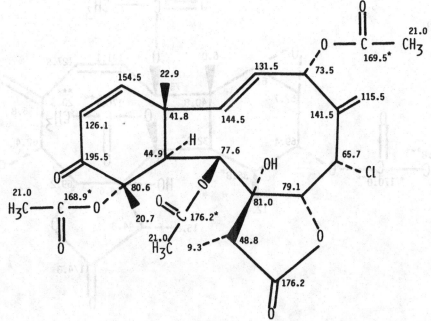

P.A. Keifer, K.L. Rinehart, Jr. and I.R. Hooper, <u>J. Org. Chem.</u>, **51** (23), 4450 (1986).

||||||||||||||||||||||||||||||||||||||||||||||||||||||||||||||||||||||||||||||||||||||||||||||||||||||||||||||||||||||||||||||||||||||||||||||

PTILOSARCEN-12-PROPIONATE

Source: *Ptilosarcus gurneyi*

Mol. formula: $C_{27}H_{35}ClO_{9}$

Mol. wt.: 538

Solvent: $CDCl_3$

R.L. Hendrickson and J.H. Cardellina, <u>Tetrahedron</u>, **42** (23), 6565 (1986).

BRIANTHEIN Z

Source: *Briareum polyanthes*

Mol. formula: $C_{26}H_{33}ClO_{10}$

Mol. wt.: 540

Solvent: $CDCl_3$

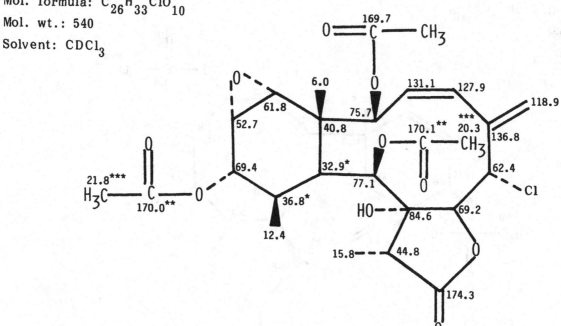

S.H. Grode, T.R. James, Jr., J.H. Cardellina and K.D. Onan, <u>J. Org. Chem.</u>, **48** (26), 5203 (1983).

JUNCEELLOLIDE A

Source: *Junceella fragilis* Ridley

Mol. formula: $C_{26}H_{33}ClO_{10}$

Mol. wt.: 540

Solvent: $CDCl_3$

J. Shin, M. Park and W. Fenical, <u>Tetrahedron</u>, **45** (6), 1633 (1989).

JUNCEELLOLIDE C

Source: *Junceella fragilis* Ridley
Mol. formula: $C_{26}H_{33}ClO_{10}$
Mol. wt.: 540
Solvent: $CDCl_3$

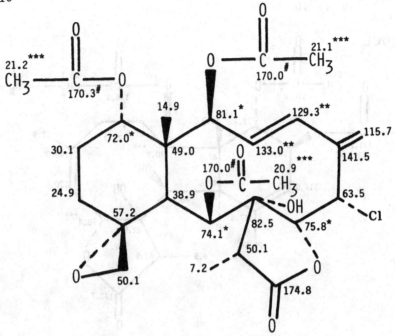

J. Shin, M. Park and W. Fenical, <u>Tetrahedron</u>, **45** (6), 1633 (1989).

STYLATULIDE

Source: *Stylatula* sp.
Mol. formula: $C_{26}H_{35}ClO_{10}$
Mol. wt.: 542
Solvent: $CDCl_3$

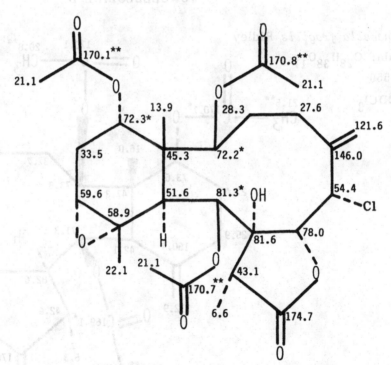

A. Clastres, A. Ahond, C. Poupat, P. Potier and S.K. Kan, <u>J. Nat. Prod.</u>, (Lloydia), **47** (1), 155 (1984).

CAVERNULINE

Source: *Cavernulina grandiflora*
Mol. formula: $C_{30}H_{44}O_{10}$
Mol. wt.: 564
Solvent: $CDCl_3$

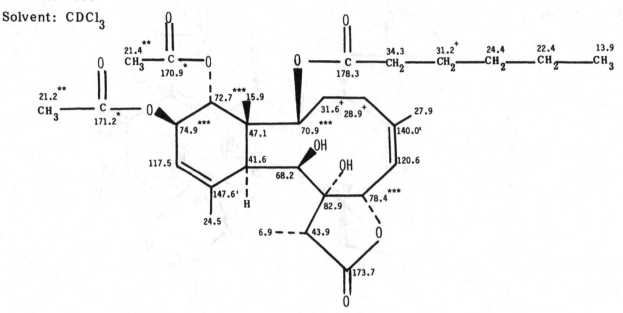

A. Clastres, P. Laboute, A. Ahond, C. Poupat and P. Potier, <u>J. Nat. Prod.</u>,(Lloydia), **47**(1), 162 (1984).

‖‖

JUNCEELLOLIDE D

Source: *Junceella fragilis* Ridley
Mol. formula: $C_{28}H_{38}O_{11}$
Mol. wt.: 550
Solvent: $CDCl_3$

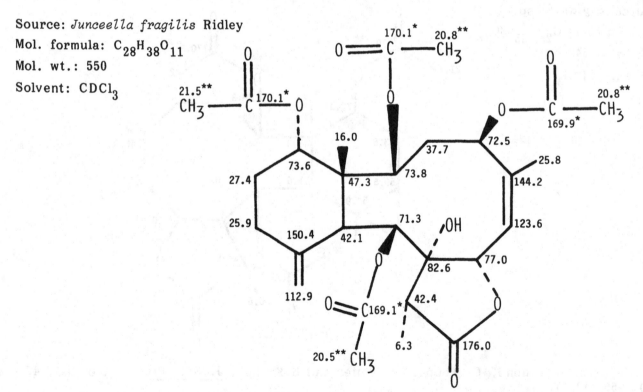

J. Shin, M. Park and W. Fenical, <u>Tetrahedron</u>, **45** (6), 1633 (1989).

BRIANTHEIN Y

Source: *Briareum polyanthes*
Mol. formula: $C_{28}H_{37}ClO_{10}$
Mol. wt.: 568
Solvent: $CDCl_3$

S.H. Grode, T.R. James, Jr., J.H. Cardellina and K.D. Onan, <u>J. Org. Chem.</u>, **48** (26), 5203 (1983).

PTILOSARCONE

Source: *Ptilosarcus gurneyi*
Mol. formula: $C_{28}H_{37}ClO_{10}$
Mol. wt.: 568
Solvent: $CDCl_3$

R.L. Hendrickson and J.H. Cardellina, <u>Tetrahedron</u>, **42** (23), 6565 (1986).

JUNCEELLIN

Source: *Junceella fragilis* Ridley

Mol. formula: $C_{28}H_{35}Cl\,O_{11}$

Mol. wt.: 582

Solvent: $CDCl_3$

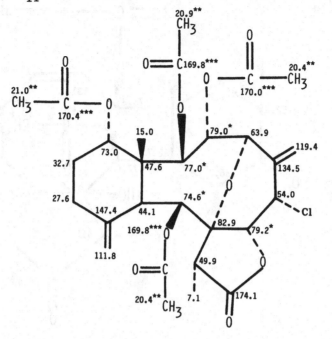

J. Shin, M. Park and W. Fenical, <u>Tetrahedron</u>, **45** (6), 1633 (1989).

PRAELOLIDE

Source: *Junceella fragilis* Ridley

Mol. formula: $C_{28}H_{35}Cl\,O_{12}$

Mol. wt.: 598

Solvent: $CDCl_3$

J. Shin, M. Park and W. Fenical, <u>Tetrahedron</u>, **45** (6), 1633 (1989).

PTEROIDINE

Source: *Pteroides laboutei*

Mol. formula: $C_{28}H_{37}ClO_{12}$

Mol. wt.: 600

Solvent: $CDCl_3$

A. Clastres, A. Ahond, C. Poupat, P. Potier and S.K. Kan, <u>J. Nat. Prod.</u>, (Lloydia), **47** (1), 155 (1984).

||||||||||||||||||||||||||||||||||||||||||||||||||||||||||||||||||||||||||||||||||||||||||||||||||||||||||||||||||||||||||||||||||||

O-DEACETYL-12-O-BENZOYL-12-PTEROIDINE

Source: *Pteroides laboutei*

Mol. formula: $C_{33}H_{39}ClO_{12}$

Mol. wt.: 662

Solvent: $CDCl_3$

A. Clastres, A. Ahond, C. Poupat, P. Potier and S.K. Kan, <u>J. Nat. Prod.</u>, (Lloydia), **47** (1), 155 (1984).

Basic skeleton:

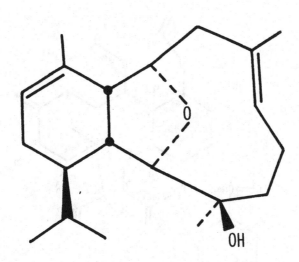

||||||||||||||||||||||||||||||||||||||||||||||||||||||||||||||||||||||||||||||||||||||||||||||||||||||||||||||||||

DITERPENE C-II

Source: *Cladiella* sp.

Mol. formula: $C_{20}H_{32}O_2$

Mol. wt.: 304

Solvent: C_6D_6

CH_3: 18.9, 20.8, 21.8, 22.2 and 27.7

CH_2: 23.2, 23.3, 37.1 and 44.7

CH: 29.3, 38.4 40.5, 47.2, 80.9, 89.9, 121.6 and 129.8

C: 76.8, 126.4 and 132.2

J.E. Hochlowski and D.J. Faulkner, Tetrahedron Lett., **21** (42), 4055 (1980).

SCLEROPHYTIN A

Source: *Sclerophytum capitalis*
Mol. formula: $C_{20}H_{32}O_3$
Mol. wt.: 320
Solvent: $CDCl_3$

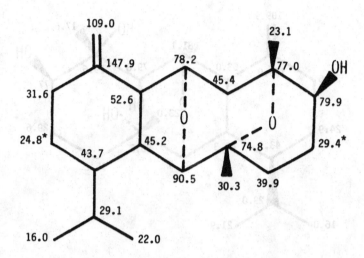

P. Sharma and M. Alam, J. Chem. Soc. Perkin. Trans. 1, (9), 2537 (1988).

||||||||||||||||||||||||||||||||||||||||||||||||||||||||||||||||||||||||||||||||||||||||||||||||||||||||||||||||||||

CLADIELLIN

Source: *Cladiella* sp.
Mol. formula: $C_{22}H_{34}O_3$
Mol. wt.: 346

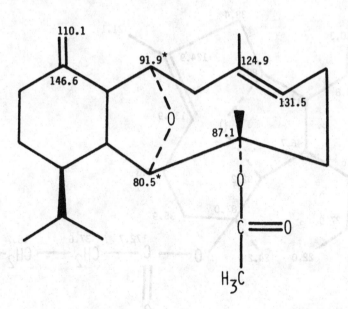

R. Kazlauskas, P.T. Murphy, R.J. Wells and P. Schonholzer, Tetrahedron Lett., 19, (52), 4643 (1977).

SCLEROPHYTIN D

Source: *Sclerophytum capitalis*
Mol. formula: $C_{20}H_{34}O_5$
Mol. wt.: 354

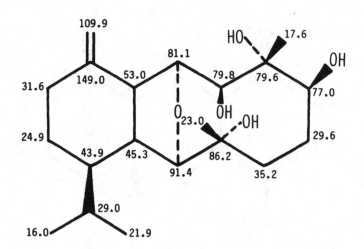

M. Alam, P. Sharma, A.S. Zektzer, G.E. Martin, X. Ji and D.V. der Helm, J. Org. Chem., 54 (8), 1896 (1989).

LITOPHYNIN A

Source: *Litophyton* sp.
Mol. formula: $C_{24}H_{38}O_3$
Mol. wt.: 374
Solvent: $CDCl_3$

M. Ochi, K. Futatsugi, H. Kotsuki, M. Ishii and K. Shibata, Chem. Lett., 28 (11), 2207 (1987).

ALCYONIN

Source: *Sinularia flexibilis*

Mol. formula: $C_{22}H_{34}O_5$

Mol. wt.: 378

Solvent: $CDCl_3$

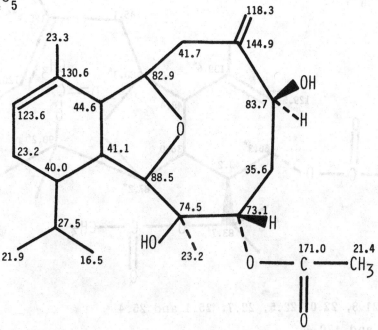

T. Kusumi, H. Uchida, M.O. Ishitsuka, H. Yamamoto and H. Kakisawa, <u>Chem. Lett.</u>, (6), 1077 (1988).

LITOPHYNIN B

Source: *Litophyton* sp.

Mol. formula: $C_{28}H_{44}O_5$

Mol. wt.: 460

Solvent: $CDCl_3$

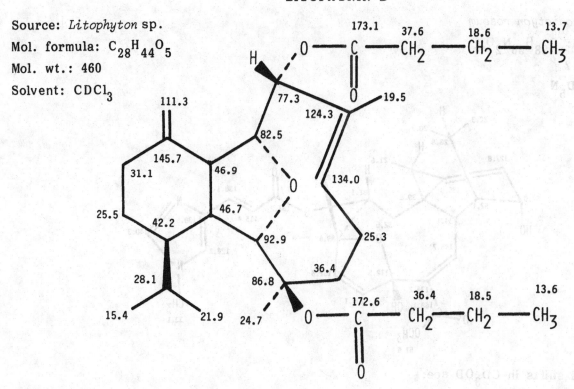

M. Ochi, K. Futatsugi, H. Kotsuki, M. Ishii and K. Shibata, <u>Chem. Lett.</u>, (11), 2207 (1987).

OPHIRIN

Source: *Muricella* sp.
Mol. formula: $C_{26}H_{38}O_7$
Mol. wt.: 462
Solvent: $CDCl_3$

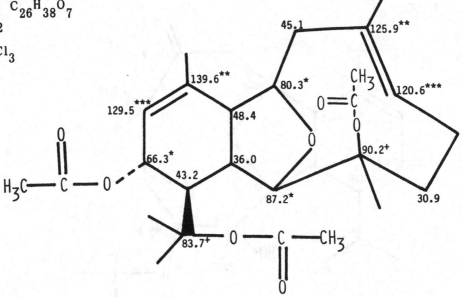

CH_3: 18.3, 21.3, 21.8, 22.0, 22.5, 22.7, 25.1 and 25.4
C=O: 167.5, 169.9 and 170.4

Y. Kashman, <u>Tetrahedron Lett.</u>, 21(9), 879 (1980).

||||||||||||||||||||||||||||||||||||||||||||||||||||||||||||||||||||||||||||||||||||||||||||||||||||||||||||||||||||||||||

SARCODICTYIN C

Source: *Sarcodictyon roseum*
Mol. formula: $C_{28}H_{36}N_2O_7$
Mol. wt.: 512
Solvent: C_5D_5N

For chemical shifts in CD_3OD see:

M.D. Ambrosio, A. Guerriero and F. Pietra, <u>Helv. Chim. Acta</u>, 71 (5), 964 (1988).

(+)-SARCODICTYIN E

Source: *Sarcodictyon roseum*
Mol. formula: $C_{28}H_{36}N_2O_7$
Mol. wt.: 512
Solvent: CD_3OD

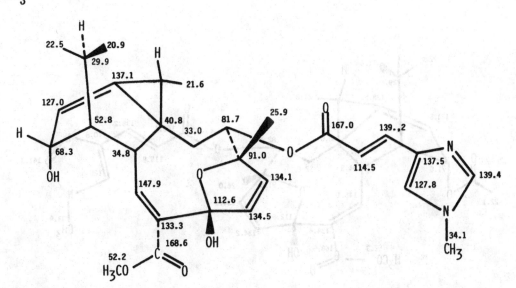

M.D. Ambrosio, A. Guerriero and F. Pietra, Helv. Chim. Acta, 71 (5), 964 (1988).

||||||||||||||||||||||||||||||||||||||||||||||||||||||||||||||||||||||||||||||||||||||||||||||||||||||||||||||||||||||||||||||||||||||||||||

(-)-SARCODICTYIN F

Source: *Sarcodictyon roseum*
Mol. formula: $C_{28}H_{36}N_2O_7$
Mol. wt.: 512
Solvent: CD_3OD

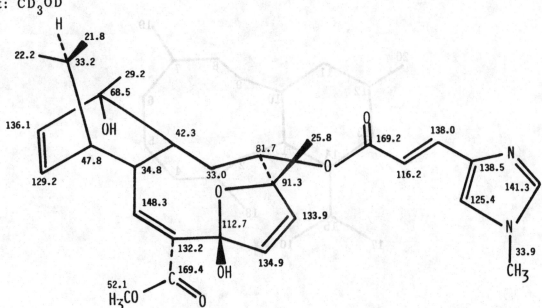

M.D. Ambrosio, A. Guerriero and F. Pietra, Helv. Chim. Acta, 71 (5), 964 (1988).

Source: *Sarcodictyon roseum*
Mol. formula: $C_{30}H_{38}N_2O_8$
Mol. wt.: 554
Solvent: CD_3OD

M.D. Ambrosio, A. Guerriero and F. Pietra, Helv. Chim. Acta, 71 (5), 964 (1988).

ASBESTININ

Basic skeleton:

ASBESTININ-III

Source: *Briareum asbestinum* (Pallas)

Mol. formula: $C_{24}H_{38}O_5$

Mol. wt.: 406

Solvent: C_6D_6

$H_3C-CH_2-CH_2-C-O$ (structure)

CH$_3$: 10.8, 13.7, 18.1, 18.5 and 18.7

CH$_2$: 18.7, 31.0, 31.6, 36.6, 44.5 and 68.1

CH: 31.3, 37.8, 37.9, 39.0, 48.8, 72.2, 73.2, 80.9, 93.2 and 126.9

C: 80.8 and 127.5

C=O: 173.3

D.B. Stierle, B. Carte, D.J. Faulkner, B. Tagle and J. Clardy, J. Am. Chem. Soc., 102 (15), 5088 (1980).

||||||||||||||||||||||||||||||||||||||||||||||||||||||||||||||||||||||||||||||||||||||||||||||||||||||||||||

ASBESTININ-I

Source: *Briareum asbestinum* (Pallas)

Mol. formula: $C_{26}H_{40}O_6$

Mol. wt.: 448

Solvent: C_6D_6

$H_3C-CH_2-CH_2-C-O$ (structure)

CH$_3$: 11.6, 13.7, 18.4, 18.8, 19.7 and 21.1

CH$_2$: 18.8, 29.7, 31.6, 36.6, 44.5 and 67.6

CH: 31.3, 37.6, 38.5, 38.5, 48.8, 72.9, 73.4, 81.0, 94.8 and 125.6

C: 79.4 and 129.5

C=O: 169.9 and 173.2

D.B. Stierle, B. Carte, D.J. Faulkner, B. Tagle and J. Clardy, J. Am. Chem. Soc., 102 (15), 5088 (1980).

Source: *Briareum asbestinum* (Pallas)

Mol. formula: $C_{26}H_{40}O_6$

Mol. wt.: 448

Solvent: C_6D_6

11.0, 13.7, 18.4, 18.8, 19.6, 21.1, 29.5, 31.7, 31.7, 34.0, 36.7, 37.1, 38.2, 38.9, 41.2, 45.5, 67.7, 73.3, 77.2, 79.4, 82.1, 92.0, 127.5, 131.8, 170.0 and 173.1

D.B. Stierle, B. Carte, D.J. Faulkner, B. Tagle and J. Clardy, J. Am. Chem. Soc., **102** (15), 5088 (1980).

ASBESTININ-IV

Source: *Briareum asbestinum* (Pallas)

Mol. formula: $C_{26}H_{38}O_7$

Mol. wt.: 462

Solvent: C_6D_6

CH_3: 11.2, 13.7, 18.2, 18.7 and 20.9

CH_2: 18.7, 36.4, 37.3, 37.9, 39.9, 41.5, 67.7 and 114.1

CH: 38.2, 48.8, 71.9, 72.5, 80.4 and 93.0

C: 77.8 and 146.8

C=O: 172.9, 172.9 and 198.7

31.4 and 31.4

D.B. Stierle, B. Carte, D.J. Faulkner, B. Tagle and J. Clardy, J. Am. Chem. Soc., **102** (15), 5088 (1980).

ASBESTININ-V

Source: *Briareum asbestinum* (Pallas)

Mol. formula: $C_{26}H_{40}O_7$

Mol. wt.: 464

Solvent: C_6D_6

11.0, 13.7, 17.8, 17.8, 18.8, 21.1, 31.5, 31.7, 36.7, 37.4, 37.4, 38.9, 39.0, 39.4, 46.0, 67.4, 72.5, 73.6, 74.0, 76.9, 83.4, 93.4, 115.5, 148.5, 171.8 and 173.1

D.B. Stierle, B. Carte, D.J. Faulkner, B. Tagle and J. Clardy, <u>J. Am. Chem. Soc.</u>, **102** (15), 5088 (1980).

||||||||||||||||||||||||||||||||||||||||||||||||||||||||||||||||||||||||||||||||||||||||||||||||||||||||||||||||||||||||||||||||||||||||||||

CUBITANE

Basic skeleton:

CUBITENE

Source: *Cubitermes umbratus* Williams
Mol. formula: $C_{20}H_{32}$
Mol. wt.: 272

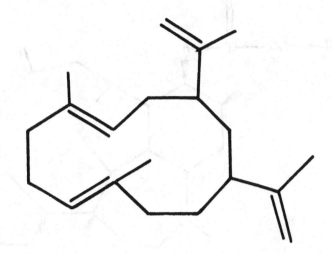

14.3, 15.1, 18.3, 22.6, 24.8, 28.0, 30.9, 31.3, 37.0, 39.9, 40.7, 40.9, 109.0, 110.7, 123.1, 127.5, 132.5, 134.4, 149.4 and 150.0

G.D. Prestwich, D.F. Wiemer, J. Meinwald and J. Clardy, J. Am. Chem. Soc., **100** (8), 2560 (1978).

IIIIIIIIIIIIIIIIIIIIIIIIIIIIIIIIIIIIIIIIIIIIIIIIIIIIIIIIIIIIIIIIIIIIIIIIIIIIIIIIIIIIIIIIIIIIIIIIIII

CALYCULONE A

Source: *Eunicea calyculata* Ellis and Solander
Mol. formula: $C_{20}H_{32}O_2$
Mol. wt.: 304
Solvent: $CDCl_3$

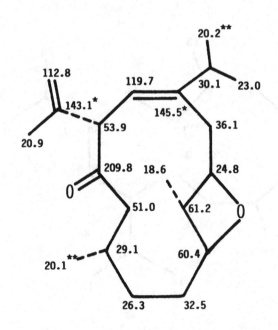

S.A. Look, W. Fenical, Z. Qi-Tai and J. Clardy, J. Org. Chem., 49(8), 1417 (1984).

CALYCULONE B

Source: *Eunicea calyculata* (Ellis and Solander)
Mol. formula: $C_{20}H_{32}O_2$
Mol. wt.: 304
Solvent: $CDCl_3$

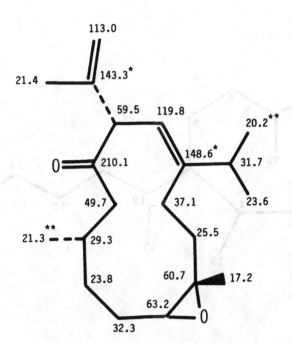

S.A. Look, W. Fenical, Z.Qi-Tai and J.Clardy, <u>J. Org. Chem.</u>, **49** (8), 1417 (1984).

CALYCULONE C

Source: *Eunicea calyculata* (Ellis and Solander)
Mol. formula: $C_{20}H_{32}O_2$
Mol. wt.: 304
Solvent: $CDCl_3$

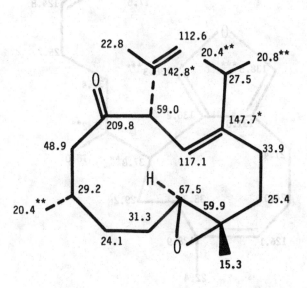

S.A. Look, W. Fenical, Z. Qi-Tai and J. Clardy, <u>J. Org. Chem.</u>, **49** (8), 1417 (1984).

CYCLONANE

Basic skeleton:

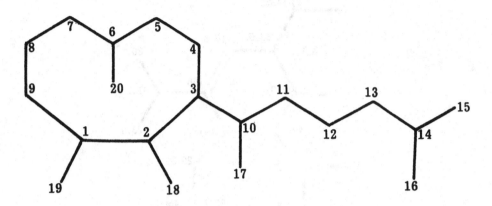

||||||||||||||||||||||||||||||||||||||||||||||||||||||||||||||||||||||||||||||||||||||||||

DICTYOFURAN C

Source: *Dictyota dichotoma*

Mol. formula: $C_{20}H_{30}O$

Mol. wt.: 286

Solvent: $CDCl_3$

25.5

129.5

17.6   124.8

25.7

37.4

O

138.5 ***   138.3 ***

35.2 **

136.2 *

131±0

21.4   37.8 **   18.0

24.0   33.3   29.2

126.1   128.9

22.4

N. Enoki, H. Shirahama, E. Osawa, S. Urano, R. Ishida and T. Matsumoto, <u>Chem. Lett.</u>, (9), 1399 (1983).

DICTYOFURAN T

Source: *Dictyota dichotoma*

Mol. formula: $C_{20}H_{30}O$

Mol. wt.: 286

Solvent: $CDCl_3$

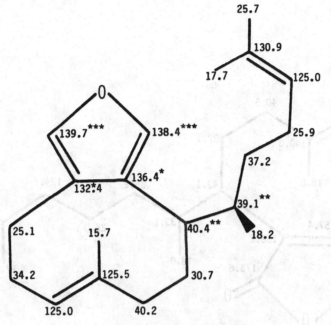

N. Enoki, H. Shirahama, E. Osawa, S. Urano, R. Ishida and T. Matsumoto, <u>Chem. Lett.</u>, (9), 1399 (1983).

||||||||||||||||||||||||||||||||||||||||||||||||||||||||||||||||||||||||||||||||||||||||||||||||||||||||||||||||||||||||||||||||||||||||||

DICTYODIAL

Source: *Dictyota crenulata* and *D. flabellata*

Mol. formula: $C_{20}H_{30}O_2$

Mol. wt.: 302

Solvent: $CDCl_3$

J. Finer, J. Clardy, W. Fenical, L. Minale, R. Riccio, J. Battaile, M. Kirkup and R.E. Moore, <u>J. Org. Chem.</u>, 44 (12), 2044 (1979).

NEODICTYOLACTONE

Source: *Pachydictyon coriaceum*

Mol. formula: $C_{20}H_{30}O_2$

Mol. wt.: 302

Solvent: $CDCl_3$

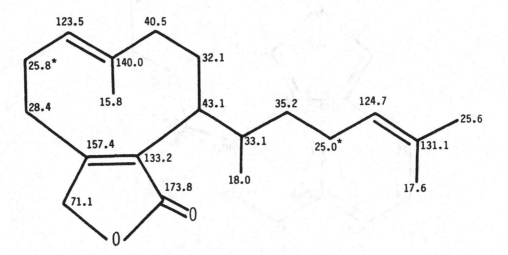

M. Ishitsuka, T. Kusumi, J. Tanaka, M. Chihara and H. Kakisawa, Chem. Lett., (1), 151 (1984).

ISODICTYOHEMIACETAL

Source: *Dictyota dichotoma*

Mol. formula: $C_{20}H_{32}O_2$

Mol. wt.: 304

Solvent: $CDCl_3$

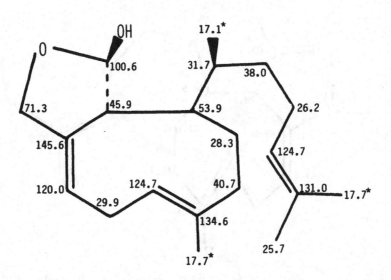

N. Enoki, R. Ishida and T. Matsumoto, Chem. Lett., 1749 (1982).

ACETYLDICTYOLAL

Source: *Dictyota dichotoma*

Mol. formula: $C_{22}H_{34}O_3$

Mol. wt.: 346

Solvent: $CDCl_3$

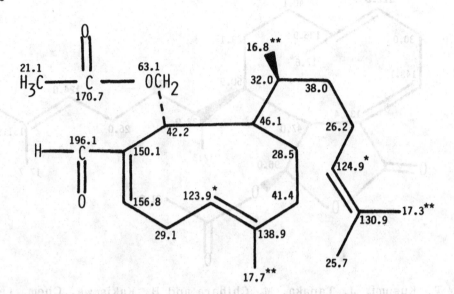

N. Enoki, R. Ishida and T. Matsumoto, <u>Chem. Lett.</u>, 1749 (1982).

||||||||||||||||||||||||||||||||||||||||||||||||||||||||||||||||||||||||||||||||||||||||||||||||||||

DITERPENE PC-VA

Source: *Pachydictyon coriaceum* (Sanadagusa)

Mol. formula: $C_{22}H_{36}O_3$

Mol. wt.: 348

Solvent: $CDCl_3$

CH_3: 17.1, 17.1, 17.6, 25.7, 54.5 and 54.7

CH_2: 25.9, 28.2, 30.4, 38.4 and 40.6

CH: 31.5, 46.9, 51.6, 107.3, 107.3, 125.0, 125.4 and 126.2

C: 130.8, 135.0 and 146.2

M. Ishitsuka, T. Kusumi, J. Tanaka and H. Kakisawa, <u>Chem. Lett.</u>, 1517 (1982).

18-ACETOXYDICTYOLACTONE

Source: *Pachydictyon coriaceum*

Mol. formula: $C_{22}H_{32}O_4$

Mol. wt.: 360

Solvent: $CDCl_3$

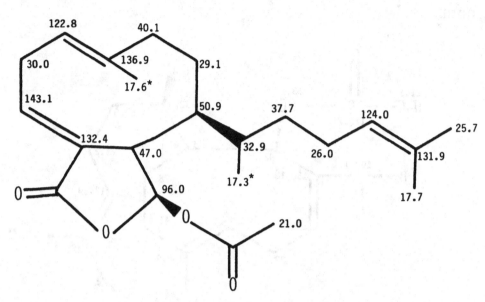

M. Ishitsuka, T. Kusumi, J. Tanaka, M. Chihara and H. Kakisawa, <u>Chem. Lett.</u>, (1), 151 (1984).

||||||||||||||||||||||||||||||||||||||||||||||||||||||||||||||||||||||||||||||||||||||||||||||||||||||||||||||||||

HYDROXYACETYLDICTYOLAL

Source: *Dictyota dichotoma*

Mol. formula: $C_{22}H_{34}O_4$

Mol. wt.: 362

Solvent: $CDCl_3$

N. Enoki, R. Ishida and T. Matsumoto, <u>Chem. Lett.</u>, 1749 (1982).

CRENULACETAL C

Source: Unidentified species of Dictyotaceae

Mol. formula: $C_{24}H_{38}O_5$

Mol. wt.: 406

Solvent: $CDCl_3$

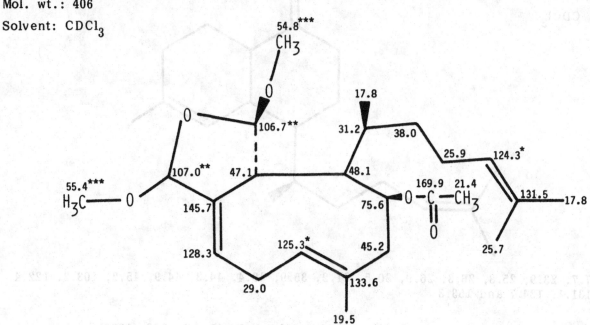

CH_3: 17.7, 21.5, 25.7, 53.5 and 54.5

CH_2: 8.1, 26.6 and 35.4

CH: 12.6, 32.1, 45.9, 73.6, 109.3 and 124.5

C: 131.5

C=O: 169.8

T. Kusumi, D. Muanza-Nkongolo, M. Goya, M. Ishitsuka, T. Iwashita and H. Kakisawa, J. Org. Chem., 51 (3), 384 (1986).

|||||||||||||||||||||||||||||||||||||||||||||||||||||||||||||||||||||||||||||||||||||||||||||||||||||||||||||||||||||||||||||||||||

DICTYODIACETAL

Source: *Dictyota dichotoma*

Mol. formula: $C_{24}H_{38}O_5$

Mol. wt.: 406

Solvent: $CDCl_3$

N. Enoki, R. Ishida and T. Matsumoto, Chem. Lett., 1749 (1982).

Basic skeleton:

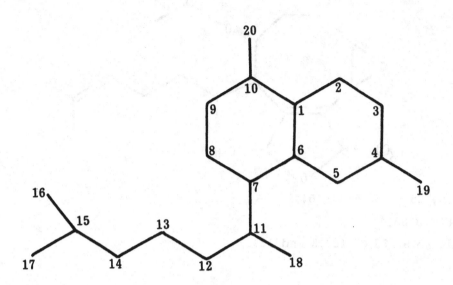

||||||||||||||||||||||||||||||||||||||||||||||||||||||||||||||||||||||||||||||||||||||||||||||||||||||||||||||||||||||||||||||||||||||

(−)-BIFLORA-4,10(19),15-TRIENE

Source: *Cubitermes umbratus* Williams and *Xenia obscuronata*

Mol. formula: $C_{20}H_{32}$

Mol. wt.: 272

Solvent: $CDCl_3$

13.3, 17.7, 23.9, 25.8, 26.3, 26.9, 30.5, 31.2, 35.9, 36.4, 44.3, 44.9, 45.2, 103.2, 122.4, 125.0, 131.1, 134.7 and 153.3

S.K. Paknikar and A.E. Greene, J. Nat. Prod., (Lloydia), 51 (2), 326 (1988).

(2S)-SERRULAT-14-ENE-2,8,20-TRIOL

Source: *Eremophila gibsonii* F.Muell.

Mol. formula: $C_{20}H_{30}O_3$

Mol. wt.: 318

Solvent: DMSO-d_6

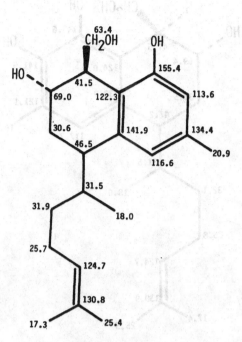

P.G. Forster, E.L. Ghisalberti, P.R. Jefferies, V.M. Poletti and N.J. Whiteside, <u>Phyto-chemistry</u>, 25(6), 1377 (1986).

||||||||||||||||||||||||||||||||||||||||||||||||||||||||||||||||||||||||||||||||||||||||||||||||||||||||||||||||

8,20-DIHYDROXYSERRULAT-14-EN-19-OIC ACID

Source: *Eremophila glabra* (R.Br.) Ostenf.

Mol. formula: $C_{20}H_{28}O_4$

Mol. wt.: 332

Solvent: $CDCl_3$

P.G. Forster, E.L. Ghisalberti, P.R. Jefferies, V.M. Poletti and N.J. Whiteside, <u>Phyto-chemistry</u>, 25(6), 1377 (1986).

(2S)-SERRULAT-14-ENE-2,7,8,20-TETRAOL

Source: *Eremophila hughesii*

Mol. formula: $C_{20}H_{30}O_4$

Mol. wt.: 334

Solvent: DMSO-d_6

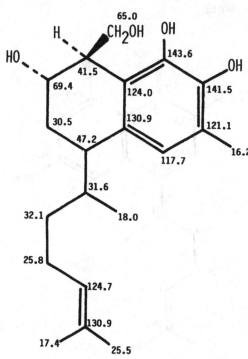

P.G. Forster, E.L. Ghisalberti, P.R. Jefferies, V.M. Poletti and N.J. Whiteside, <u>Phyto-chemistry</u>, 25(6), 1377 (1986).

||||||||||||||||||||||||||||||||||||||||||||||||||||||||||||||||||||||||||||||||||||||||||||||||||||||||||||||||

KALIHINOL C

Source: *Acanthella* sp.

Mol. formula: $C_{22}H_{32}N_2O_2$

Mol. wt.: 356

Solvent: $CDCl_3$

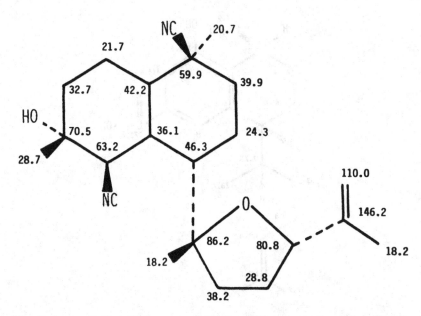

C.W.J. Chang, A. Patra, J.A. Baker and P.J. Scheuer, <u>J. Am. Chem. Soc.</u>, 109(20), 6119 (1987).

KALIHINOL Y

Source: *Acanthella* sp.
Mol. formula: $C_{21}H_{32}ClNO_2$
Mol. wt.: 365
Solvent: $CDCl_3$

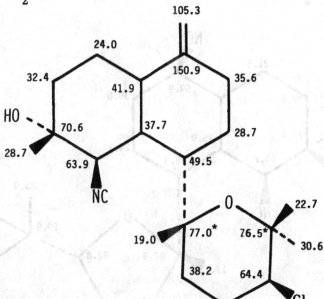

C.W.J. Chang, A. Patra, J.A. Baker and P.J. Scheuer, <u>J. Am. Chem. Soc.</u>, **109(20)**, 6119 (1987).

7,8,16,17-TETRAHYDROXYSERRULATAN-19-OIC ACID

Source: *Eremophila drummondii*
Mol. formula: $C_{20}H_{30}O_6$
Mol. wt.: 366
Solvent: $(CD_3)_2CO$

P.G. Forster, E.L. Ghisalberti, P.R. Jefferies, V.M. Poletti and N.J. Whiteside, <u>Phytochemistry</u>, **25(6)**, 1377 (1986).

KALIHINOL F

Source: *Acanthella* sp.
Mol. formula: $C_{23}H_{33}N_3O_2$
Mol. wt.: 383
Solvent: $CDCl_3$

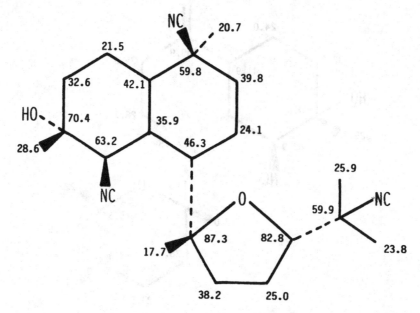

C.W.J. Chang, A. Patra, J.A. Baker and P.J. Scheuer, <u>J. Am. Chem. Soc.</u>, **109** (20), 6119 (1987).

KALIHINOL A

Source: *Acanthella* sp.
Mol. formula: $C_{22}H_{33}ClN_2O_2$
Mol. wt.: 392
Solvent: $CDCl_3$

C.W.J. Chang, A. Patra, J.A. Baker and P.J. Scheuer, <u>J. Am. Chem. Soc.</u>, **109** (20), 6119 (1987).

588

KALIHINOL B

Source: *Acanthella* sp.
Mol. formula: $C_{22}H_{33}ClN_2O_2$
Mol. wt.: 392
Solvent: $CDCl_3$

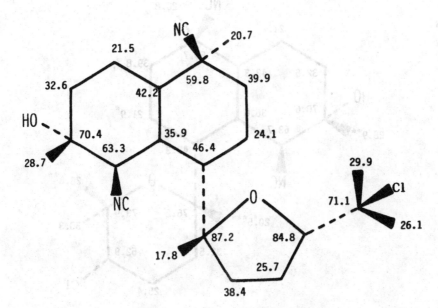

C.W.J. Chang, A. Patra, J.A. Baker and P.J. Scheuer, <u>J. Am. Chem. Soc.</u>, **109** (20), 6119 (1987).

KALIHINOL D

Source: *Acanthella* sp.
Mol. formula: $C_{22}H_{33}ClN_2O_2$
Mol. wt.: 392
Solvent: $CDCl_3$

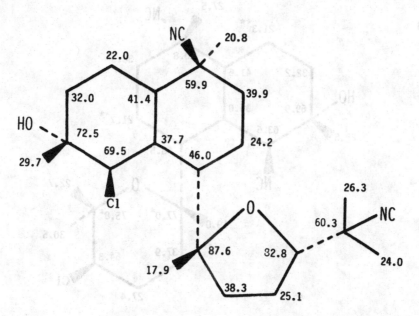

C.W.J. Chang, A. Patra, J.A. Baker and P.J. Scheuer, <u>J. Am. Chem. Soc.</u>, **109** (20), 6119 (1987).

KALIHINOL E

Source: *Acanthella* sp.
Mol. formula: $C_{22}H_{33}ClN_2O_2$
Mol. wt.: 392
Solvent: $CDCl_3$

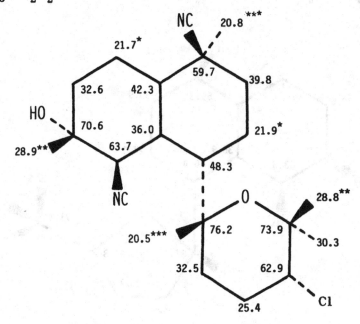

C.W.J. Chang, A. Patra, J.A. Baker and P.J. Scheuer, <u>J. Am. Chem. Soc.</u>, **109** (20), 6119 (1987).

KALIHINOL Z

Source: *Acanthella* sp.
Mol. formula: $C_{22}H_{33}ClN_2O_2$
Mol. wt.: 392
Solvent: $CDCl_3$

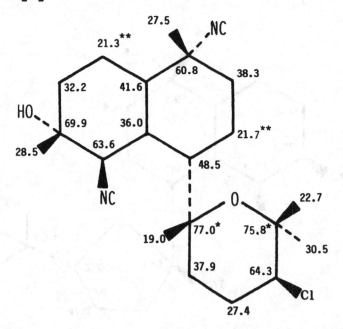

C.W.J. Chang, A. Patra, J.A. Baker and P.J. Scheuer, <u>J. Am. Chem. Soc.</u>, **109**(20), 6119 (1987).

KALIHINOL G

Source: *Acanthella* sp.
Mol. formula: $C_{23}H_{33}N_3O_2S$
Mol. wt.: 415
Solvent: $CDCl_3$

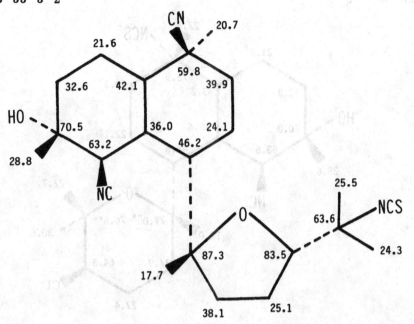

C.W.J. Chang, A. Patra, J.A. Baker and P.J. Scheuer, <u>J. Am. Chem. Soc.</u>, **109(20)**, 6119 (1987).

||||||||||||||||||||||||||||||||||||||||||||||||||||||||||||||||||||||||||||||||||||||||||||||||||||||||||||||

KALIHINOL H

Source: *Acanthella* sp.
Mol. formula: $C_{23}H_{33}N_3O_2S$
Mol. wt.: 415
Solvent: $CDCl_3$

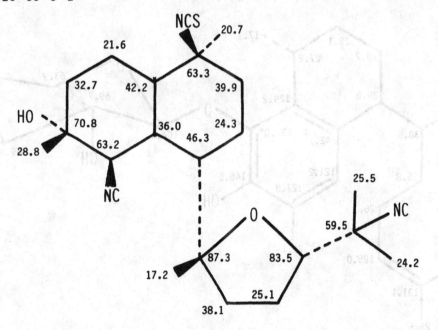

C.W.J. Chang, A. Patra, J.A. Baker and P.J. Scheuer, <u>J. Am. Chem. Soc.</u>, **109(20)**, 6119 (1987).

KALIHINOL X

Source: *Acanthella* sp.
Mol. formula: $C_{22}H_{33}CIN_2O_2S$
Mol. wt.: 424
Solvent: $CDCl_3$

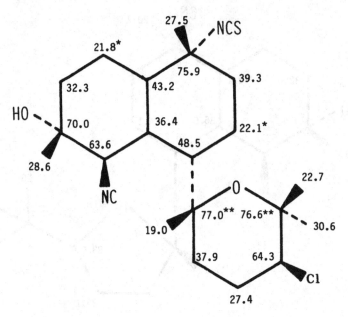

C.W.J. Chang, A. Patra, J.A. Baker and P.J. Scheuer, J. Am. Chem. Soc., **109**(20), 6119 (1987).

||||||||||||||||||||||||||||||||||||||||||||||||||||||||||||||||||||||||||||||||||||||||||||||||||||||||||||||||||||||||||||||||||||||||

SECOPSEUDOPTEROSIN A

Source: *Pseudopterogorgia kallos*
Mol. formula: $C_{25}H_{38}O_6$
Mol. wt.: 434
Solvent: $CDCl_3$

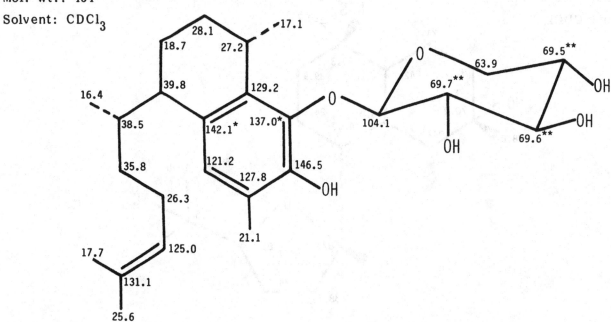

S.A. Look and W. Fenical, Tetrahedron, **43** (15), 3363 (1987).

SECOPSEUDOPTEROSIN C

Source: *Pseudopterogorgia kallos*
Mol. formula: $C_{27}H_{40}O_7$
Mol. wt.: 476
Solvent: $CDCl_3$

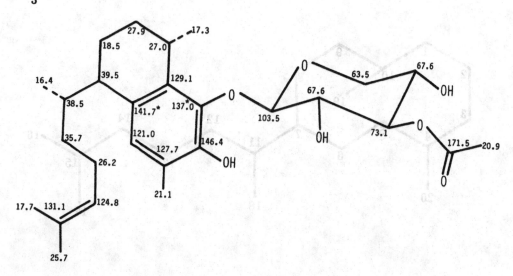

S.A. Look and W. Fenical, <u>Tetrahedron</u>, **43** (15), 3363 (1987).

||||||||||||||||||||||||||||||||||||||||||||||||||||||||||||||||||||||||||||||||||||||||||||||||||||||

SECOPSEUDOPTEROSIN D

Source: *Pseudopterogorgia kallos*
Mol. formula: $C_{27}H_{40}O_7$
Mol. wt.: 476
Solvent: $CDCl_3$

S.A. Look and W. Fenical, <u>Tetrahedron</u>, **43** (15), 3363 (1987).

Basic skeleton:

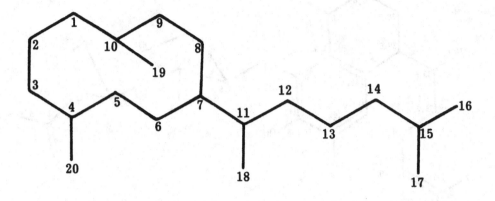

||||||||||||||||||||||||||||||||||||||||||||||||||||||||||||||||||||||||||||||||

PACHYALDEHYDE

Source: *Pachydictyon coriaceum*
Mol. formula: $C_{19}H_{30}O$
Mol. wt.: 274
Solvent: $CDCl_3$

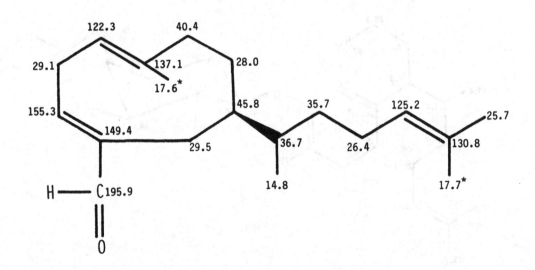

M. Ishitsuka, T. Kusumi, H. Kakisawa, Y. Nagai, Y. Kawakami and T. Sato, <u>J. Chem. Soc. Chem. Commun.</u>, (14), 906 (1984).

DILOPHOLONE

Source: *Dilophus prolificans*
Mol. formula: $C_{20}H_{30}O_2$
Mol. wt.: 302
Solvent: $CDCl_3$

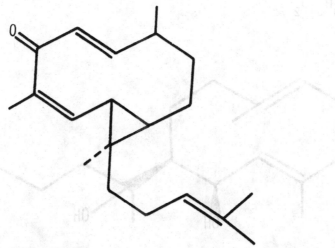

CH_2: 42.3 and 47.9

CH: 36.5, 124.4, 125.3, 130.8 and 159.5

C: 71.7, 131.0 and 141.6

C=O: 202.8

13.3, 17.6, 19.6, 20.4, 24.5, 25.1, 25.6, 26.9 and 28.4

R. Kazlauskas, P.T. Murphy, R.J. Wells and J.F. Blount, <u>Tetrahedron Lett.</u>, (43), 4155 (1978).

||||||||||||||||||||||||||||||||||||||||||||||||||||||||||||||||||||||||||||||||||||||||||||||||||||||||||||||||||||||||||||||||||

EPOXYDILOPHONE

Source: *Dilophus prolificans*
Mol. formula: $C_{20}H_{30}O_2$
Mol. wt.: 302
Solvent: $CDCl_3$

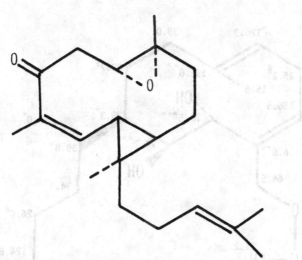

CH_2: 43.6

CH: 58.2, 124.3 and 125.6

C: 60.5, 130.9 and 137.4

C=O: 208.3

17.5, 18.1, 19.7, 19.7, 24.9, 25.3, 25.6, 26.4, 29.7, 33.0, 36.1 and 36.7

R. Kazlauskas, P.T. Murphy, R.J. Wells and J.F. Blount, <u>Tetrahedron Lett.</u>, (43), 4155 (1978).

HYDROXYDILOPHOL

Source: *Dictyota masonii* Setchell and Gardner
Mol. formula: $C_{20}H_{34}O_2$
Mol. wt.: 306
Solvent: $CDCl_3$

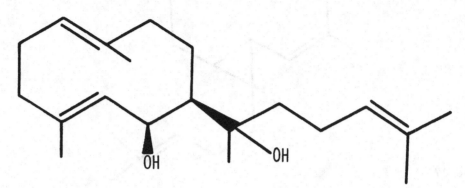

CH: 68.8 and 124.4
C: 76.6, 131.7 and 133.3
16.6, 17.7, 23.5, 24.4, 24.7, 25.7 and 41.5

H.H. Sun and W. Fenical, J. Org. Chem., 44 (8), 1354 (1979).

||||||||||||||||||||||||||||||||||||||||||||||||||||||||||||||||||||||||||||||||||||||||||||||||||

ACETOXYPACHYDIOL

Source: *pachydictyon coriaceum*
Mol. formula: $C_{22}H_{36}O_4$
Mol. wt.: 364
Solvent: CD_2Cl_2

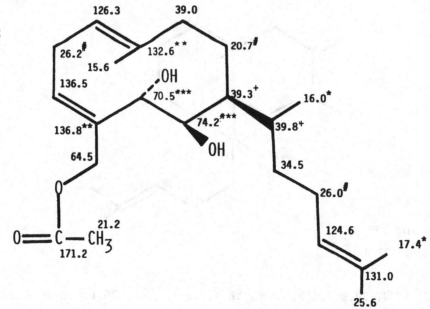

M. Ishitsuka, T. Kusumi, H. Kakisawa, Y. Kawakami, Y. Nagai and T. Sato, Tetrahedron Lett., 27(23), 2639 (1986).

ACETOXYDILOPHOLONE

Source: *Dilophus prolificans*

Mol. formula: $C_{22}H_{32}O_4$

Mol. wt.: 360

Solvent: $CDCl_3$

CH: 39.9, 79.9, 123.9, 126.0, 138.3 and 150.0

C: 73.4, 131.4 and 135.8

C=O: 169.7 and 200.0

17.5, 18.1, 19.7, 21.0, 24.9, 25.6, 25.7, 31.0, 31.4, 33.9 and 35.7

R. Kazlauskas, P.T. Murphy, R.J. Wells and J.F. Blount, Tetrahedron Lett., (43), 4155 (1978).

||||||||||||||||||||||||||||||||||||||||||||||||||||||||||||||||||||||||||||||||||||||||||||||||||||||||||||||||||||||||||||||||||||||||

EPI-ACETOXYDILOPHOLONE

Source: *Dilophus prolificans*

Mol. formula: $C_{22}H_{32}O_4$

Mol. wt.: 360

Solvent: $CDCl_3$

CH_2: 42.1

CH: 35.4, 82.1, 124.1 128.7, 130.1 and 155.6

C: 73.1, 131.2 and 142.5

C=O: 169.6 and 201.9

13.4, 17.7, 20.5, 21.0, 23.8, 25.1, 25.1, 25.6, 26.6 and 28.4

R. Kazlauskas, P.T. Murphy, R.J. Wells and J.F. Blount, Tetrahedron Lett., (43), 4155 (1978).

SACCULATANE

Basic skeleton:

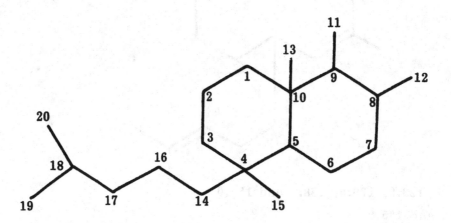

||||||||||||||||||||||||||||||||||||||||||||||||||||||||||||||||||||||||||||||||||||||||||||||||||||||||||||||||||

SACCULATAL

Source: *Pellia endiviaefolia*
Mol. formula: $C_{20}H_{30}O_2$
Mol. wt.: 302
Solvent: $CDCl_3$

Y. Asakawa, M. Toyota and T. Takemoto, <u>Phytochemistry</u>, 19(8), 1799 (1980).

SACCULAPLAGIN

Source: *Plagiochila acanthophylla*
Mol. formula: $C_{24}H_{38}O_7$
Mol. wt.: 438
Solvent: $CDCl_3$

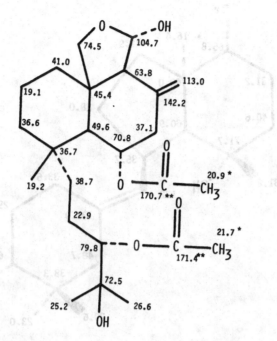

T. Hashimoto, M. Tori and Y. Asakawa, <u>Tetrahedron Lett.</u>, 28(50), 6293 (1987).

||||||||||||||||||||||||||||||||||||||||||||||||||||||||||||||||||||||||||||||||||||||||||||||||||||

IRIEOL

Basic skeleton:

IRIEOL

Source: *Laurencia irieii*

Mol. formula: $C_{20}H_{32}Br_2O$

Mol. wt.: 448

Solvent: $CDCl_3$

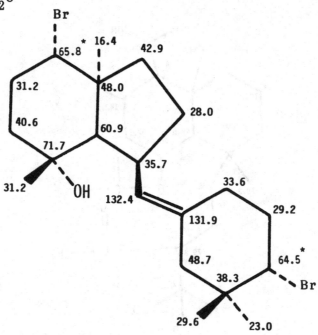

B.M. Howard and W. Fenical, J. Org. Chem., **43** (23), 4401 (1978).

IRIEDIOL

Source: *Laurencia irieii*

Mol. formula: $C_{20}H_{32}Br_2O_2$

Mol. wt.: 464

Solvent: $CDCl_3$

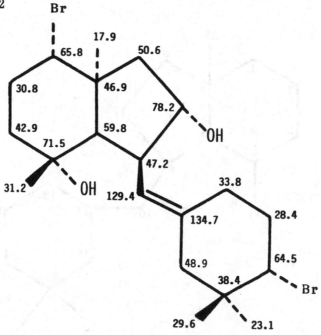

B.M. Howard and W. Fenical, J. Org. Chem., **43** (23), 4401 (1978).

IRIEOL A

Source: *Laurencia irieii*
Mol. formula: $C_{20}H_{32}Br_2O_2$
Mol. wt.: 464
Solvent: $CDCl_3$

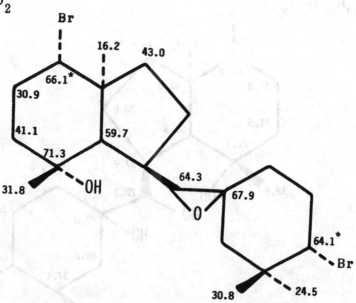

24.8, 25.7, 29.4, 34.9, 36.1, 37.1 and 40.0

B.M. Howard and W. Fenical, <u>J. Org. Chem.</u>, 43 (23), 4401 (1978).

|||||||||||||||||||||||||||||||||||||||||||||||||||||||||||||||||||||||||||||||||||||||||||||||||||||||||||||||||||||||||

IRIEOL C

Source: *Laurencia irieii*
Mol. formula: $C_{20}H_{34}Br_2O_2$
Mol. wt.: 466
Solvent: $CDCl_3$

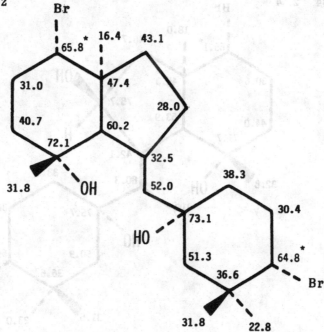

B.M. Howard and W. Fenical, <u>J. Org. Chem.</u>, 43 (23), 4401 (1978).

IRIEOL E

Source: *Laurencia irieii*
Mol. formula: $C_{20}H_{34}Br_2O_3$
Mol. wt.: 482
Solvent: $CDCl_3$

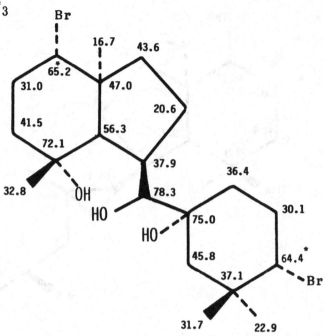

B.M. Howard and W. Fenical, J. Org. Chem., **43** (23), 4401 (1978).

||||||||||||||||||||||||||||||||||||||||||||||||||||||||||||||||||||||||||||||||||||||||||||||||||||||||||||||||||

IRIEOL F

Source: *Laurencia irieii*
Mol. formula: $C_{20}H_{34}Br_2O_4$
Mol. wt.: 498
Solvent: $CDCl_3$

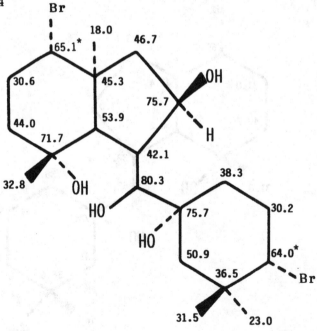

B.M. Howard and W. Fenical, J. Org. Chem., **43** (23), 4401 (1978).

Source: *Laurencia irieii*
Mol. formula: $C_{22}H_{36}Br_2O_5$
Mol. wt.: 540
Solvent: $CDCl_3$

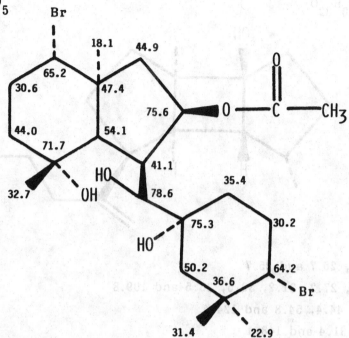

B.M. Howard and W. Fenical, J. Org. Chem., 43 (23), 4401 (1978).

SPATANE

Basic skeleton:

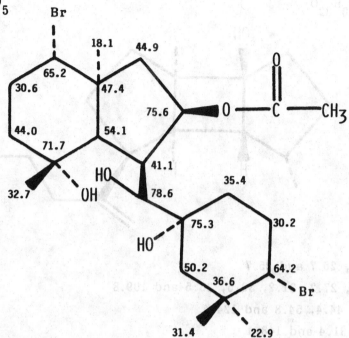

DITERPENE DO-I

Source: *Dilophus okamurai* Dawson
Mol. formula: $C_{20}H_{32}O$
Mol. wt.: 288
Solvent: $CDCl_3$

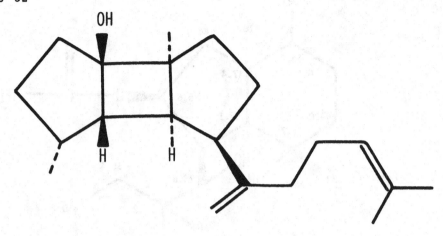

CH_3: 17.5, 17.7, 20.7 and 25.7
CH_2: 15.1, 26.4, 27.2, 31.2, 33.5, 36.5 and 109.8
CH: 34.6, 43.7, 44.4, 54.8 and 124.4
C: 47.9, 93.7, 131.4 and 148.4

K. Kurata, M. Suzuki, K. Shiraishi and K. Taniguchi, <u>Phytochemistry</u>, 27(5), 1321 (1988).

||||||||||||||||||||||||||||||||||||||||||||||||||||||||||||||||||||||||||||||||||||||||||||||||||||||||||||||||||||||||||||

DITERPENE DO-II

Source: *Dilophus okamurai* Dawson
Mol. formula: $C_{20}H_{32}O$
Mol. wt.: 288
Solvent: $CDCl_3$

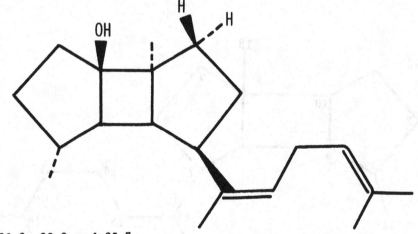

CH_3: 14.0, 17.7, 20.3, 23.9 and 25.7
CH_2: 26.8, 29.5, 33.2, 34.6 and 36.9
CH: 35.6, 42.4, 43.9, 50.0, 123.6 and 125.4
C: 47.9, 82.2, 131.0 and 134.8

K. Kurata, M. Suzuki, K. Shiraishi and K. Taniguchi, <u>Phytochemistry</u>, 27(5), 1321 (1988).

DITERPENE DO-III (DIACETATE)

Source: *Dilophus okamurai* Dawson

Mol. formula: $C_{24}H_{36}O_4$

Mol. wt.: 388

Solvent: $CDCl_3$

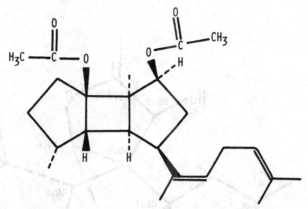

CH_3: 13.7, 13.8, 17.7, 21.2, 21.3, 23.7 and 25.7

CH_2: 26.9, 33.2, 34.6 and 36.1

CH: 34.6, 41.3, 43.1, 46.4, 77.9, 123.3 and 126.6

C: 51.2, 88.4, 131.3 and 133.3

C=O: 170.1 and 170.2

K. Kurata, M. Suzuki, K. Shiraishi and K. Taniguchi, Phytochemistry, 27 (5), 1321 (1988).

||||||||||||||||||||||||||||||||||||||||||||||||||||||||||||||||||||||||||||||||||||||||||||||||||||||||||||||||||||||||||||||||||||||||

GINKGOLIDE

Basic skeleton:

GINKGOLIDE A

Source: *Ginkgo biloba*

Mol. formula: $C_{20}H_{24}O_9$

Mol. wt.: 408

Solvent: DMSO-d_6

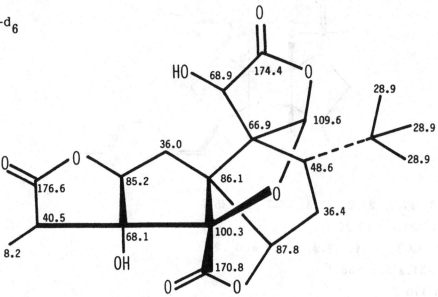

C. Roumestand, B. Perly, D. Hosford and P. Braquet, <u>Tetrahedron</u>, **45** (7), 1975 (1989).

||||||||||||||||||||||||||||||||||||||||||||||||||||||||||||||||||||||||||||||||||||||||||||||||||||||||||||||||||||||||||||||

GINKGOLIDE B

Source: *Ginkgo biloba*

Mol. formula: $C_{20}H_{24}O_{10}$

Mol. wt.: 424

Solvent: DMSO-d_6

C. Roumestand, B. Perly, D. Hosford and P. Braquet, <u>Tetrahedron</u>, **45** (7), 1975 (1989).

GINKGOLIDE J

Source: *Ginkgo biloba*

Mol. formula: $C_{20}H_{24}O_{10}$

Mol. wt.: 424

Solvent: DMSO-d_6

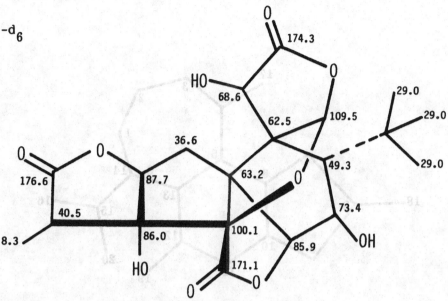

C. Roumestand, B. Perly, D. Hosford and P. Braquet, <u>Tetrahedron</u>, **45** (7), 1975 (1989).

GINKGOLIDE C

Source: *Ginkgo biloba*

Mol. formula: $C_{20}H_{24}O_{11}$

Mol. wt.: 440

Solvent: DMSO-d_6

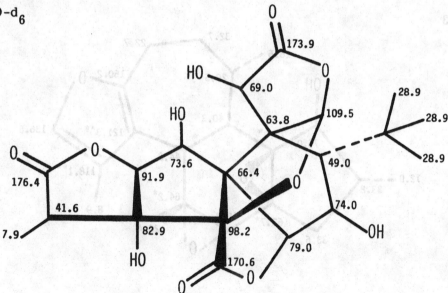

C. Roumestand, B. Perly, D. Hosford and P. Braquet, <u>Tetrahedron</u>, **45** (7), 1975 (1989).

CROTOFOLANE

Basic skeleton:

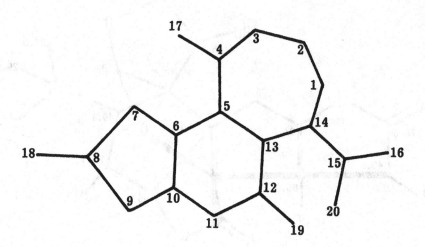

||||||||||||||||||||||||||||||||||||||||||||||||||||||||||||||||||||||||||||||||||||||||||||||||||||

CROTOXIDE A

Source: *Croton dichogamus* Pax.

Mol. formula: $C_{20}H_{26}O_4$

Mol. wt.: 330

Solvent: $CDCl_3$

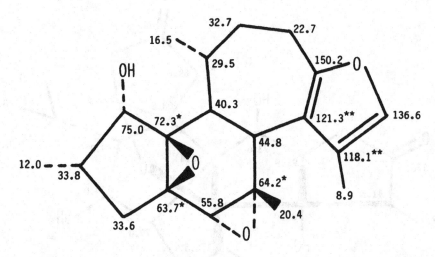

M.K. Jogia, R.J. Andersen, L. Parkanyi, J. Clardy, H.T. Dublin and A.R.E. Sinclair, <u>J. Org. Chem.</u>, **54** (7), 1654 (1989).

CROTOFOLIN E

Source: *Croton corylifolius* L.
Mol. formula: $C_{20}H_{22}O_5$
Mol. wt.: 342
Solvent: $CDCl_3$

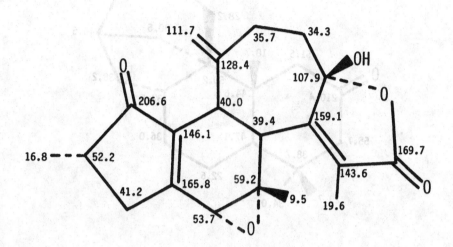

B.A. Burke, W.R. Chan, K.O. Pascoe, M. Jamaica, J.F. Blount and P.S. Manchand, Tetrahedron Lett., (36), 3345 (1979).

||||||||||||||||||||||||||||||||||||||||||||||||||||||||||||||||||||||||||||||||||||||||||||||||||||||||||||||||||||||||||||||||||

STEMODANE

Basic skeleton:

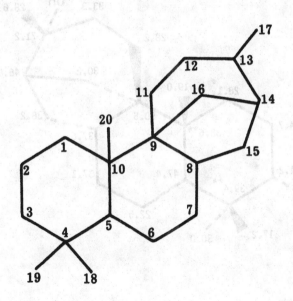

STEMODINONE

Source: *Stemodia maritima* L.

Mol. formula: $C_{20}H_{32}O_2$

Mol. wt.: 304

Solvent: C_5D_5N

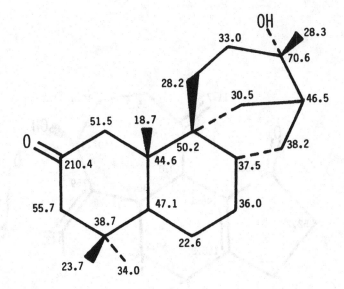

C.D. Hufford, <u>J. Nat. Prod.</u>, (Lloydia), 51 (2), 367 (1988).

MARITIMOL

Source: *Stemodia maritima* L.

Mol. formula: $C_{20}H_{34}O_2$

Mol. wt.: 306

Solvent: C_5D_5N

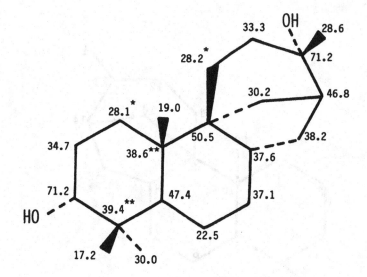

C.D. Hufford, <u>J. Nat. Prod.</u>, (Lloydia), 51 (2), 367 (1988).

610

STEMODIN

Source: *Stemodia maritima* L.

Mol. formula: $C_{20}H_{34}O_2$

Mol. wt.: 306

Solvent: C_5D_5N

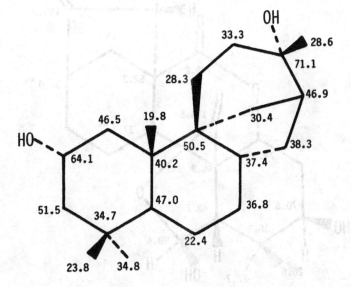

C.D. Hufford, J. Nat. Prod., (Lloydia), 51 (2), 367 (1988).

||||||||||||||||||||||||||||||||||||||||||||||||||||||||||||||||||||||||||||||||||||||||||||||||||||||||||||||||||||||||||||||||||

ENMEIN

Basic skeleton:

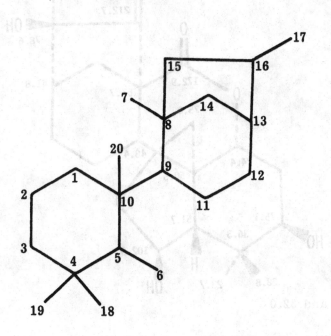

EMEMOGIN

Source: *Rabdosia trichocarpa* Kudo

Mol. formula: $C_{20}H_{26}O_7$

Mol. wt.: 378

Solvent: C_5D_5N

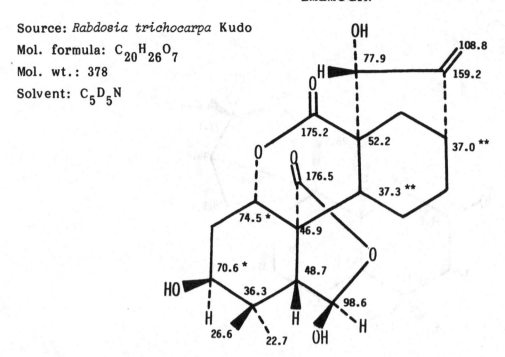

CH$_2$: 17.8, 32.5, 32.6 and 34.1

Y. Takeda, K.-I. Takeda and T. Fujita, J. Chem. Soc. Perkin Trans. I, (4), 689 (1986).

TRICHORABDONIN

Source: *Rabdosia trichocarpa* Kudo

Mol. formula: $C_{20}H_{28}O_7$

Mol. wt.: 380

Solvent: C_5D_5N

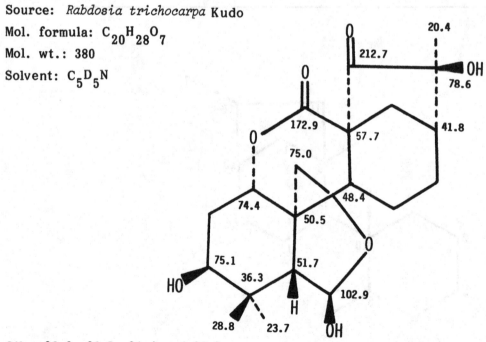

CH$_2$: 20.0, 22.5, 31.4 and 32.0

Y. Takeda, K.-I. Takeda and T. Fujita, J. Chem. Soc. Perkin Trans. I, (4), 689 (1986).

RABDOLASIONAL

Source: *Rabdosia lasiocarpa* (Hayata) Hara

Mol. formula: $C_{22}H_{30}O_7$

Mol. wt.: 406

Solvent: $CDCl_3$

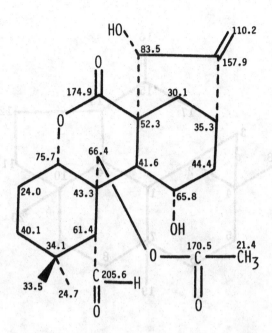

Y. Takeda, T. Fujita and C.-C. Chen, <u>Chem. Lett.</u>, 833 (1982).

||||||||||||||||||||||||||||||||||||||||||||||||||||||||||||||||||||||||||||||||||||||||||||||||||||||

CARPALASIONIN

Source: *Rabdosia lasiocarpa* (Hayata) Hara

Mol. formula: $C_{22}H_{28}O_8$

Mol. wt.: 420

Solvent: C_5D_5N

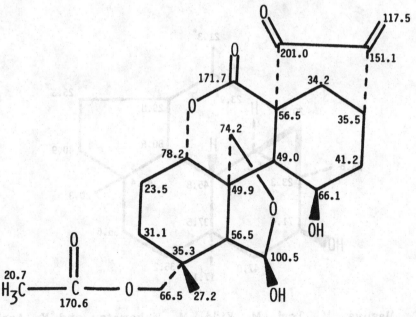

Y. Takeda, T. Fujita and C.-C. Chen, <u>Chem. Lett.</u>, 833 (1982).

Basic skeleton:

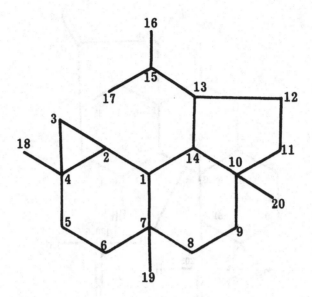

||||||||||||||||||||||||||||||||||||||||||||||||||||||||||||||||||||||||||||||||||||||||||||||||||

(+)-13-EPI-NEOVERRUCOSAN-5β-OL

Source: *Plagiochila stephensoniana* Mitten.

Mol. formula: $C_{20}H_{34}O$

Mol. wt.: 290

Solvent: $CDCl_3$

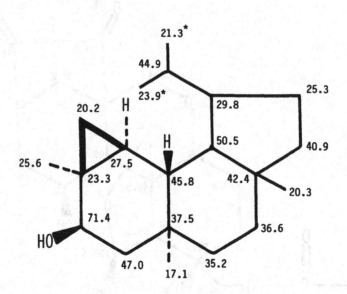

Y. Fukuyama, T. Masuya, M. Tori, M. Kido, M. Wakamatsu and Y. Asakawa, <u>Phytochemistry</u>, 27(6), 1797 (1988).

Source: *Mylia verrucosa* Lindb.
Mol. formula: $C_{22}H_{36}O_3$
Mol. wt.: 348
Solvent: $CDCl_3$

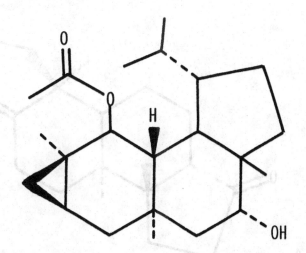

CH_3: 15.1, 19.1, 21.2, 22.2 and 23.4
CH_2: 18.3, 21.8, 32.9, 42.3 and 43.9
CH: 25.2, 29.1, 40.3, 43.0, 46.3, 73.4 and 73.5
C: 23.4, 36.8 and 48.5

D. Takaoka, J. Chem. Soc. Perkin Trans. I, (11), 2711 (1979).

||||||||||||||||||||||||||||||||||||||||||||||||||||||||||||||||||||||||||||||||||||||||||||||||||||||||||||||||||||||||||||||||||

BARBACENANE

Basic skeleton:

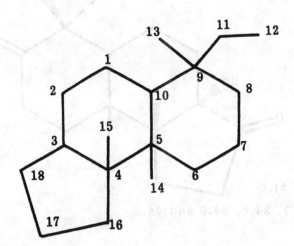

3-DEOXYBARBACENIC ACID

Source: *Barbacenia flava* Martius ex Schultes

Mol. formula: $C_{18}H_{26}O_4$

Mol. wt.: 306

Solvent: $CDCl_3$

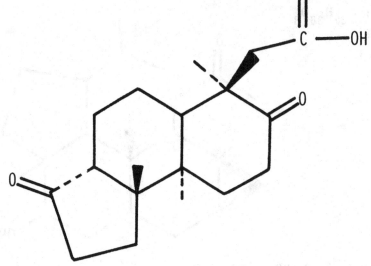

CH_3: 16.6, 26.4 and 29.4

CH_2: 19.6, 20.5, 28.4, 29.4, 36.5, 45.1 and 45.6

CH: 42.8 and 56.8

C: 38.7, 38.7 and 46.1

C=O: 172.6, 218.4 and 218.4

A.C. Pinto, M.D.C. Frechiani and A.L. Pereira, <u>Phytochemistry</u>, **27** (12), 3917 (1988).

|||||||||||||||||||||||||||||||||||||||||||||||||||||||||||||||||||||||||||||||||||||||||||||||||||||||||||||||

BARBACENIC ACID (ACETATE)

Source: *Barbacenia flava* Martius ex Schultes

Mol. formula: $C_{19}H_{28}O_5$

Mol. wt.: 336

Solvent: $CDCl_3$

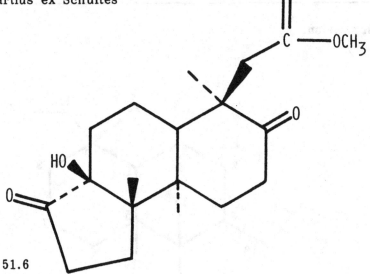

CH_3: 16.0, 16.5, 23.0 and 51.6

CH_2: 20.3, 27.3, 27.9, 29.2, 34.4, 34.9 and 44.1

CH: 41.9

C: 38.8, 48.0, 49.0 and 75.8

C=O: 172.4, 215.5 and 216.7

A.C. Pinto, M.D.C. Frechiani and A.L. Pereira, <u>Phytochemistry</u>, **27** (12), 3917 (1988).

TRINERVERANE

Basic skeleton:

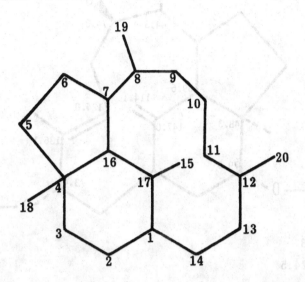

‖‖‖

TRINERVI-2β,3α,9α-TRIOL-9-O-ACETATE

Source: *Trinervitermes soldiers*
Mol. formula: $C_{22}H_{34}O_4$
Mol. wt.: 362
Solvent: $CDCl_3$

117.5
151.3
51.7
69.6
171.0
21.4
CH₃
57.6
127.9
22.8
20.3
46.7
27.7
21.6
72.3
73.5
134.6
HO
HO
H

CH_3: 24.8, 30.3, 32.2, 32.2, 32.6 and 36.4

G.D. Prestwich, S.P. Tanis, F.G. Pilkiewicz, I. Miura and K. Nakanishi, <u>J. Am. Chem. Soc.</u>, 98 (19), 6062 (1976).

Source: *Nasutitermes* sp.

Mol. formula: $C_{29}H_{44}O_6$

Mol. wt.: 488

Solvent: $CDCl_3$

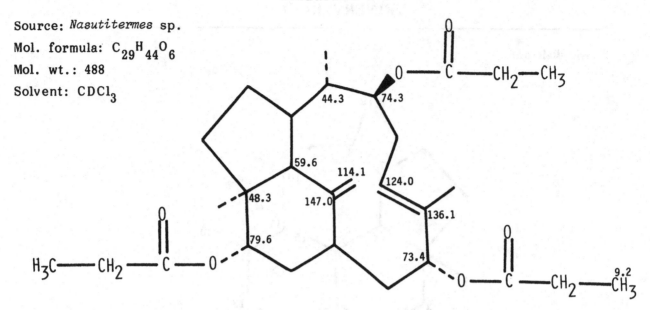

CH_3: 11.7, 11.9 and 19.8

C=O: 173.6, 174.4 and 174.5

27.9, 28.0, 28.4, 30.1, 33.6, 36.3, 36.5, 37.5 and 38.5

G.D. Prestwich, S.G. Spanton, S.H. Goh and Y.P. Tho, <u>Tetrahedron Lett.</u>, 22(17), 1563 (1981).

|||||||||||||||||||||||||||||||||||||||||||||||||||||||||||||||||||||||||||||||||||||||||||||||||||||

(8,19)β-EPOXY-2β,3α,7α,9α,14α,17-HEXAHYDROXY-17-METHYLTRINERVITENE 17-O-ACETOXY-2,3,9,14-O-TETRAPROPIONATE

Source: *Hospitalitermes umbrinus*

Mol. formula: $C_{35}H_{52}O_{12}$

Mol. wt.: 664

Solvent: $CDCl_3$

S.H. Goh, C.H. Chauh, J.C. Beloeil and N. Morellet, <u>Tetrahedron Lett.</u>, 29 (1), 113 (1988).

VISCIDANE

Basic skeleton:

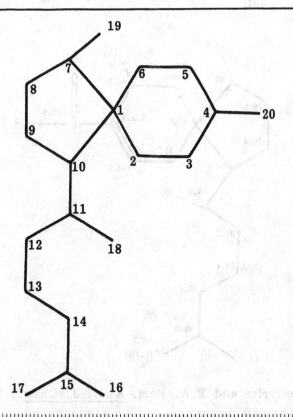

||||||||||||||||||||||||||||||||||||||||||||||||||||||||||||||||||||||||||||||||||||||||||||||||||||||||||||||||||||||||

DITERPENE EV-II

Source: *Eremophila viscida*

Mol. formula: $C_{20}H_{32}O_3$

Mol. wt.: 320

Solvent: $CDCl_3$

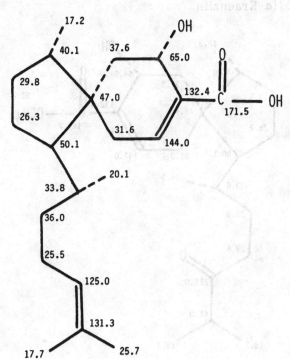

E.L. Ghisalberti, P.R. Jefferies and T.A. Mori, <u>Aust. J. Chem.</u>, 37(3), 635 (1984).

Source: *Eremophila viscida* Endl.

Mol. formula: $C_{20}H_{34}O_4$

Mol. wt.: 338

Solvent: $CDCl_3$

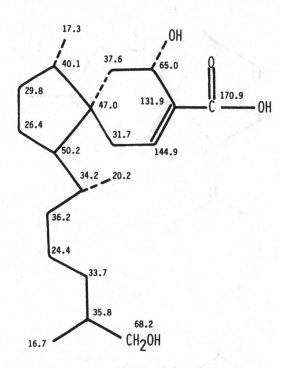

E.L. Ghisalberti, P.R. Jefferies and T.A. Mori, <u>Aust. J. Chem.</u>, 37(3), 635 (1984).

||||||||||||||||||||||||||||||||||||||||||||||||||||||||||||||||||||||||||||||||||||||||||||||||||||||||||||||||||||||||||||||||||||||

(1R,4S,5S,9S,1'R)-1-(1',5'-DIMETHYL-4-OXOHEXYL)-9-HYDROXY-4-METHYL-
SPIRO[4.5]DEC-7-ENE-8-CARBOXYLIC ACID (METHYL ESTER)

Source: *Eremophila cuneifolia* Kraenzlin

Mol. formula: $C_{21}H_{34}O_4$

Mol. wt.: 350

Solvent: $CDCl_3$

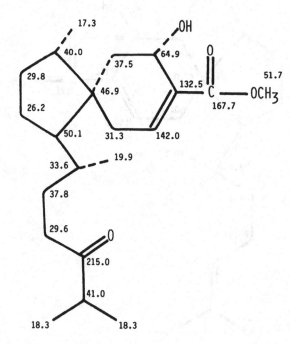

E.L. Ghisalberti, P.R. Jefferies and T.A. Mori, <u>Aust. J. Chem.</u>, 37 (3), 635 (1984).

DITERPENE EE-II (METHYL ESTER)

Source: *Eremophila exotrachys* Kraenzlin

Mol. formula: $C_{21}H_{34}O_4$

Mol. wt.: 350

Solvent: $CDCl_3$

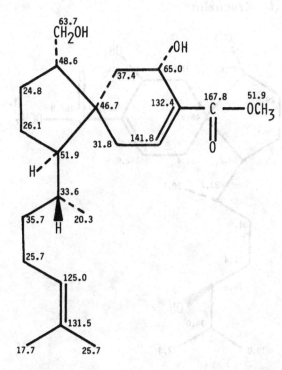

P.G. Forster, E.L. Ghisalberti and P.R. Jefferies, <u>Aust. J. Chem.</u>, 39 (12), 2111 (1986).

||||||||||||||||||||||||||||||||||||||||||||||||||||||||||||||||||||||||||||||||||||||||||||||||||||||||

DITERPENE EV-I (METHYL ESTER)

Source: *Eremophila viscida* Endl

Mol. formula: $C_{21}H_{36}O_4$

Mol. wt.: 352

Solvent: $CDCl_3$

CH_3: 16.7, 17.3, 20.1 and 51.7

CH_2: 24.4, 26.4, 29.8, 31.4, 33.7, 36.2, 37.6 and 68.2

CH: 34.2, 35.8, 40.0, 50.2, 64.9 and 141.9

C: 46.9 and 131.9

C=O: 167.8

E.L. Ghisalberti, C.H. Hocart, P.R. Jefferies, G.M. Proudfoot, B.W. Skelton and A.H. White, <u>Aust. J. Chem.</u>, 36(5), 993 (1983).

(1R,4S,5S,9S,1'R,4'R)-9-HYDROXY-1-(4'-HYDROXY-1',5'-DIMETHYLHEXYL)-4-METHYLSPIRO-[4.5]DEC-7-ENE-8-CARBOXYLIC ACID (METHYL ESTER)

Source: *Eremophila cuneifolia* Kraenzlin

Mol. formula: $C_{21}H_{36}O_4$

Mol. wt.: 352

Solvent: $CDCl_3$

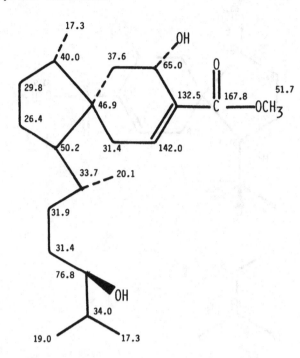

E.L. Ghisalberti, P.R. Jefferies and T.A. Mori, Aust. J. Chem., 37 (3), 635 (1984).

MISCELLANEOUS DITERPENES

DICTYOTETRAENE

Source: *Dictyota* sp.

Mol. formula: $C_{20}H_{28}$

Mol. wt.: 268

Solvent: C_6D_6

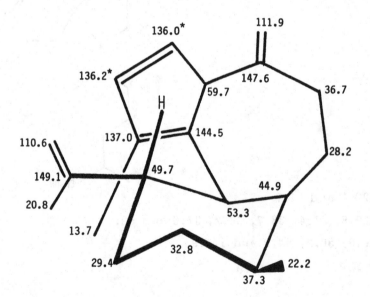

B. Dematte, A. Guerriero and F. Pietra, J. Chem. Soc. Chem. Commun., (7), 391 (1985).

CUBUGENE

Source: *Cubitermes ugandensis* (Fuller)

Mol. formula: $C_{20}H_{32}$

Mol. wt.: 272

Solvent: C_6D_6

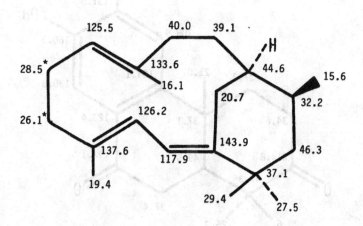

M.S. Tempesta, J.K. Pawlak, T. Iwashita, Y. Naya, K. Nakanishi and G.D. Prestwich, <u>J. Org. Chem.</u>, **49** (11), 2077 (1984).

ACROSTALIDIC ACID

Source: *Acrostalagmus* sp.

Mol. formula: $C_{16}H_{22}O_4$

Mol. wt.: 278

Solvent: $CDCl_3$

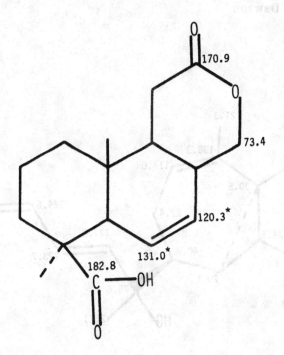

10.7, 18.9, 28.0, 30.6, 35.2, 35.2, 35.5, 37.0, 43.1, 46.1 and 54.7

M. Sato and H. Kakisawa, <u>J. Chem. Soc. Perkin Trans. I</u>, (22), 2407 (1976).

NIMBIONE

Source: *Azadirachta indica* A.Juss

Mol. formula: $C_{18}H_{22}O_3$

Mol. wt.: 286

Solvent: $CDCl_3$

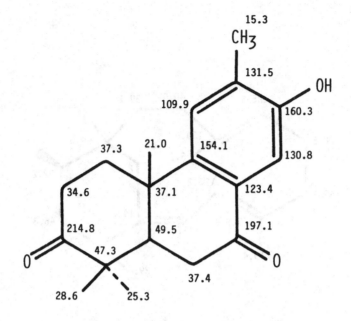

I. Ara, B.S. Siddiqui, S. Faizi and S. Siddiqui, <u>Phytochemistry</u>, 27(6), 1801 (1988).

DITERPENE DO-III

Source: *Dilophus okamurai* Dawson

Mol. formula: $C_{20}H_{30}O$

Mol. wt.: 286

Solvent: $CDCl_3$

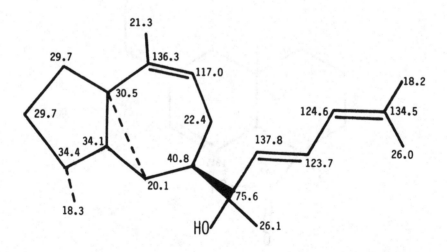

K. Kurata, K. Shiraishi, T. Takato, K. Taniguchi and M. Suzuki, <u>Chem. Lett.</u>, (10), 1629 (1988).

EPOXYDICTYMENE

Source: *Dictyota dichotoma*

Mol. formula: $C_{20}H_{32}O$

Mol. wt.: 288

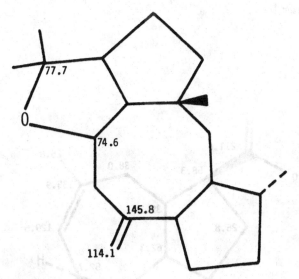

CH$_3$: 19.1, 24.0, 24.3 and 29.2

CH$_2$: 20.2, 34.1, 34.6, 41.1, 44.4 and 48.3

CH: 43.2, 43.8, 50.8, 57.4 and 59.5

N. Enoki, A. Furusaki, K. Suehiro, R. Ishida and T. Matsumoto, <u>Tetrahedron Lett.</u>, 24 (40), 4341 (1983).

|||||||||||||||||||||||||||||||||||||||||||||||||||||||||||||||||||||||||||||||||||||||||||||||||||||||||||||||||||||||||||||||||||||||

(4E,7E,11E)-11-ISOPROPYL-1,4,8-TRIMETHYL-1,2,3,3a,6,9- 10,12a-OCTAHYDROCYCLOPENTACYCLOUNDECEN-1-OL

Source: *Cespitularia* sp.

Mol. formula: $C_{20}H_{32}O$

Mol. wt.: 288

Solvent: $CDCl_3$

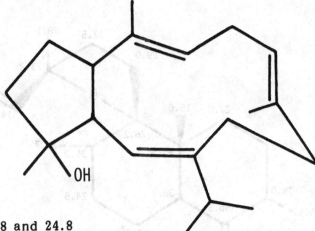

CH$_3$: 15.5, 17.8, 21.8, 24.8 and 24.8

CH$_2$: 24.8, 27.5, 29.2, 37.4 and 41.2

CH: 31.7, 54.9, 57.7, 122.8, 122.8 and 124.0

C: 80.4, 131.9, 132.2 and 147.0

B.F. Bowden, J.C. Coll, J.M. Gulbis, M.F. Mackay and R.H. Willis, <u>Aust. J. Chem.</u>, 39(5), 803 (1986).

18-OXO-3-VIRGENE

Source: *Nicotiana tabacum* L.

Mol. formula: $C_{20}H_{32}O$

Mol.wt.: 288

Solvent: $CDCl_3$

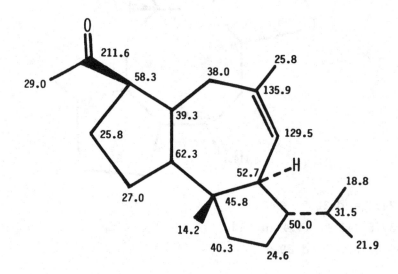

R. Uegaki, T. Fujimori, N. Ueda and A. Ohnishi, <u>Phytochemistry</u>, 26(11), 3029 (1987).

ROTALIN A

Source: *Mycale rotalis* (Bowerbank).

Mol. formula: $C_{20}H_{32}O$

Mol. wt.: 288

Solvent: C_6D_6

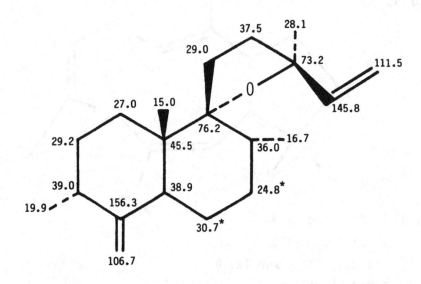

G. Corriero, A. Madaio, L. Mayol, V. Piccialli and D. Sica, <u>Tetrahedron</u>, 45 (1), 277 (1989).

SPHAEROPYRANE

Source: *Sphaerococcus coronopifolius*
Mol. formula: $C_{20}H_{32}O$
Mol. wt.: 288

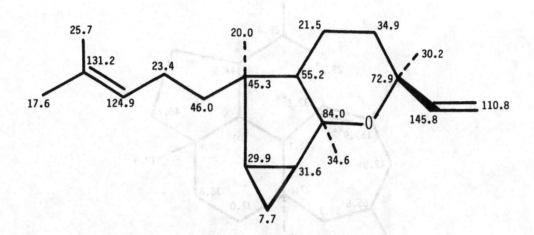

F. Cafieri, L. De Napoli, E. Fattorusso and C. Santacroce, <u>Phytochemistry</u>, 27 (2), 621 (1988).

||||||||||||||||||||||||||||||||||||||||||||||||||||||||||||||||||||||||||||||||||||||||||||||||||||||||||

WENTILACTONE B

Source: *Aspergillus wentii* Wehmer
Mol. formula: $C_{16}H_{18}O_5$
Mol. wt.: 290
Solvent: $CDCl_3$ - DMSO-d_6 (1:1)

J.W. Dorner, R.J. Cole, J.P.Springer, R.H. Cox, H. Cutler and D.T. Wicklow, <u>Phytochemistry</u>, 19(6), 1157 (1980).

627

Source: *Schistochila nobilis* (Hook.) Trev.

Mol. formula: $C_{20}H_{34}O$

Mol. wt.: 290

Solvent: $CDCl_3$

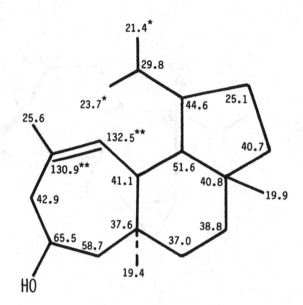

Y. Asakawa, T. Masuya, M. Tori and Y. Fukuyama, Phytochemistry, **27** (11), 3509 (1988).

||||||||||||||||||||||||||||||||||||||||||||||||||||||||||||||||||||||||||||||||||||||||||||||||||||||||||||||||||||||||||||||||||||||||||

PRESPHAEROL

Source: *Sphaerococcus coronopifolius* (Good. et Woodw.) C. Ag.

Mol. formula: $C_{20}H_{34}O$

Mol. wt.: 290

CH_3: 14.7, 22.8, 23.4, 23.5 and 30.6

CH_2: 22.4, 26.9, 27.9, 30.5, 37.0 and 38.1

CH: 28.8, 35.8, 47.8, 50.3, 59.3 and 126.0

C: 45.2, 75.3 and 132.5

F. Cafieri, L. De Napoli, E. Fattorusso, M. Piattelli and S. Sciuto, Tetrahedron Lett., (11), 963 (1979).

AETHIOPINONE

Source: *Salvia aethiopis* L.
Mol. formula: $C_{20}H_{24}O_2$
Mol. wt.: 296

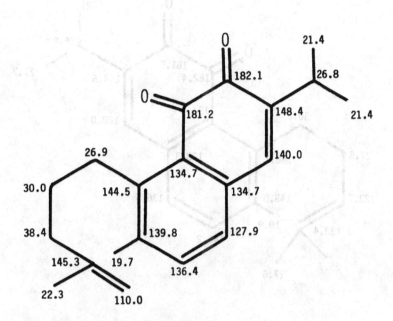

M.T. Boya and S. Valverde, <u>Phytochemistry</u>, 20(6), 1367 (1981).

||||||||||||||||||||||||||||||||||||||||||||||||||||||||||||||||||||||||||||||||||||||||||||||||||||||||||

CURCUSONE B

Source: *Jatropha curcus*
Mol. formula: $C_{20}H_{24}O_2$
Mol. wt.: 296
Solvent: $CDCl_3$

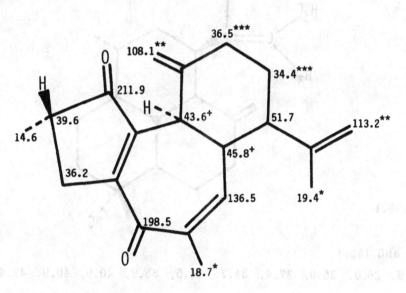

C: 140.8, 146.8, 148.6, 148.8 and 158.4

W. Naengchomnong, Y. Thebtaranonth, P. Wiriyachitra, K.T. Okamoto and J. Clardy, <u>Tetrahedron Lett.</u>, 27 (22), 2439 (1986).

SAPRORTHOQUINONE

Source: *Salvia prionitis* Hance
Mol. formula: $C_{20}H_{24}O_2$
Mol. wt.: 296
Solvent: $CDCl_3$

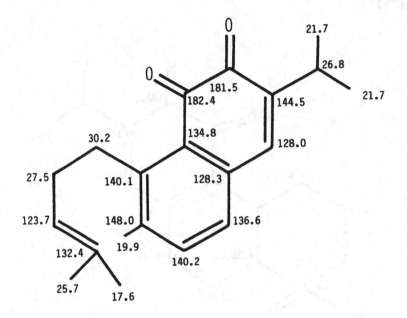

L.-Z. Lin, G. Blasko and G.A. Cordell, <u>Phytochemistry</u>, 28 (1), 177(1989).

||||||||||||||||||||||||||||||||||||||||||||||||||||||||||||||||||||||||||||||||||||||||||||||||||||||||||||||||||||||||||||

7-ISOCYANO-11(20),14-EPIAMPHILECTADIENE

Source: *Adocia* sp.
Mol. formula: $C_{21}H_{31}N$
Mol. wt.: 297
Solvent: $CDCl_3$

CH_2: 60.8 and 108.1

CH: 126.4

C: 130.3, 148.5 and 152.1

17.7, 20.0, 20.6, 26.0, 26.0, 27.4, 31.7, 34.5, 35.9, 40.9, 40.9, 42.4, 48.2, 49.7 and 49.9

R. Kazlauskas, P.T. Murphy, R.J. Wells and J.F. Blount, <u>Tetrahedron Lett.</u>, 21 (3), 315 (1980).

8,9,11,14-DIDEHYDROVOUACAPEN-5α-OL

Source: *Caesalpinia pulcherrima* Swartz

Mol. formula: $C_{20}H_{26}O_2$

Mol. wt.: 298

Solvent: $CDCl_3$

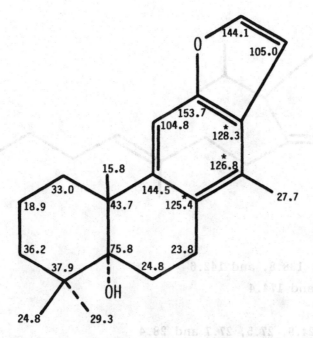

D.D. McPherson, C.-T. Che, G.A. Cordell, D.D. Soejarto, J.M. Pezzuto and H.H.S. Fong, <u>Phytochemistry</u>, 25 (1), 167 (1986).

MARGOLONE

Source: *Azadirachta indica* A. Juss.

Mol. formula: $C_{19}H_{24}O_3$

Mol. wt.: 300

Solvent: $CDCl_3$

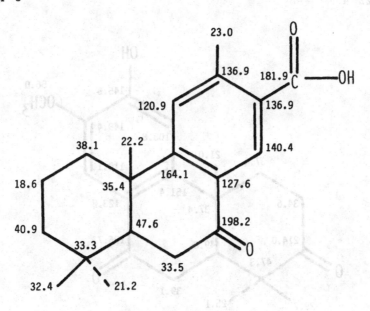

I. Ara, B.S. Siddiqui, S. Faizi and S. Siddiqui, <u>J. Chem. Soc. Perkin Trans. 1</u>, (2), 343 (1989).

TAONIANONE

Source: *Taonia australasica*

Mol. formula: $C_{20}H_{28}O_2$

Mol. wt.: 300

Solvent: $CDCl_3$

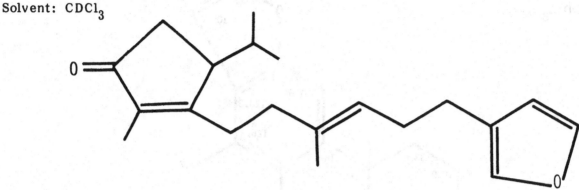

CH_2: 34.8 and 37.0

CH: 45.9, 110.9, 124.7, 138.8, and 142.6

C: 124.7, 134.5, 137.1 and 174.4

C=O: 209.3

8.0, 14.7, 16.0, 21.9, 24.9, 27.5, 27.7 and 28.4

P.T. Murphy, G. March and R.J. Wells, <u>Tetrahedron Lett.</u>, 22(16), 1555 (1981).

||||||||||||||||||||||||||||||||||||||||||||||||||||||||||||||||||||||||||||||||||||||||||||||||||||||||||||||||||||||||||||||

NIMBIONONE

Source: *Azadirachta indica* A.Juss

Mol. formula: $C_{18}H_{22}O_4$

Mol. wt.: 302

Solvent: $CDCl_3$

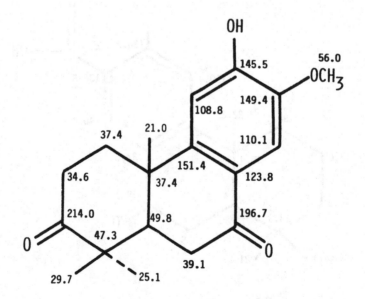

S. Siddiqui, I. Ara, S. Faizi, T. Mahmood and B.S. Siddiqui, <u>Phytochemistry</u>, 27(12), 3903 (1988).

PACHYLACTONE

Source: *Pachydictyon coriaceum*
Mol. formula: $C_{20}H_{30}O_2$
Mol. wt.: 302

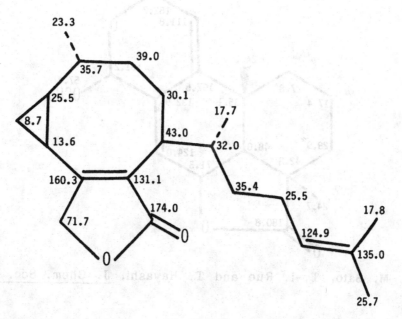

M. Ishitsuka, T. Kusumi, H. Kakisawa, Y. Kawakami, Y. Nagai and T. Sato, <u>Tetrahedron Lett.</u>, **24** (46), 5117 (1983).

WENTILACTONE A

Source: *Aspergillus wentii* Wehmer
Mol. formula: $C_{16}H_{16}O_6$
Mol. wt.: 304
Solvent: DMSO-d_6

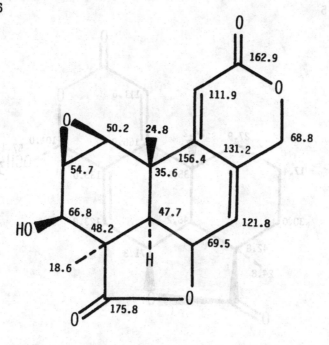

J.W. Dorner, R.J.Cole, J.P. Springer, R.H. Cox, H.Cutler and D.T. Wicklow, <u>Phytochemistry</u>, **19**(6), 1157 (1980).

Source: *Acrostalagmus* sp.

Mol. formula: $C_{17}H_{20}O_5$

Mol. wt.: 304

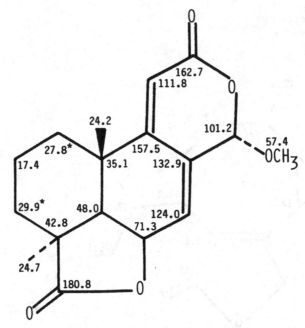

1. H. Kakisawa, M. Sato, T.-I. Ruo and T. Hayashi, J. Chem. Soc. Chem. Commun., (20), 802 (1973).

2. G.A. Ellesteo, R.H. Evans, Jr., M.P. Kunstmann, J.E. Lancaster and G.O. Morton, J. Am. Chem. Soc., 92, 5483 (1970).

PR-1387

Source: *Oidiodendron truncatum*

Mol. formula: $C_{17}H_{20}O_5$

Mol. wt.: 304

Solvent: $CDCl_3$

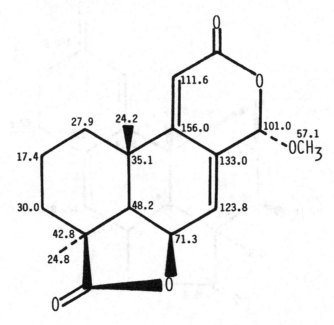

N.R. Andersen, P.R. Rasmussen, C.P. Falshaw and T.J. King, Tetrahedron Lett., 25 (4), 469 (1984).

AMBLIOL A

Source: *Dysidea amblia* (de Laubenfels)
Mol. formula: $C_{20}H_{32}O_2$
Mol. wt.: 304
Solvent: C_6D_6

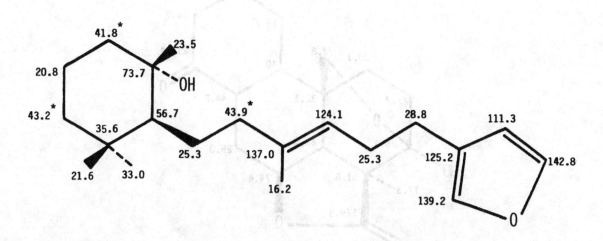

1. V.J. Paul and W. Fenical, <u>Phytochemistry</u>, 24 (10), 2239 (1985).

2. R.P. Walker and D.J. Faulkner, <u>J. Org. Chem.</u>, 46 (6), 1098 (1981).

DICTYOXEPIN

Source: *Dictyota acutiloba*
Mol. formula: $C_{20}H_{32}O_2$
Mol. wt.: 304

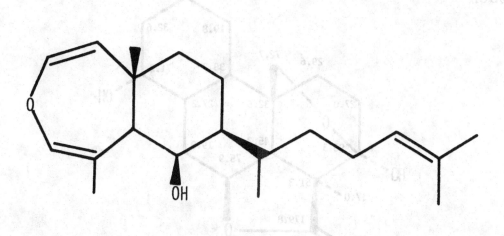

75.0, 114.1, 119.5, 125.4, 131.0, 139.4 and 140.7

H.H. Sun, S.M. Waraszkiewicz, K.L. Erickson, J. Finer and J. Clardy, <u>J. Am. Chem. Soc.</u>, 99 (10), 3516 (1977).

HUMIRIANTHENOLIDE A

Source: *Humirianthera rupestris* Ducle

Mol. formula: $C_{17}H_{22}O_5$

Mol. wt.: 306

Solvent: $CDCl_3$

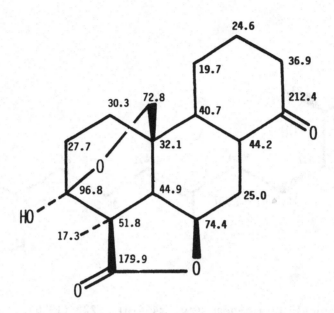

M.D.G.B. Zoghbi, N.F. Roque and H.E. Gottlieb, <u>Phytochemistry</u>, **20** (7), 1669 (1981).

HUMIRIANTHENOLIDE B

Source: *Humirianthera rupestris* Ducle

Mol. formula: $C_{17}H_{24}O_5$

Mol. wt.: 308

Solvent: $CDCl_3$

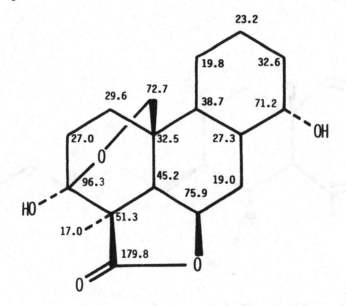

For chemical shifts in DMSO-d$_6$ see:

M.D.G.B. Zoghbi, N.F. Roque and H.E. Gottlieb, <u>Phytochemistry</u>, **20** (7), 1669 (1981).

AEGYPTINONE A

Source: *Salvia aegyptiaca* L.
Mol. formula: $C_{20}H_{22}O_3$
Mol. wt.: 310
Solvent: $CDCl_3$

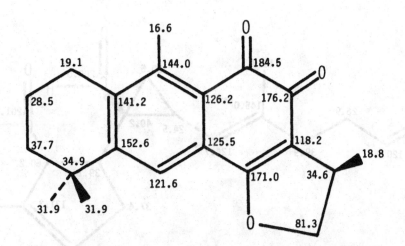

N.N. Sabri, A.A. Abou-Donia, N.M. Ghazy, A.M. Assad, A.M. El-Lakany, D.R. Sanson, H. Gracz, C.L. Barnes, E.O. Schlemper and M.S. Tempesta, J. Org. Chem., **54** (17), 4097 (1989).

HALIMEDALACTONE

Source: *Halimeda* sp.
Mol. formula: $C_{20}H_{26}O_3$
Mol. wt.: 314
Solvent: $CDCl_3$

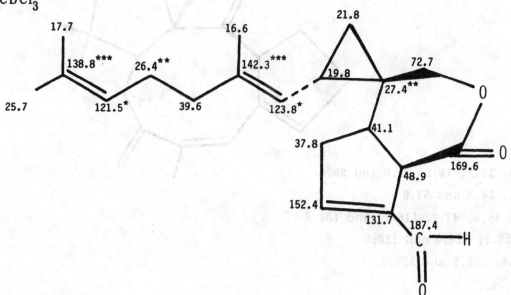

V.J. Paul and W. Fenical, Tetrahedron, **40**(16), 3053 (1984).

HALIMEDATRIAL

Source: *Halimeda* sp.

Mol. formula: $C_{20}H_{26}O_3$

Mol. wt.: 314

Solvent: $(CD_3)_2CO$

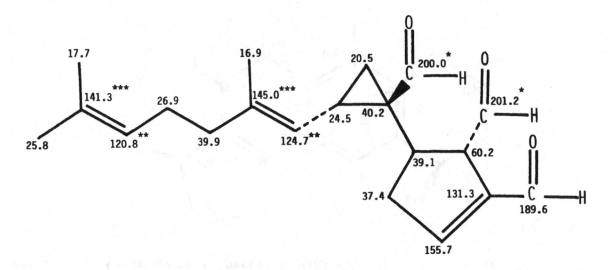

V.J. Paul and W. Fenical, <u>Tetrahedron</u>, **40**(16), 3053 (1984).

||||||||||||||||||||||||||||||||||||||||||||||||||||||||||||||||||||||||||||||||||||||||||||||||||||||||||||||

JATROPHATRIONE

Source: *Jatropha macrorhiza* Benth.

Mol. formula: $C_{20}H_{26}O_3$

Mol. wt.: 314

Solvent: $CDCl_3$

CH_3: 13.0, 19.2, 19.2, 21.6 and 25.6

CH_2: 34.1, 34.5 and 51.0

CH: 36.6, 46.8, 47.2, 116.7, and 131.8

C: 35.2, 59.1, 124.5 and 126.8

C=O: 193.8, 196.2 and 198.8

S.J. Torrance, R.M. Wiedhopf, J.R. Cole, S.K. Arora, R.B. Bates, W.A. Beavers and R.S. Cutler, <u>J. Org. Chem.</u>, **41**(10), 1855 (1976).

BIS-NORDITERPENOID HS-V

Source: *Halimeda scabra*
Mol. formula: $C_{20}H_{26}O_3$
Mol. wt.: 314
Solvent: C_6D_6

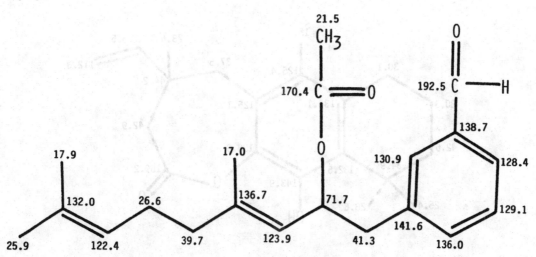

V.J. Paul and W. Fenical, *Tetrahedron*, 40(16), 3053 (1984).

||||||||||||||||||||||||||||||||||||||||||||||||||||||||||||||||||||||||||||||||||||||||||||||||||||||||||||||||||||||||||

STRICTIC ACID

Source: *Conyza stricta* Willd.
Mol. formula: $C_{20}H_{26}O_3$
Mol. wt.: 314
Solvent: $CDCl_3$

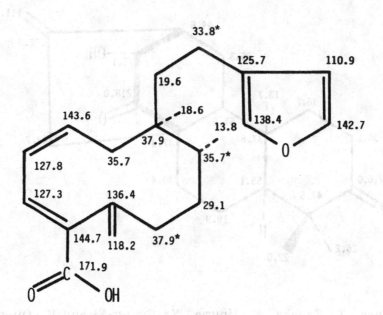

S. Tandon and R.P. Rastogi, *Phytochemistry*, 18(3), 494 (1979).

VELLOZIOLIDE

Source: *Vellozia Candida* Mikan

Mol. formula: $C_{20}H_{26}O_3$

Mol. wt.: 314

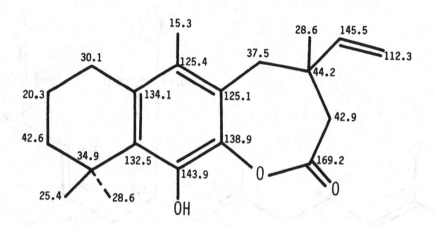

A.C. Pinto, M.L.A. Goncalves, R.B. Filho, A. Neszmelyi and G. Lukacs, <u>J. Chem. Soc. Chem. Commun.</u>, (5), 293 (1982).

ANTIQUORIN

Source: *Euphorbia antiquorum* L.

Mol. formula: $C_{20}H_{28}O_3$

Mol. wt.: 316

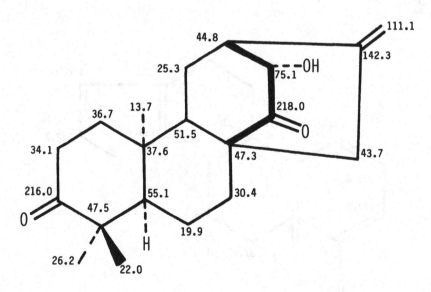

M. Zhi-Da, M. Mizuno, T. Tanaka, M. Iinuma, X. Guang-Yi and H. Qing, <u>Phytochemistry</u>, **28** (2), 553 (1989).

GALEOPSINOLONE

Source: *Galeopsis angustifolia* Hoffm.
Mol. formula: $C_{20}H_{28}O_3$
Mol. wt.: 316
Solvent: $CDCl_3$

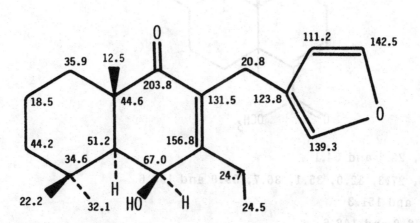

L. Perez-Sirvent, B. Rodriguez, G. Savona and O. Servettaz, <u>Phytochemistry</u>, 22(2), 527 (1983).

||||||||||||||||||||||||||||||||||||||||||||||||||||||||||||||||||||||||||||||||||||||||||||||||||

TRICYCLOSOLIDAGOLACTONE

Source: *Solidago altissima* L.
Mol. formula: $C_{20}H_{28}O_3$
Mol. wt.: 316
Solvent: $CDCl_3$

CH_3: 16.4, 20.3, 21.2 and 23.8
CH_2: 20.7, 23.2, 37.2, 37.8 and 75.3
CH: 32.1, 38.8, 49.2, 116.7 and 125.9
C: 34.4, 43.8, 76.5, 139.5 and 173.8
C=O: 174.6

S. Yamamura, M. Ito, M. Niwa, I. Hasegawa, S. Ohba and Y. Saito, <u>Tetrahedron Lett.</u>, 22 (8), 739 (1981).

ERYTHROXA-3,15-DIEN-18-OIC ACID (METHYL ESTER)

Source: *Helichrysum refluxum* N.E. Br.

Mol. formula: $C_{21}H_{32}O_2$

Mol. wt.: 316

Solvent: $CDCl_3$

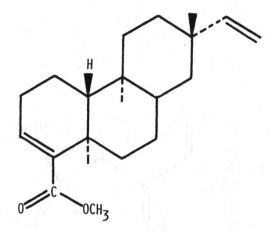

CH_3: 12.4, 21.5, 23.2 and 54.1

CH_2: 16.9, 25.8, 27.3, 32.0, 35.1, 36.7, 39.0 and 108.6

CH: 42.4, 136.7 and 151.3

C: 36.4, 36.8, 38.0 and 142.5

C=O: 167.9

F. Bohlmann, L. Hartono and J. Jakupovic, <u>Phytochemistry</u>, **24** (3), 611 (1985).

||||||||||||||||||||||||||||||||||||||||||||||||||||||||||||||||||||||||||||||||||||||||||||||||||||||||||||||||||||

DITERPENE H-I

Source: *Higginsia* sp.

Mol. formula: $C_{20}H_{30}O_3$

Mol. wt.: 318

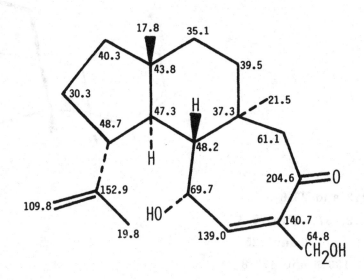

M.P. Cassidy, E.L. Ghisalberti, P.R. Jefferies, B.W. Skelton and A.H. White, <u>Aust. J. Chem.</u>, **38** (8), 1187 (1985).

GALANAL A

Source: *Alpinia galanga* (L.) Willd.
Mol. formula: $C_{20}H_{30}O_3$
Mol. wt.: 318
Solvent: $CDCl_3$

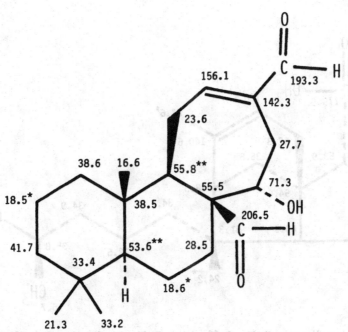

H. Morita and H. Itokawa, <u>Planta Med.</u>, **54**(2), 117 (1988).

GALANAL B

Source: *Alpinia galanga* (L.) Willd.
Mol. formula: $C_{20}H_{30}O_3$
Mol. wt.: 318
Solvent: $CDCl_3$

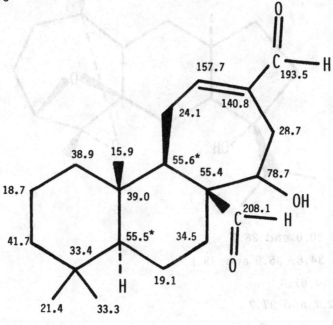

H. Morita and H. Itokawa, <u>Planta Med.</u>, **54**(2), 117 (1988).

Source: *Eremophila georgii* Diels

Mol. formula: $C_{20}H_{32}O_3$

Mol. wt.: 320

Solvent: $CDCl_3$

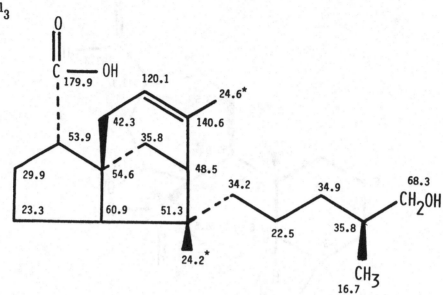

P.G. Forster, E.L. Ghisalberti, P.R. Jefferies, B.W. Skelton and A.H. White, Tetrahedron, 42(1), 215 (1986).

DICTYOXETANE

Source: *Dictyota dichotoma*

Mol. formula: $C_{20}H_{32}O_3$

Mol. wt.: 320

Solvent: $CDCl_3$

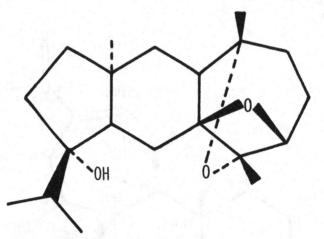

CH_3: 16.1, 17.5, 18.5, 20.0 and 26.7

CH_2: 23.4, 24.6, 27.6, 34.6, 35.9 and 39.1

CH: 36.6, 48.0, 52.6 and 81.3

C: 42.5, 80.1, 80.6, 82.7 and 97.2

C.B. Rao, K.C. Pullaiah, R.K. Surapaneni, B.W. Sullivan, K.F. Albizati, D.J. Faulkner, H. Cun-Heng and J. Clardy, J. Org. Chem., 51 (14), 2736 (1986).

DITERPENE EC-III

Source: *Eremophila cuneifolia* Kraenzlin
Mol. formula: $C_{20}H_{32}O_3$
Mol. wt.: 320

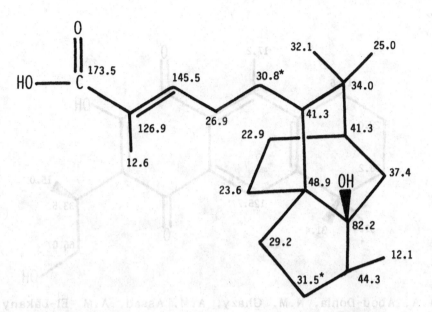

K.D. Croft, E.L. Ghisalberti, P.R. Jefferies, T.A. Mori, B.W. Skelton and A.H. White, Aust. J. Chem., 37(4), 785 (1984).

||||||||||||||||||||||||||||||||||||||||||||||||||||||||||||||||||||||||||||||||||||||||||||||||||||

PYGMAEOCINE E

Source: *Pygmaeopremna herbacea* (Roxb.) Moldenke
Mol. formula: $C_{20}H_{20}O_4$
Mol. wt.: 324
Solvent: C_5D_5N

Q. Meng, N. Zhu and W. Chen, Phytochemistry, 27(4), 1151 (1988).

Source: *Salvia aegyptiaca* L.

Mol. formula: $C_{20}H_{24}O_4$

Mol. wt.: 328

Solvent: $CDCl_3$

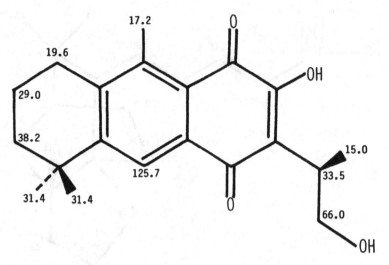

N.N. Sabri, A.A. Abou-Donia, N.M. Ghazy, A.M. Assad, A.M. El-Lakany, D.R. Sanson, H. Gracz, C.L. Barnes, E.O. Schlemper and M.S. Tempesta, J. Org. Chem., 54 (17), 4097 (1989).

|||||||||||||||||||||||||||||||||||||||||||||||||||||||||||||||||||||||||||||||||||||||||||||||||||||||||||||||

SIRUTEKKONE

Source: *Premna herbacea* Roxb. (syn. *Pygmacopremna herbacea* Moldenke)

Mol. formula: $C_{20}H_{24}O_4$

Mol. wt.: 328

Solvent: $CDCl_3$

G. Sandhya, K. Rajagopalan and N. Chandrakumar, Phytochemistry, 27 (7), 2249 (1988).

SEMPERVIROL (ACETATE)

Source: *Podocarpus neriifolius* D.Don ex Lamb.

Mol. formula: $C_{22}H_{32}O_2$

Mol. wt.: 328

Solvent: $CDCl_3$

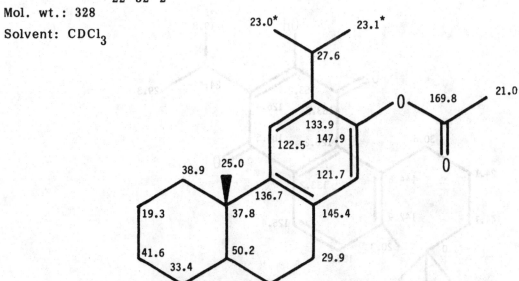

R.C. Cambie, R.E. Cox, K.D. Croft and D. Sidwell, <u>Phytochemistry</u>, 22(5), 1163 (1983).

||||||||||||||||||||||||||||||||||||||||||||||||||||||||||||||||||||||||||||||||||||||||||||||||||||||||||||||||

CITLALITRIONE

Source: *Jatropha dioica* (Hook) Mc Vaugh

Mol. formula: $C_{20}H_{26}O_4$

Mol. wt.: 330

Solvent: C_6D_6

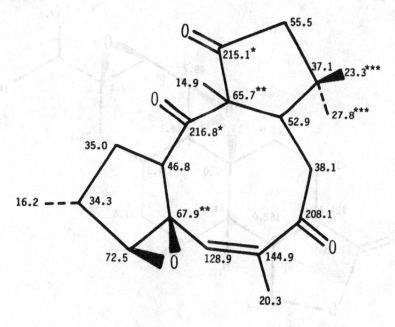

A.M. Villarreal, X.A. Dominguez, H.J. Williams, A.I. Scott and J. Reibenspies, <u>J. Nat. Prod.</u>, (Lloydia), **51** (4), 749 (1988).

4-HYDROXYSAPRIPARQUINONE

Source: *Salvia prionitis* Hance

Mol. formula: $C_{20}H_{26}O_4$

Mol. wt.: 330

Solvent: $CDCl_3$

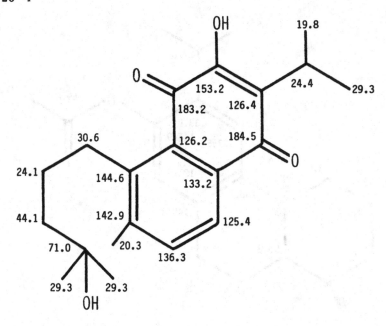

L.-Z. Lin, G. Blasko and G.A. Cordell, <u>Phytochemistry</u>, 28 (1), 177 (1989).

STEMOLIDE

Source: *Stemodia maritima*

Mol. formula: $C_{20}H_{26}O_4$

Mol. wt.: 330

Solvent: $CDCl_3$

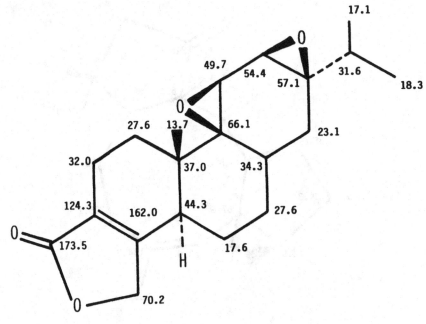

P.S. Manchand and J.F. Blount, <u>Tetrahedron Lett.</u>, (29), 2489 (1976).

GRACILIN E

Source: *Spongionella gracilis*

Mol. formula: $C_{21}H_{32}O_3$

Mol. wt.: 332

Solvent: $CDCl_3$

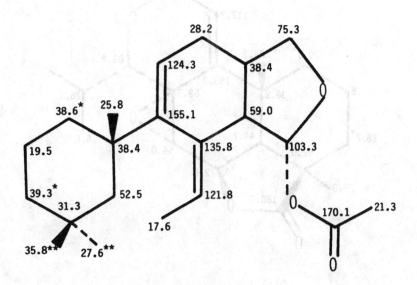

L. Mayol, V. Piccialli and D. Sica, <u>Tetrahedron</u>, **42** (19), 5369 (1986).

||||||||||||||||||||||||||||||||||||||||||||||||||||||||||||||||||||||||||||||||||||||||||||||||||||||||||||||||||||||||||||||||

SALVICANARIC ACID

Source: *Salvia canariensis* L.

Mol. formula: $C_{19}H_{26}O_5$

Mol. wt.: 334

Solvent: $CDCl_3$

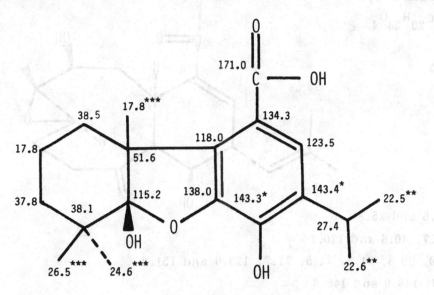

A.G. Gonzalez, J.R. Herrera, J.G. Luis, A.G. Ravelo and A. Perales, <u>J. Nat. Prod.</u>, (Lloydia), 50(3), 341 (1987).

NAGILACTONE G

Source: *Decussocarpus rospigiliosii* (Pilger) De Laubenfels

Mol. formula: $C_{19}H_{24}O_5$

Mol. wt.: 332

Solvent: $CDCl_3$

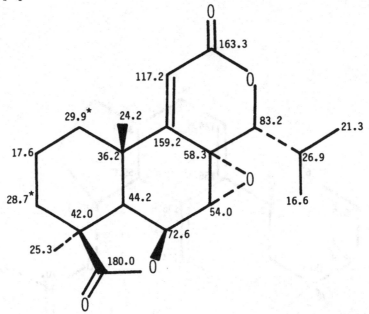

J.M. Amaro-Luis and D. Carroz U., <u>J. Nat. Prod.</u>, (Lloydia), **51** (6), 1249 (1988).

||||||||||||||||||||||||||||||||||||||||||||||||||||||||||||||||||||||||||||||||||||||

(1R*,5S*,6S*,7R*,10E,12R*,13S*)5,6-EPOXY-7,13-DIHYDROXY-2-METHYLENE-12-(2-METHYLPROP-1-ENYL)-6-METHYLBICYCLO[7,4,0]TRIDEC-10-ENE-10-CARBALDEHYDE

Source: *Efflatounaria* sp.

Mol. formula: $C_{20}H_{28}O_4$

Mol. wt.: 332

Solvent: $CDCl_3$

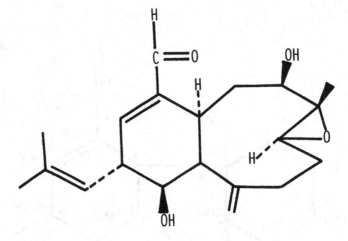

CH_3: 13.2, 18.6 and 25.8

CH_2: 25.8, 26.7, 40.6 and 120.7

CH: 32.7, 44.0, 59.5, 61.5, 71.5, 77.7, 123.0 and 151.2

C: 62.2, 137.6, 144.9 and 145.6

C=O: 193.6

K.P. Burns, G. Englert, R. Kazlauskas, P.T. Murphy, P. Schonholzer and R.J. Wells, <u>Aust. J. Chem.</u>, **36** (1), 171 (1983).

APHIDICOLIN

Source: *Cephalosporium aphidicola* and *Nigrospora sphaerica*

Mol. formula: $C_{20}H_{34}O_4$

Mol. wt.: 338

Solvent: C_5D_5N

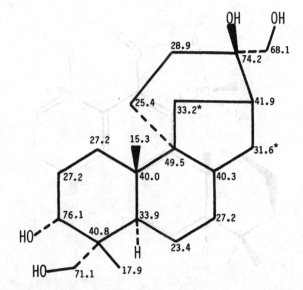

J. Ipsen and J.P. Rosazza, J. Nat. Prod., (Lloydia), **47** (3), 497 (1984).

For chemical shifts in (CD_3CO_2D) see:

M.R. Adams and J.D. Bu'Lock, J. Chem. Soc. Chem. Commun., (10), 389 (1975).

KEMPENE-2

Source: *Nasutitermes kempae*

Mol. formula: $C_{22}H_{30}O_3$

Mol. wt.: 342

Solvent: $CDCl_3$

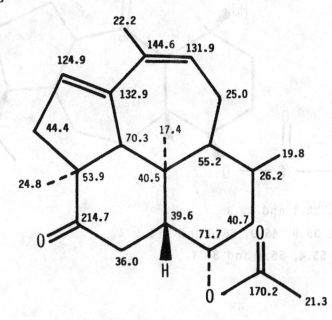

G.D. Prestwich, B.A. Solheim, J. Clardy, F.G. Pilkiewicz, I. Miura, S.P. Tanis and K. Nakanishi, J. Am. Chem. Soc., **99** (24), 8082 (1977).

BETOLIDE

Source: *Betonica officinalis* L. (syn. *Stachys officinalis*)

Mol. formula: $C_{20}H_{24}O_5$

Mol. wt.: 344

Solvent: $CDCl_3$

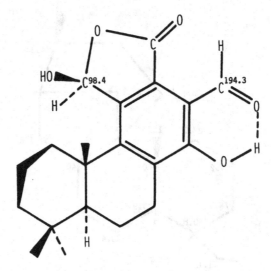

Benzene ring C signals: 112.2, 124.5, 133.4, 136.6, 155.3 and 161.5

Other signals at: 16.9, 18.3, 20.9, 24.4, 32.6, 33.1, 35.6, 40.4 and 49.1

V.V. Tkachev, G.K. Nikonov, L.O. Atovmyan, A. Ya. Kobzar and T.V. Zinchenko, <u>Chem. Nat. Compds.</u>, **23** (6), 673 (1987); <u>Khim. Prir. Soedin.</u>, **23** (6), 811 (1987).

||||||||||||||||||||||||||||||||||||||||||||||||||||||||||||||||||||||||||||||||||||||||||||||||||||||||||||

CHROMOPHYCADIOL MONOACETATE

Source: *Dictyota* sp.

Mol. formula: $C_{22}H_{36}O_3$

Mol. wt.: 348

Solvent: $CDCl_3$

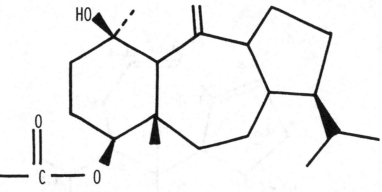

CH_3: 15.4, 15.5, 21.5, 22.1 and 22.3

CH_2: 23.6, 26.8, 28.8, 39.9, 45.9, 66.0 and 113.3

CH: 30.1, 46.6, 52.3, 53.4, 55.4 and 83.4

C: 40.8, 72.6 and 149.1

C=O: 171.1

J. Clardy, G.V. Duyne, A.G. Gonzalez, E. Manta, J.D. Martin, C. Perez, J.L. Ravelo and G.K. Schulte, <u>J. Chem. Soc. Chem. Commun.</u>, (10), 767 (1987).

α-DICTALEDIOL MONOACETATE

Source: *Dictyota* sp.

Mol. formula: $C_{22}H_{36}O_3$

Mol. wt.: 348

Solvent: $CDCl_3$

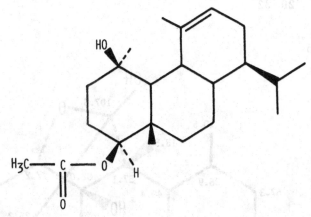

CH$_3$: 20.8, 21.3, 21.4, 22.3 and 27.7

CH$_2$: 17.0, 24.8, 26.3, 32.3, and 37.5

CH: 23.0, 32.3, 39.7, 45.5, 47.4, 78.1 and 122.4

C: 40.0, 73.6 and 137.4

C=O: 170.6

A.G. Gonzalez, J.D. Martin, B. Gonzalez, J.L. Ravelo, C. Perez, S. Rafii and J. Clardy, J. Chem. Soc. Chem. Commun., (10), 669 (1984).

||||||||||||||||||||||||||||||||||||||||||||||||||||||||||||||||||||||||||||||||||||||||||||||||||||||||

CHRYSOTHAME

Source: *Chrysothamnus paniculatus* (Gray) Hall

Mol. formula: $C_{20}H_{32}O_5$

Mol. wt.: 352

Solvent: $CDCl_3$

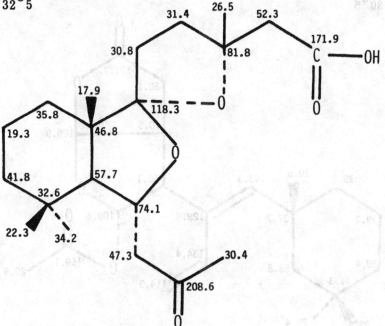

J.J. Hoffmann, S.P. McLaughlin, S.D. Jolad, K.H. Schram, M.S. Tempesta and R.B. Bates, J. Org. Chem., 47 (9), 1725 (1982).

CINNCASSIOL D$_1$

Source: *Cinnamomum cassia* Blume
Mol. formula: $C_{20}H_{32}O_5$
Mol. wt.: 352
Solvent: C_5D_5N

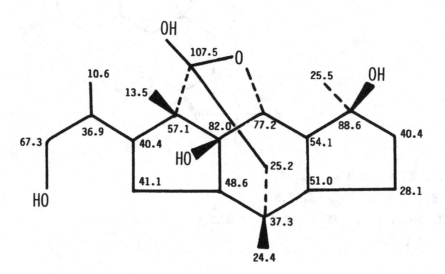

T. Nohara, Y. Kashiwada, K. Murakami, T. Tomimatsu, M. Kido, A. Yagi and I. Nishioka, Chem. Pharm. Bull. (Tokyo), 29(9), 2451 (1981).

SPONGIONELLIN

Source: *Spongionella gracilis*
Mol. formula: $C_{21}H_{30}O_5$
Mol. wt.: 362
Solvent: $CDCl_3$

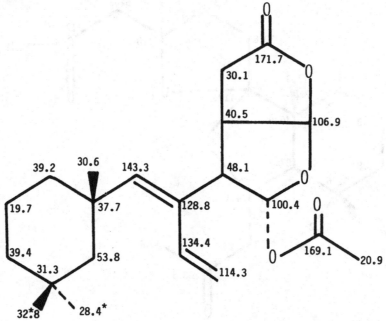

L. Mayol, V. Piccialli and D. Sica, Tetrahedron, 42 (19), 5369 (1986).

CINNCASSIOL C$_2$

Source: *Cinnamomum cassia* Blume

Mol. formula: C$_{20}$H$_{28}$O$_6$

Mol. wt.: 364

Solvent: CD$_3$OD

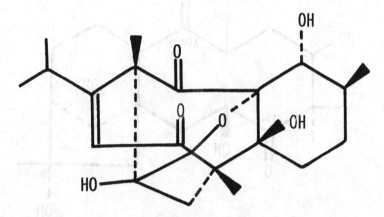

13.3, 18.8, 21.6, 22.4, 22.9, 28.4, 29.2, 31.5, 43.5, 54.2 and 60.6

-C-O-: 72.9, 79.1 and 91.1

O-C-O-: 104.8

C=C: 125.8 and 163.5

Y. Kashiwada, T. Nohara, T. Tomimatsu and I. Nishioka, <u>Chem. Pharm. Bull.</u> (Tokyo), 29(9), 2686 (1981).

||||||||||||||||||||||||||||||||||||||||||||||||||||||||||||||||||||||||||||||||||||||||||||||||||||||||||||

CINNCASSIOL D$_2$

Source: *Cinnamomum cassia* Blume

Mol. formula: C$_{20}$H$_{32}$O$_6$

Mol. wt.: 368

Solvent: C$_5$D$_5$N

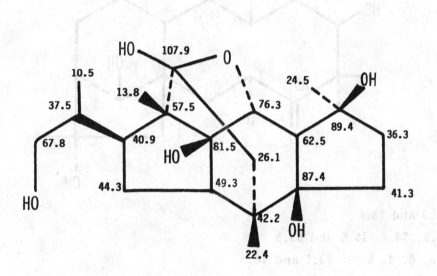

T. Nohara, Y. Kashiwada, K. Murakami, T. Tomimatsu, M. Kido, A. Yagi and I. Nishioka, <u>Chem. Pharm. Bull.</u> (Tokyo), 29(9), 2451 (1981).

GRAYANOL A

Source: *Leucothoe grayana*

Mol. formula: $C_{20}H_{32}O_6$

Mol. wt.: 368

Solvent: C_5D_5N

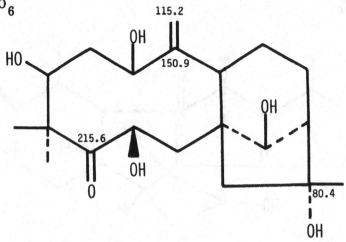

CH$_3$: 16.4, 24.0 and 25.9

CH$_2$: 26.6, 27.2, 35.4, 45.3 and 57.8

CH: 51.7, 53.9, 67.9, 70.7, 79.2 and 85.7

C: 53.3 and 54.3

T. Kaiya, N. Shirai and J. Sakakibara, J. Chem. Soc. Chem. Commun., (1), 22 (1981).

GRAYANOL B

Source: *Leucothoe grayana*

Mol. formula: $C_{20}H_{32}O_6$

Mol. wt.: 368

Solvent: C_5D_5N

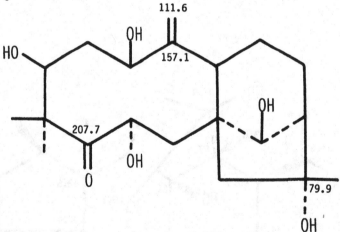

CH$_3$: 16.2, 23.5 and 25.6

CH$_2$: 25.6, 27.9, 38.4, 45.5 and 53.5

CH: 50.7, 53.5, 67.4, 69.8, 73.1 and 78.9

C: 53.9 and 58.0

T. Kaiya, N. Shirai and J. Sakakibara, J. Chem. Soc. Chem. Commun., (1), 22 (1981).

BROMOCORODIENOL

Source: *Sphaerococcus coronopifolius*

Mol. formula: $C_{20}H_{33}BrO$

Mol. wt.: 369

Solvent: $CDCl_3$

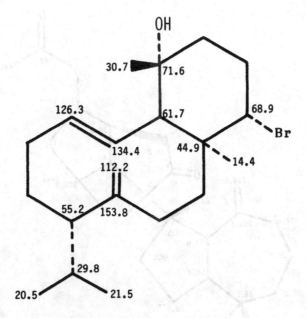

CH$_2$: 25.6, 27.8, 30.8, 31.6, 39.5 and 41.1

F. Cafieri, E. Fattorusso and C. Santacroce, <u>Tetrahedron Lett.</u>, **25** (29), 3141 (1984).

||||||||||||||||||||||||||||||||||||||||||||||||||||||||||||||||||||||||||||||||||||||||||||||||||||||||||||||||||||||

PSEUDOPTEROLIDE

Source: *Pseudopterogorgia acerosa* (Pallas)

Mol. formula: $C_{21}H_{22}O_6$

Mol. wt.: 370

Solvent: $CDCl_3$

CH$_3$: 20.8, 21.4 and 51.4

CH$_2$: 28.1, 112.5 and 114.8

CH: 42.6, 48.9, 52.6, 59.9, 79.7, 111.5 and 151.3

C: 114.8, 129.0, 140.5, 145.3, 150.4, 160.3 and 163.8

C=O: 171.9

M.M. Bandurraga, W. Fenical, S.F. Donovan and J. Clardy, <u>J. Am. Chem. Soc.</u>, **104** (23), 6463 (1982).

APLYVIOLENE

Source: *Aplysilla polyrhaphis*

Mol. formula: $C_{22}H_{32}O_5$

Mol. wt.: 376

Solvent: $CDCl_3$

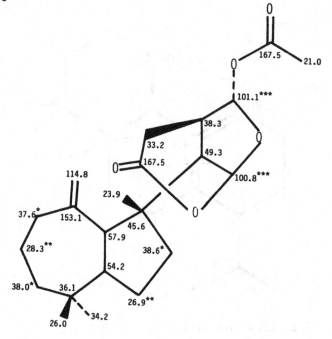

S.C. Bobzin and D.J. Faulkner, J. Org. Chem., **54** (16), 3902 (1989).

DENDRILLOLIDE A

Source: *Dendrilla* sp.

Mol. formula: $C_{22}H_{32}O_5$

Mol. wt.: 376

Solvent: C_6D_6

CH_3: 20.7, 24.1, 25.8 and 34.5

CH_2: 27.1, 28.7, 28.8, 37.7, 37.7, 38.8 and 113.3

CH: 41.9, 54.5, 54.9, 55.7, 97.1 and 105.0

C: 36.0, 46.7 and 153.7

C=O: 169.1 and 175.2

B. Sullivan and D.J. Faulkner, J. Org. Chem., **49** (17), 3204 (1984).

HYPOESTOXIDE

Source: *Hypoestes rosea* R. Br
Mol. formula: $C_{22}H_{32}O_5$
Mol. wt.: 376

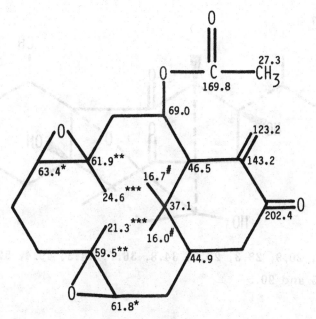

CH_2: 23.9, 31.4, 36.1, 42.2 and 42.8

A.A. Adesomoju and J.I. Okogun, <u>J. Nat. Prod.</u>, (Lloydia), **47** (2), 308 (1984).

||||||||||||||||||||||||||||||||||||||||||||||||||||||||||||||||||||||||||||||||||||||||||||||||||||||||||||||||||||||||||||||

DITERPENE AM-XX

Source: *Archidoris montereyensis*
Mol. formula: $C_{23}H_{38}O_4$
Mol. wt.: 378
Solvent: $CDCl_3$

CH_3: 15.6, 15.7, 18.5, 18.7 and 33.4
CH_2: 21.1, 21.7, 22.7, 40.0, 41.9, 63.6 and 65.2
CH: 54.4, 56.6, 62.8, 70.5 and 124.3
C: 33.2, 36.7, 37.5 and 128.6
C=O: 173.3

K. Gustafson and R.J. Andersen, <u>Tetrahedron</u>, 41(6), 1101 (1985).

CINNCASSIOL C₃

Source: *Cinnamomum cassia* Blume

Mol. formula: $C_{20}H_{30}O_7$

Mol. wt.: 382

Solvent: CD_3OD

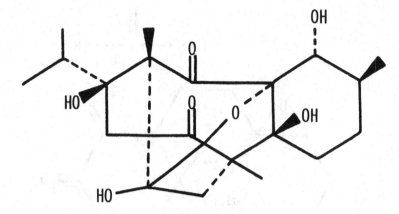

11.9, 17.8, 17.8, 18.7, 20.9, 28.3, 28.6, 34.8, 36.7, 44.3, 49.4, 52.4 and 60.5

-C-O: 71.6, 77.4, 80.5 and 90.5

O-C-O: 103.7

C=O: 222.1 and 222.9

Y. Kashiwada, T. Nohara, T. Tomimatsu and I. Nishioka, <u>Chem. Pharm. Bull.</u> (Tokyo), **29(9)**, 2686 (1981).

||||||||||||||||||||||||||||||||||||||||||||||||||||||||||||||||||||||||||||||||||||||||||||||||||||||||||||||||||||||||||||||||||||||||||||||||||

BROMOTETRASPHAEROL

Source: *Sphaerococcus coronopifolius*

Mol. formula: $C_{20}H_{33}BrO_2$

Mol. wt.: 385

Solvent: $CDCl_3$

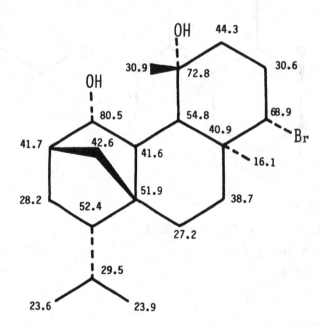

F. Cafieri, E. Fattorusso, L. Mayol and C. Santacroce, <u>Tetrahedron</u>, **42(15)**, 4273 (1986).

Source: *Coleus fredericii* G. Tayl.

Mol. formula: $C_{22}H_{28}O_6$

Mol. wt.: 388

Solvent: $CDCl_3$

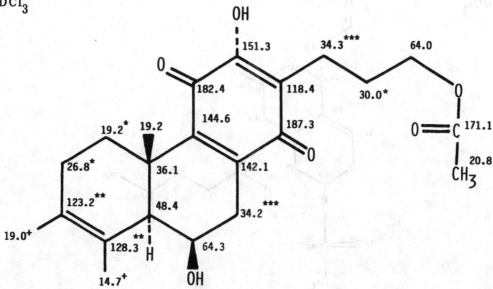

Z.-Y. Zhu, H. Nayeshiro, R. Prewo, P. Ruedi and C.H. Eugster, <u>Helv. Chim. Acta</u>, 71(3), 577 (1988).

||||||||||||||||||||||||||||||||||||||||||||||||||||||||||||||||||||||||||||||||||||||||||||||||||||

(1R*, 4S*, 4aR*, 6S*)-6-ACETOXY-1-(3-METHYLBUT-2-ENYL)-4-(1-METHYLENE-5-OXOHEXYL)-4,4a,5,6-TETRAHYDRO-1H, 3H-PYRANO[3,4-c] PYRAN-3-ONE

Source: *Efflatounaria* sp.

Mol. formula: $C_{22}H_{30}O_6$

Mol. wt.: 390

Solvent: $CDCl_3$

CH_3: 17.9, 20.8, 25.6 and 29.7

CH_2: 21.1, 30.9, 32.1, 32.6, 42.4 and 114.6

CH: 26.7, 54.0, 77.6, 88.5, 117.8 and 136.7

C: 110.4, 135.1 and 143.2

C=O: 169.1, 170.3 and 208.1

B.F. Bowden, J.C.Coll, V.A. Patrick, D.M. Tapiolas and A.H. White, <u>Aust. J. Chem.</u>, 36(11), 2279 (1983).

ALCYONOLIDE

Source: *Alcyonium* sp.

Mol. formula: $C_{22}H_{30}O_6$

Mol. wt.: 390

Solvent: $CDCl_3$

M. Kobayashi, T. Yasuzawa, Y. Kobayashi, Y. Kyogoku and I. Kitagawa, <u>Tetrahedron Lett.</u>, **22** (44), 4445 (1981).

1-ACETOXY-12-HYDROXY-2,3,6,7-BISEPOXYSMALLANTHA-<u>10E</u>,14(21)-DIENE

Source: *Heliopsis helianthoides* (L.) Sweet

Mol. formula: $C_{23}H_{38}O_5$

Mol. wt.: 394

Solvent: $CDCl_3$

J. Jakupovic, A. Schuster, T.V. Chau-Thi, F. Bohlmann and X.A. Dominguez, <u>Phytochemistry</u>, **27** (7), 2235 (1988).

LATRUNCULIN B

Source: *Latrunculia magnifica* (Keller)

Mol. formula: $C_{20}H_{29}NO_5S$

Mol. wt.: 395

Solvent: $CDCl_3$

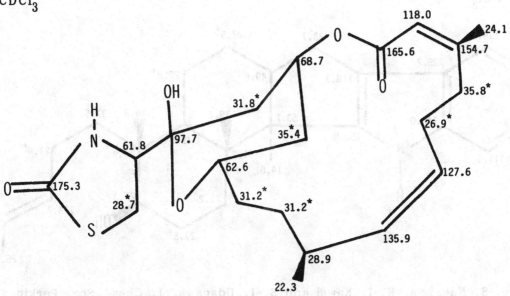

A. Groweiss, U. Shmueli and Y. Kashman, <u>J. Org. Chem.</u>, 48 (20), 3512 (1983).

EMINDOLE DA

Source: *Emericella desertorum* Samson and Mouchacca

Mol. formula: $C_{28}H_{39}NO$

Mol. wt.: 405

Solvent: $CDCl_3$

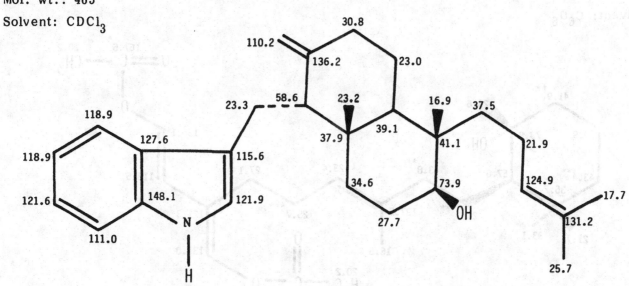

K. Nozawa, S. Nakajima, K.-I. Kawai and S.-I. Udagawa, <u>J. Chem. Soc. Perkin Trans. 1</u>, (9), 2607 (1988).

EMINDOLE SB

Source: *Emericella striata* Malloch and Cain

Mol. formula: $C_{28}H_{39}NO$

Mol. wt.: 405

Solvent: $CDCl_3$

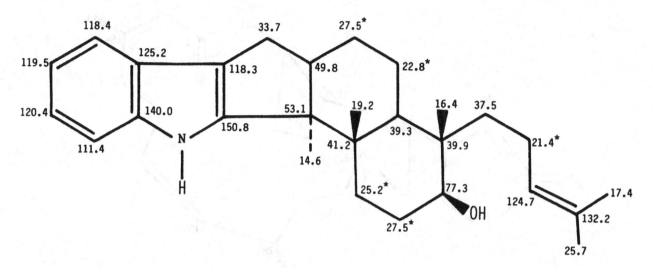

K. Nozawa, S. Nakajima, K.-I. Kawai and S.-I. Udagawa, <u>J. Chem. Soc. Perkin Trans. 1</u>, (9), 2607 (1988).

||||||||||||||||||||||||||||||||||||||||||||||||||||||||||||||||||||||||||||||||||||||||||||||||||||||||||||||||

DITERPENOID CB-IV

Source: *Caulerpa brownii* (C. Agardh) Endl.

Mol. formula: $C_{24}H_{38}O_5$

Mol. wt.: 406

Solvent: C_6D_6

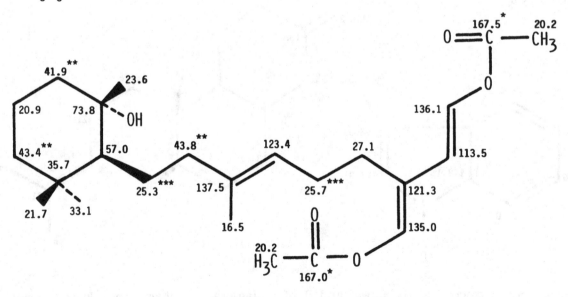

V.J. Paul and W. Fenical, <u>Phytochemistry</u>, 24(10), 2239 (1985).

BRIANTHEIN W

Source: *Briareum polyanthes*

Mol. formula: $C_{24}H_{32}O_6$

Mol. wt.: 416

Solvent: C_6D_6

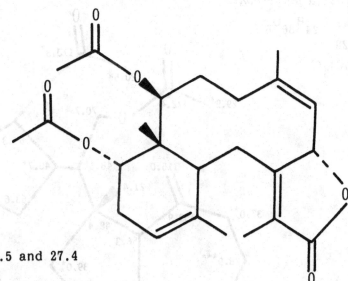

CH_3: 9.5, 14.5, 20.7, 20.8, 21.5 and 27.4

CH_2: 26.4, 29.2, 33.7 and 53.4

CH: 37.6, 72.4, 74.5, 80.4, 116.8 and 123.6

C: 41.5, 124.7, 136.9, 143.2 and 159.5

C=O: 170.2, 170.4 and 173.5

J.H. Cardellina, T.R. James, Jr., M.H.M. Chen and J. Clardy, <u>J. Org. Chem.</u>, 49 (18), 3398 (1984).

||||||||||||||||||||||||||||||||||||||||||||||||||||||||||||||||||||||||||||||||||||||||||||||||||||||||||||

DEHYDROXYPAXILLINE

Source: *Emericella striata* Malloch and Cain

Mol. formula: $C_{27}H_{33}NO_3$

Mol. wt.: 419

Solvent: $CDCl_3$

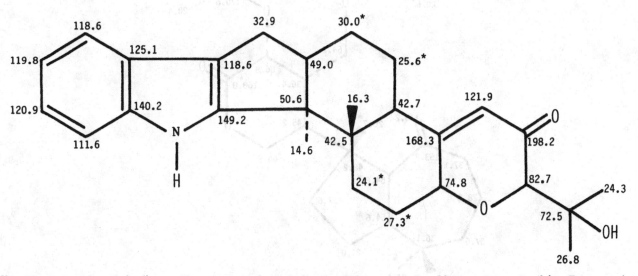

K. Nozawa, S. Nakajima, K.-I. Kawai and S.-I. Udagawa, <u>J. Chem. Soc. Perkin Trans. 1</u>, (9), 2607 (1988).

665

POLYRHAPHIN A

Source: *Aplysilla polyrhaphis*
Mol. formula: $C_{24}H_{36}O_6$
Mol. wt.: 420
Solvent: $CDCl_3$

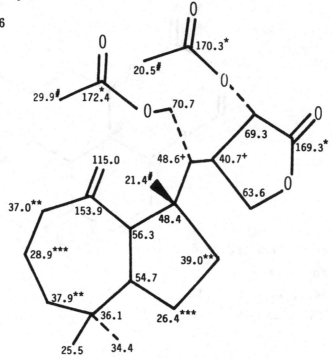

S.C. Bobzin and D.J. Faulkner, <u>J. Org. Chem.</u>, **54** (16), 3902 (1989).

SHAHAMIN C

Source: *Aplysilla polyrhaphis*
Mol. formula: $C_{24}H_{36}O_6$
Mol. wt.: 420
Solvent: $CDCl_3$

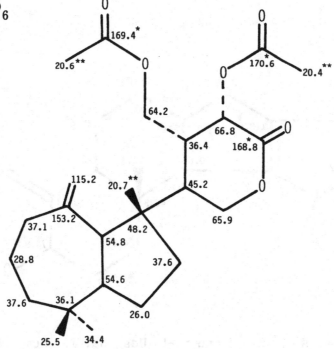

S.C. Bobzin and D.J. Faulkner, <u>J. Org. Chem.</u>, **54** (16), 3902 (1989).

666

VERRUCOSIN A

Source: *Doris verrucosa* Cuvier
Mol. formula: $C_{25}H_{40}O_5$
Mol. wt.: 420
Solvent: $CDCl_3$

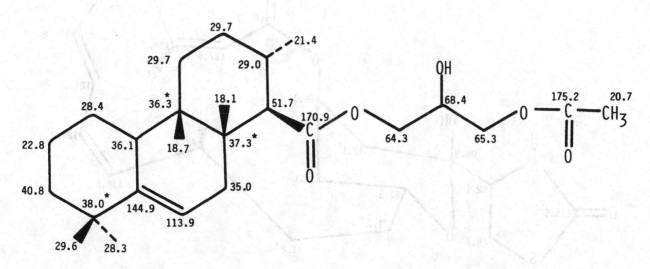

G. Cimino, M. Gavagnin, G. Sodano, R. Puliti, C.A. Mattia and L. Mazzarella, Tetrahedron, 44(8), 2301 (1988).

VERRUCOSIN B

Source: *Doris verrucosa* Cuvier
Mol. formula: $C_{25}H_{40}O_5$
Mol. wt.: 420
Solvent: $CDCl_3$

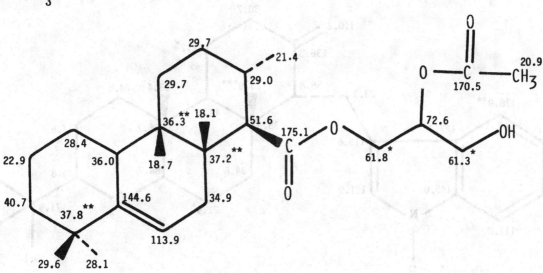

G. Cimino, M. Gavagnin, G. Sodano, R. Puliti, C.A. Mattia and L. Mazzarella, Tetrahedron, 44(8), 2301 (1988).

LATRUNCULIN A

Source: *Latrunculia magnifica* (Keller)

Mol. formula: $C_{22}H_{31}NO_5S$

Mol. wt.: 421

Solvent: $CDCl_3$

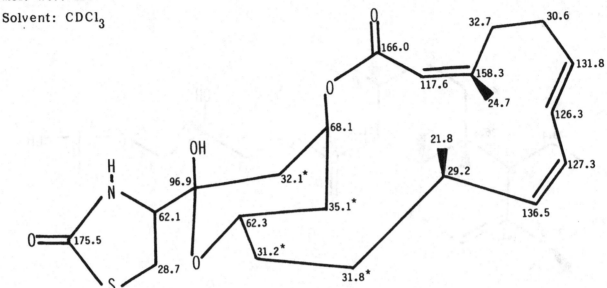

A. Groweiss, U. Shmueli and Y. Kashman, <u>J. Org. Chem.</u>, **48** (20), 3512 (1983).

EMINDOLE DB

Source: *Emericella desertorum* Samson and Mouchacca

Mol. formula: $C_{28}H_{39}NO_2$

Mol. wt.: 421

Solvent: $CDCl_3$

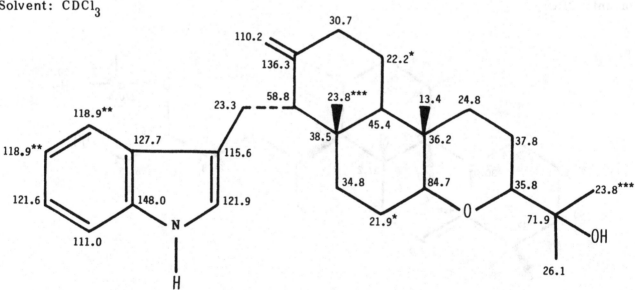

K. Nozawa, S. Nakajima, K.-I. Kawai and S.-I. Udagawa, <u>J. Chem. Soc. Perkin Trans. 1</u>, (9), 2607 (1988).

PASPALINE

Source: *Emericella striata* Malloch and Cain

Mol. formula: $C_{28}H_{39}NO_2$

Mol. wt.: 421

Solvent: $CDCl_3$

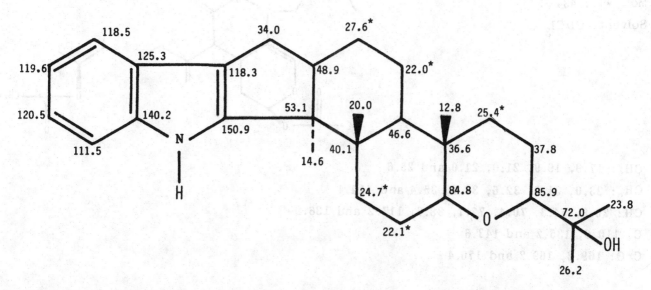

K. Nozawa, S. Nakajima, K.-I. Kawai and S.-I. Udagawa, J. Chem. Soc. Perkin Trans. 1, (9), 2607 (1988).

|||||||||||||||||||||||||||||||||||||||||||||||||||||||||||||||||||||||||||||||||||||||||||||||||||||||||||||||||

[2-(3-METHYLBUT-2-ENYL)-5-(5-CINNAMOLOXY-2-OXO-1,5-DIMETHYLHEX-3-ENYL)-3-METHYL-2,5-DIHYDROFURAN]

Source: *Artemisia pallens*

Mol. formula: $C_{27}H_{34}O_4$

Mol. wt.: 422

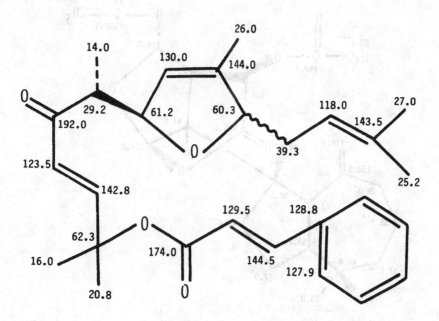

A. Chandra, L.N. Misra and R.S. Thakur, Tetrahedron Lett., 28 (50), 6377 (1987).

(1R*, 4S*, 4aR*, 6S*)-6-ACETOXY-4-[(5R*)-5-ACETOXY-1-METHYLENEHEXYL]-1-(3-METHYLBUT-2-ENYL)-4,4a,5,6-TETRAHYDRO-1H,3H-PYRANO [3,4-c]-PYRAN-3-ONE

Source: *Efflatounaria* sp.

Mol. formula: $C_{24}H_{34}O_7$

Mol. wt.: 434

Solvent: $CDCl_3$

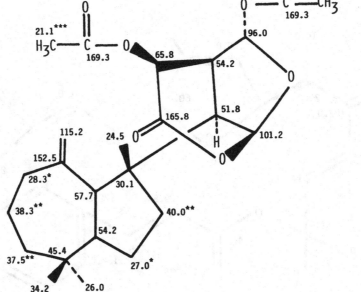

CH₃: 17.9, 19.9, 21.0, 21.0 and 25.6

CH₂: 23.0, 31.0, 32.6, 32.6, 35.4 and 114.4

CH: 26.8, 54.3, 70.4, 77.7, 88.6, 117.2 and 136.9

C: 110.6, 135.2 and 143.6

C=O: 169.2, 169.2 and 170.4

B.F. Bowden, J.C. Coll, V.A. Patrick, D.M. Tapiolas and A.H. White, Aust. J. Chem., **36**(11), 2279 (1983).

|||||||||||||||||||||||||||||||||||||||||||||||||||||||||||||||||||||||||||||||||||||||||||||||||||

MACFARLANDIN E

Source: *Chromodoris macfarlandi*

Mol. formula: $C_{24}H_{34}O_7$

Mol. wt.: 434

Solvent: $CDCl_3$

T.F. Molinski, D.J. Faulkner, He Cun-heng, G.D.V Duyne and J. Clardy, J. Org. Chem., **51**(24), 4564 (1986).

1-BROMO-8-KETOAMBLIOL A ACETATE

Source: *Dendrilla* sp.

Mol. formula: $C_{22}H_{31}BrO_4$

Mol. wt.: 439

Solvent: $CDCl_3$

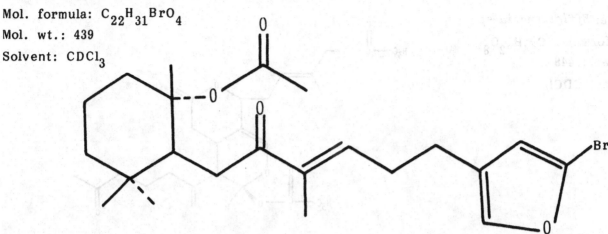

CH_3: 11.8, 19.7, 21.9, 22.5 and 32.4

CH_2: 19.5, 23.9, 28.8, 33.9, 37.4 and 40.4

CH: 50.4, 112.1, 139.5 and 140.5

C: 35.0, 86.2, 122.1, 126.8 and 137.7

C=O: 169.7 and 201.2

B. Sullivan and D.J. Faulkner, <u>J. Org. Chem.</u>, **49** (17), 3204 (1984).

|||||||||||||||||||||||||||||||||||||||||||||||||||||||||||||||||||||||||||||||||||||||||||||||||||||||||||||||||||||||||

CYSTOSEIROL B

Source: *Cystoseira mediterranea*

Mol. formula: $C_{27}H_{38}O_5$

Mol. wt.: 442

Solvent: $CDCl_3$

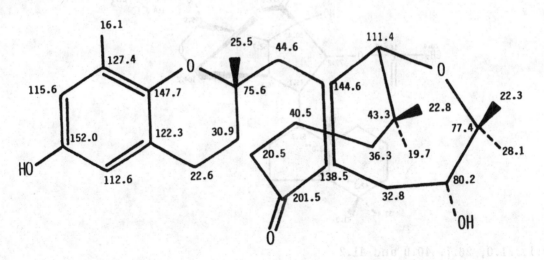

C. Francisco, B. Banaigs, M. Rakba, J. Teste and A. Cave, <u>J. Org. Chem.</u>, **51**(14), 2707 (1986).

(1R*,4aS*,5S*,6R*)-6-ACETOXY-1[(2E)-4-ACETOXY-3-METHYLBUT-2-ENYL]-5-(1-METHYLENE-5-OXOHEXYL)-4,4a,5,6-TETRAHYDRO-1H, 3H-PYRANO[3,4-C]PYRAN-3-ONE

Source: *Efflatounaria* sp.

Mol. formula: $C_{24}H_{32}O_8$

Mol. wt.: 448

Solvent: $CDCl_3$

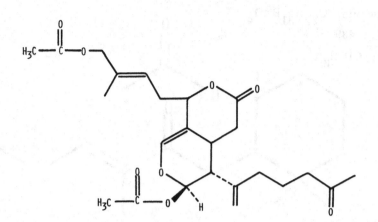

CH_3: 14.2, 20.7, 20.7 and 29.7

CH_2: 21.2, 34.3, 34.8, 34.8, 42.5, 69.1 and 113.4

CH: 30.9, 48.3, 78.5, 92.5, 121.8 and 137.5

C: 109.7, 133.9 and 144.6

C=O: 168.8, 169.3, 170.4 and 207.0

B.F. Bowden, J.C. Coll, V.A. Patrick, D.M. Tapiolas and A.H. White, <u>Aust. J. Chem.</u>, **36** (11), 2279 (1983).

||||||||||||||||||||||||||||||||||||||||||||||||||||||||||||||||||||||||||||||||||||||||||||||||||||||||||||||||||

CHROMODOROLIDE A

Source: *Chromodoris cavae*

Mol. formula: $C_{24}H_{34}O_8$

Mol. wt.: 450

Solvent: C_6D_6

CH_2: 20.1, 21.0, 26.7, 40.0 and 41.2

E.J. Dumdei, E.D. de Silva and R.J. Andersen, <u>J. Am. Chem. Soc.</u>, 111 (7), 2712 (1989).

POLYRHAPHIN B

Source: *Aplysilla polyrhaphis*
Mol. formula: $C_{25}H_{38}O_7$
Mol. wt.: 450
Solvent: $CDCl_3$

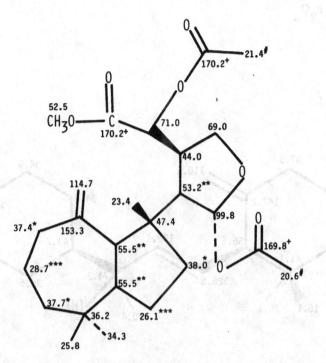

S.C. Bobzin and D.J. Faulkner, <u>J. Org. Chem.</u>, 54 (16), 3902 (1989).

||||||||||||||||||||||||||||||||||||||||||||||||||||||||||||||||||||||||||||||||||||||||||||||||||||||

STRICTAKETAL

Source: *Cystoseira stricta* (Mont.) Sauv.
Mol. formula: $C_{28}H_{40}O_5$
Mol. wt.: 456
Solvent: $CDCl_3$

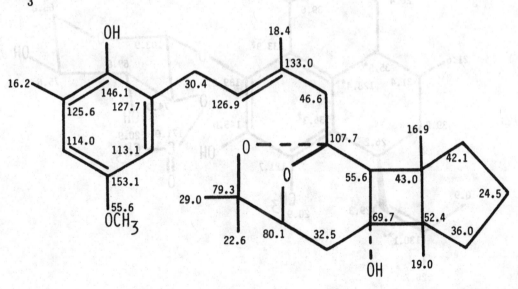

V. Amico, F. Cunsolo, M. Piattelli, G. Ruberto and L. Mayol, <u>J. Nat. Prod.</u>, (Lloydia), 50 (3), 449 (1987).

OBTUSADIOL

Source: *Laurencia obtusa* (Huds.) Lamouroux

Mol. formula: $C_{20}H_{32}Br_2O_2$

Mol. wt.: 464

Solvent: $CDCl_3$

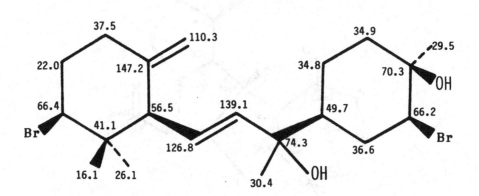

B.M. Howard and W. Fenical, <u>Tetrahedron Lett.</u>, (28), 2453 (1978).

||||||||||||||||||||||||||||||||||||||||||||||||||||||||||||||||||||||||||||||||||||||||||||||||||||||||||||

PSEUDOPTEROSIN B

Source: *Pseudopterogorgia elisabethae*

Mol. formula: $C_{27}H_{38}O_7$

Mol. wt.: 474

Solvent: $CDCl_3$

S.A. Look, W. Fenical, G.K. Matsumoto and J. Clardy, <u>J. Org. Chem.</u>, 51(25), 5140 (1986).

PSEUDOPTEROSIN C

Source: *Pseudopterogorgia elisabethae*
Mol. formula: $C_{27}H_{38}O_7$
Mol. wt.: 474
Solvent: $CDCl_3$

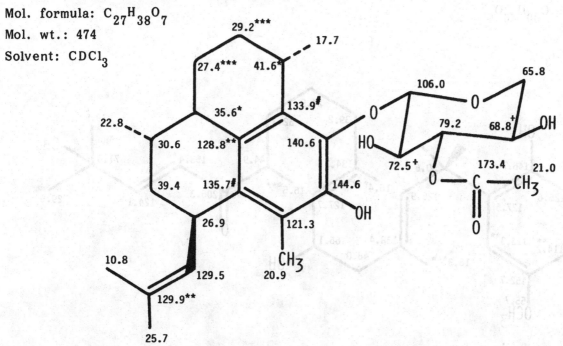

S.A. Look, W. Fenical, G.K. Matsumoto and J. Clardy, <u>J. Org. Chem.</u>, 51(26), 5140 (1986).

PSEUDOPTEROSIN D

Source: *Pseudopterogorgia elisabethae*
Mol. formula: $C_{27}H_{38}O_7$
Mol. wt.: 474
Solvent: $CDCl_3$

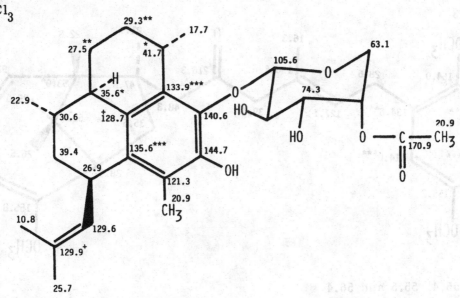

S.A. Look, W. Fenical, G.K. Matsumoto and J. Clardy, <u>J. Org. Chem.</u>, 51(26), 5140 (1986).

DITERPENE CE-III

Source: *Cystoseira elegans*

Mol. formula: $C_{30}H_{46}O_5$

Mol. wt.: 486

Solvent: $CDCl_3$

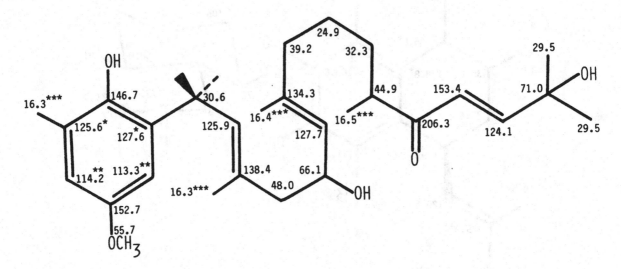

B. Banaigs, C. Francisco, E. Gonzalez, L. Codomier and W. Fenical, <u>Tetrahedron Lett.</u>, 23 (32), 3271 (1982).

|||||||||||||||||||||||||||||||||||||||||||||||||||||||||||||||||||||||||||||||||||||||||||||||||||||||||||||||||||||||||||||||||||||||||

MEDITERRANEOL B

Source: *Cystoseira mediterranea*

Mol. formula: $C_{31}H_{44}O_5$

Mol. wt.: 496

Solvent: $CDCl_3$

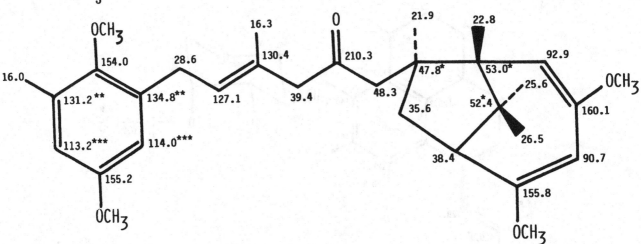

OCH_3: 55.3, 55.4, 55.5 and 56.4

C. Francisco, B. Banaigs, J. Teste and A. Cave, <u>J. Org. Chem.</u>, 51(7), 1115 (1986).

CINNCASSIOL D₁ GLUCOSIDE

Source: *Cinnamomum cassia* Blume

Mol. formula: $C_{26}H_{42}O_{10}$

Mol. wt.: 514

Solvent: C_5D_5N

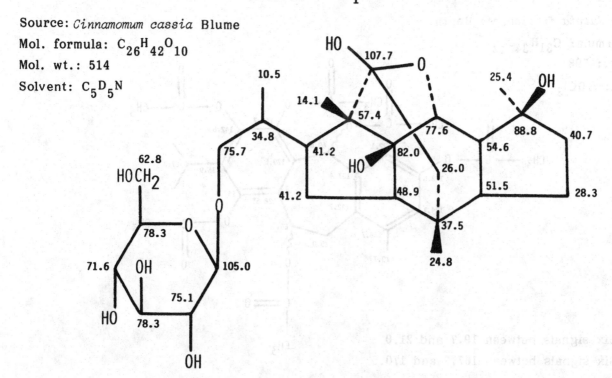

T. Nohara, Y. Kashiwada, K. Murakami, T. Tomimatsu, M. Kido, A. Yagi and I. Nishioka, Chem.Pharm.Bull. (Tokyo), 29(9), 2451 (1981).

‖‖

17-ACETOXYINGENOL-5,20-DIACETATE-3-ANGELATE

Source: *Euphorbia kamerunica*

Mol. formula: $C_{31}H_{40}O_{10}$

Mol. wt.: 572

Solvent: $CDCl_3$

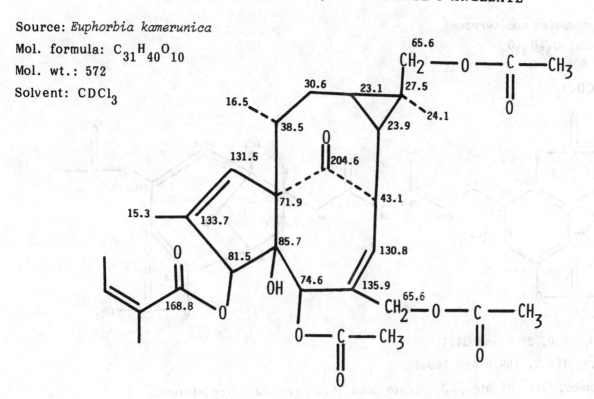

J.D. Connolly, C.O. Fakunle and D.S. Rycroft, Tetrahedron Lett., 25 (34), 3773 (1984).

DEMETHYL FRUTICULIN A (HEXA ACETATE)

Source: *Salvia fruticulosa* Benth.

Mol. formula: $C_{31}H_{34}O_{12}$

Mol. wt.: 598

Solvent: $CDCl_3$

CH_3: Six signals between 19.7 and 21.0

C=O: Six signals between 167.3 and 170.5

L. Rodriguez-Hahn, B. Esquivel, C. Sanchez, L. Estebanes, J. Cardenas, M. Soriano-Garcia, R. Toscano and T.P. Ramamoorthy, Phytochemistry, 28(2), 567 (1989).

MEDITERRANEOL A (TETRA ACETATE)

Source: *Cystoseira mediterranea*

Mol. formula: $C_{35}H_{44}O_9$

Mol. wt.: 608

Solvent: $CDCl_3$

CH_3: 20.0, 20.0, 20.9 and 21.1.

C=O: 168.6, 168.7, 168.8 and 169.4.

C. Francisco, B. Banaigs, J. Teste and A. Cave, J. Org.-Chem., 51(7), 1115 (1986).

DIHYPOESTOXIDE

Source: *Hypoestes rosea* R. Br

Mol. formula: $C_{44}H_{64}O_{10}$

Mol. wt.: 752

CH_2: 22.1, 23.7, 24.2, 26.7, 31.2, 32.4, 32.5, 35.8, 37.4, 40.6, 43.2 and 43.9

A.A. Adesomoju and J. I. Okogun, <u>J. Nat. Prod.</u>, (Lloydia), **47** (2), 308 (1984).

DIHYPOESTOXIDE

Source: *Hypoestes rosea* R. Br

Mol. formula: $C_{44}H_{64}O_{10}$

Mol. wt.: 752

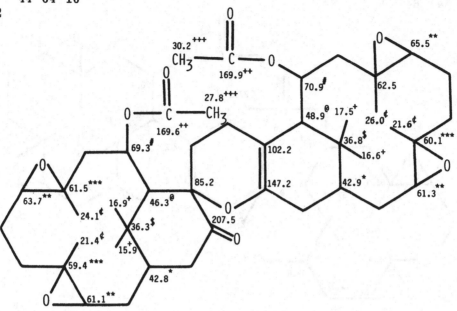

CH_2: 22.1, 23.7, 24.2, 26.7, 31.2, 32.4, 32.5, 35.8, 37.4, 40.6, 43.2 and 43.9

A.A. Adesomoju and J. I. Okogun, <u>J. Nat. Prod.</u>, (Lloydia), **47** (2), 308 (1984).

NAME INDEX

SOURCE INDEX

703

713

717

COMPOUND TYPE INDEX

DITERPINES 3

LINEAR 3

17-Acetoxythymifodioic acid (dimethyl ester) 19

(2*E*, 5*E*, 10*E*)-1,7-Dihydroxy-13-keto-3,7,11,15-tetramethyl hexadeca-2,6,10,14-tetraene 9

(2*E*,10*E*)-1,6-Dihydroxy-7-methylene-13-keto-3,11,15-trimethyl hexadeca-2,10,14-triene 10

Dimethyl-(14*S*)-(2*E*, 6*E*,10*Z*)-6-formyl-2, 10,14-trimethyl hexadeca-2,6,10- trienedioic acid (dimethyl ester) 14

Dimethyl(2*E*,6*Z*)-2-[(3'*Z*,7'*E*)-9'-hydroxy-4',8'-dimethylnona-3',7'-dienyl]-6-methylocta-2,6-dienedioate 15

Dimethyl (2*E*,6*Z*)-2-[(3'*Z*,7'*Z*)-9'- hydroxy-4',8'-dimethylnona-3',7'-dienyl]-6-methylocta-2,6-dienedioic acid (methyl ester) 15

Dimethyl(2*E*,6*E*,10*Z*,14*E*)-6-methoxy-carbonyl-2,10,14-trimethylhexadeca-2,6,10,14- tetraenedioate (trimethyl ester) 18

Dimethyl(2*Z*,6*E*,10*Z*,14*E*)-7-methoxy-carbonyl-3,11,15-trimethylhexadeca-2,6,10,14-tetraenedioic acid (dimethyl ester) 17

Dimethyl(2*E*,6*E*,10*Z*,14*E*)-6-methoxycarbonyl-2,10,14-trimethyl-hexadeca-2,6,10,14-tetraenedioic acid (dimethyl ester) 17

Dimethyl(14*S*)-(2*E*,6*E*,10*Z*)-6-methoxy-carbonyl-2,10,14-trimethylhexadeca-2,6,10-trienedioic acid (dimethyl ester) 18

Diterpene CE-III 21

Diterpene PD-III 20

Diterpene PF-I 19

Diterpene UA-I 20

Diterpene UF-IX 8

Diterpenoid UF-VIII (acetate) 14

Eleganediol 7

Eleganolone 6

Epoxyeleganolone 10

(5*E*,9*E*)-13,14-Epoxy-6,10,14-trimethylpentadeca-5,9, dien-2-one 4

Flexilin 9

Geranylgeraniol 5

Halimedatrial-IV 21

(2*E*,10*E*)-1-Hydroxy-6,13-diketo-7-methylene-3,11,15-trimethylhexadeca-2,10,14-triene 8

3ε-Hydroxy-4ε,9-dimethyl- 6*E*,9*E*-dodecadienedioic acid (methyl ester) 6

12-(*S*)-Hydroxygeranylgeraniol 7

13-Hydroxygeranyllinalool-3,13-O-β-diglucoside 22

13-Hydroxygeranyllinalool-13-O-β-D-(6'-O-β-L-fucopyranosyl)-glucopyranoside 22

(*E*,*E*,*E*)-3-Hydroxymethyl-7,11-dimethyl-2,6,10-hexadecatrien-1,14,15-triol 13

(2*E*,6*E*,10*E*)-12-(S)-Hydroxy-3,7,11,15-tetramethyl-hexadeca-2,6,10,14-tetranoic acid 11

Marislin 12

13-Methoxygeranyllinalool 11

15-Methoxy-3,7,11,15-tetramethylhexadeca-1,6,(*E*),10(*E*), 13(*E*)-tetraene 12

Phytol 5

β-Springene 3

(2*E*,6*E*,10*E*,13*E*)-3,7,11,15-Tetramethylhexadeca 2,6,10,13,15-pentenol 4

Thymifodioic acid 13

Trifarin 16

Udoteal 18

DIFURAN-TERPENOID 23

Cacospongienone A 23

Cacospongienone B 24

Cacospongione A 25

Dihydrofurospongin-2 24

Furospongin-1 25

OXEPANE 26

Montanol 26

Tomentanol 27

Tomentol 27

BISABOLANE ISOPRENOLOGUE 28

Diterpene EF-1 (methyl ester) 29

ABIETANE 286

Seco-CEMBRANE 479

ISO-CEMBRANE 480

753

Spongionellin 654

C₂₁H₃₁N

7-Isocyano-11(20), 14-epiamphilectadiene 630

C₂₁H₃₂ClNO₂

Kalihinol Y 587

C₂₁H₃₂O₂

Diterpene MT-Ia 264

Diterpene MT-II 264

Diterpene XA-I 58

Erythroxa-3,15-dien-18-oic acid (methyl ester) 642

Isopimaran-19-oic acid (methyl ester) 265

Methylcommunate 58

C₂₁H₃₂O₃

Diterpene GC-XIb (methyl ester) 71

Diterpene SL-V 344

Gracilin E 649

ent-15α-Hydroxyatis-16-en-19-oic acid (methyl ester) 442

15-Methoxyabietic acid 306

12-Methoxy-7α,11-dihydroxy-dehydroabietane 306

Tetrahydrofuran 28

Tetrahydropyran 29

C₂₁H₃₂O₄

Dendrillol-III 419

Diterpene BD-II (methyl ester) 485

Diterpene BD-III (methyl ester) 485

Diterpene BD-IV (methyl ester) 486

C₂₁H₃₂O₆

Dendrillol-IV 420

C₂₁H₃₄O₂

ent-Cleroda-4 (18), 13Z-dien-15-oic acid (methyl ester)

155

C₂₁H₃₄O₃

ent-Atisan-16α-ol-18-oic methyl ester 442

4β, 18-Epoxy-*ent*-clerod-13Z-en-15-oic acid (methyl

ester) 165

C₂₁H₃₄O₄

(1R, 4S, 5S, 9S, 1'R)-1-(1',5'-Dimethyl-4-oxohexyl)-9-

hydroxy-4-methylspiro [4.5] dec-7-ene-8- carboxylic acid

(methyl ester) 620

Diterpene EE-II (methyl ester) 621

19-Hydroxy-15-oxo-seco-beyeran- 16-oic acid (methyl

ester) 391

ent- Lab-8(17), 14- dien-13,16-diol-18-oic acid methyl

ester 86

3α-Methoxy-4β-hydroxy-5β, 10β-*cis*-17α,20α-cleroda-13

(14)-ene-15,16-olide 184

Methyl-2β, 3β-dihydroxylabda-8(17),13E-dien-15-oate 86

C₂₁H₃₆O₂

13-Methoxygeranyllinalool 11

ent-16α-Methoxy-kauran-17-ol 342

15-Methoxy-3,7,11,15-tetramethylhexadeca-1,6(E), 10(E),

13(E)-tetraene 12

C₂₁H₃₆O₃

Discoidic acid (methyl ester) 75

Diterpene BR-I 169

Gomeric acid (methyl ester) 76

Havardic acid B (methyl ester) 76

3β-Hydroxylabd-7-en-15-oic acid (methyl ester) 77

4β-Hydroxymethyllabd-7-en-15-oic acid (methyl ester) 77

C₂₁H₃₆O₄

Diterpene BR-III 185

Diterpene EV-I (methyl ester) 621

Havardic acid C (methyl ester) 88

3α-Hydroxygomeric acid (methyl ester) 89

Dimethyl (14S), (2E,6E,10Z)-6-formyl-2,10,14- trimethyl-

hexadeca-2,6,10-trien dioic acid (dimethyl ester) 14

Dimethyl (2E, 6Z)-2-[3'Z, 7'E)-9'-hydroxy-4',8'-dimethyl-

nona-3',7'-dienyl]-6- methylocta-2,6-dienedioate 15

Dimethyl (2E,6Z)-2-[3'Z,7'Z]-9'-hydroxy- 4',8'-dimethyl

nona-3',7'-dienyl]-6-methylocta-2,6-dienedioic acid

(methyl ester) 15

Diterpene HA-II 105

Evilliosin 105

Havardic acid D (methyl ester) 106

Previtexilactone 106

Stolonidiol monoacetate 500

Vitexilactone 107

C22H34O6

7β-Acetoxy-6β, 9α-dihydroxy-8,13-epoxy-labd-14-ene-11-

one 112

9-Deoxyforskolin 113

Diterpene HA-XXII 138

Diterpene HA-XX 138

Inflexarabdonin F 358

C22H34O7

Forskolin 118

C22H36Br2O5

Irieol D 603

C22H36O3

(1S*, 3E, 7E, 11S*, 12S*)-9-Acetoxydolabella-3,7-dien-12-

ol 497

19-Acetoxymanoyl oxide 84

(+)-Acetoxyodontoschismenol 498

Acetyldictyol C 540

Chromophycadiol monoacetate 652

α-Dictalediol monoacetate 653

(-)-2β, 9α-Dihydroxyverrucosane 615

Diterpene PC-Va 581

Jhanol acetate 85

C22H36O4

(1S*,3E,11S*,12S*)-9-Acetoxy-7,8-epoxydolabella-3-en-

12-ol 499

15-Acetoxy-7-labden-17-oic acid 94

Acetoxypachydiol 596

6-Acetyl-isoandalusol 94

3-Acetylvarodiol 95

12-Acetylvarodiol 95

10-Deacetoxybarbilycopodin 500

8,17H, 7,8-Dehydropinifolic acid (methyl ester) 96

Hamachilobene E 96

Haplociliatic acid (dimethyl ester) 198

Havardic acid A (methyl ester) 97

Jhanidiol-18-monoacetate 97

Methyl-15-formyloxy-7-labden-17-oate 98

Methyl-7-labden-15,17-dioate 98

6-O-Acetyl-austroinulin 99

C22H36O5

Sclerophytin E 481

C22H36O6

Sclerophytin C 482

C22H38O3

Methyl-15-methoxy-7-labden-17-oate 87

C22H38O4

Methyl-15-hydroxy-7α-methoxy-8-labden-17-oate 99

C23H26O7

12β-O-Acetyl-7-O-formyl-7-O-desacetylcoleon Z 411

C23H28O8

Salvinorin 224

$C_{27}H_{38}O_7$

Pseudopterosin B 674

Pseudopterosin C 675

Pseudopterosin D 675

$C_{27}H_{40}O_7$

Secopseudopterosin C 593

Secopseudopterosin D 593

$C_{27}H_{40}O_8$

7-Oxo-8-en-acamptoic acid (methyl ester) 129

$C_{27}H_{40}O_9$

Diterpene CB-VIII 429

$C_{27}H_{42}O_5$

ent-19-(3-Hydroxy-3-methylglutaroyloxy) beyer-15-ene

  (methyl ester) 390

$C_{27}H_{42}O_7$

Acamptoic acid (methyl ester) 126

$C_{27}H_{42}O_8$

Diterpene GS-III 130

7α-Hydroxy-8-en-acamptoic acid (methyl ester) 130

$C_{27}H_{43}N_5O$

Agelasimine A 143

$C_{27}H_{45}N_5O$

Agelasimine B 143

$C_{27}H_{46}O_8$

Microlepin 372

$C_{28}H_{34}O_{11}$

Teusalvin E (tetraacetate) 243

$C_{28}H_{35}ClO_{11}$

Junceellin 564

$C_{28}H_{35}ClO_{12}$

Praelolide 564

$C_{28}H_{36}N_2O_7$

Sarcodictyin C 570

(+)-Sarcodictyin E 571

(-)-Sarcodictyin F 571

$C_{28}H_{36}O_{11}$

Teupyreinidin 244

$C_{28}H_{37}ClO_{10}$

Brianthein Y 563

Ptilosarcone 563

$C_{28}H_{37}ClO_{12}$

Pteroidine 565

$C_{28}H_{38}O_9$

Halimedatrial-IV 21

Xenicin 533

$C_{28}H_{38}O_{11}$

Junceellolide D 562

$C_{28}H_{39}NO$

Emindole DA 663

Emindole SB 664

$C_{28}H_{39}NO_2$

Emindole DB 668

Paspaline 669

$C_{28}H_{40}O_5$

Strictaketal 673

$C_{28}H_{40}O_6$

Margotianin 445

$C_{28}H_{42}O_5$

Diterpene CE-III 21

$C_{28}H_{42}O_7$

2β-Isobutoxy-18ε,19ε -diacetoxy-4ε, 17α, 19α, 20β-(-)-

  cleroda-13 (16), 14-diene-18,19-oxide 236

771

C$_{38}$H$_{60}$O$_{18}$

Rebaudioside B 337

Stevioside 378

C$_{38}$H$_{62}$O$_{15}$

Hookeroside C 284

C$_{39}$H$_{64}$O$_{14}$

Hookeroside A 283

C$_{43}$H$_{64}$O$_4$

Foliosate 282

C$_{43}$H$_{70}$O$_{18}$

Hookeroside D 285

C$_{44}$H$_{64}$O$_{10}$

Dihypoestoxide 679

C$_{44}$H$_{70}$O$_{23}$

Rebaudioside A 379